T0133560

CILIA
Development and Disease

CILIA
Development and Disease

Editor

Paraskevi Goggolidou

School of Biomedical Science and Physiology
Faculty of Science and Engineering
University of Wolverhampton
UK

and

Centre for Nephrology
UCL Medical Campus
Royal Free Hospital
London
UK

CRC Press
Taylor & Francis Group
Boca Raton London New York

CRC Press is an imprint of the
Taylor & Francis Group, an **informa** business

A SCIENCE PUBLISHERS BOOK

CRC Press
Taylor & Francis Group
6000 Broken Sound Parkway NW, Suite 300
Boca Raton, FL 33487-2742

First issued in paperback 2020

© 2018 by Taylor & Francis Group, LLC
CRC Press is an imprint of Taylor & Francis Group, an Informa business

No claim to original U.S. Government works

ISBN-13: 978-1-4987-0368-0 (hbk)
ISBN-13: 978-0-367-78145-3 (pbk)

This book contains information obtained from authentic and highly regarded sources. Reasonable efforts have been made to publish reliable data and information, but the author and publisher cannot assume responsibility for the validity of all materials or the consequences of their use. The authors and publishers have attempted to trace the copyright holders of all material reproduced in this publication and apologize to copyright holders if permission to publish in this form has not been obtained. If any copyright material has not been acknowledged please write and let us know so we may rectify in any future reprint.

Except as permitted under U.S. Copyright Law, no part of this book may be reprinted, reproduced, transmitted, or utilized in any form by any electronic, mechanical, or other means, now known or hereafter invented, including photocopying, microfilming, and recording, or in any information storage or retrieval system, without written permission from the publishers.

For permission to photocopy or use material electronically from this work, please access www.copyright.com (http://www.copyright.com/) or contact the Copyright Clearance Center, Inc. (CCC), 222 Rosewood Drive, Danvers, MA 01923, 978-750-8400. CCC is a not-for-profit organization that provides licenses and registration for a variety of users. For organizations that have been granted a photocopy license by the CCC, a separate system of payment has been arranged.

Trademark Notice: Product or corporate names may be trademarks or registered trademarks, and are used only for identification and explanation without intent to infringe.

Library of Congress Cataloging-in-Publication Data

Names: Goggolidou, Paraskevi, editor.
Title: Cilia : development and disease / editor, Paraskevi Goggolidou, School of Biomedical Science and Physiology, Faculty of Science and Engineering, University of Wolverhampton, UK and Centre for Nephrology, UCL Medical Campus, Royal Free Hospital, London, UK.
Description: Boca Raton, FL : CRC Press, Taylor & Francis Group, [2018] | "A science publishers book." | Includes bibliographical references and index.
Identifiers: LCCN 2017048143 | ISBN 9781498703680 (hardback : alk. paper)
Subjects: LCSH: Cilia and ciliary motion--Diseases.
Classification: LCC QP310.C5 C558 2018 | DDC 571.6/7--dc23
LC record available at https://lccn.loc.gov/2017048143

Visit the Taylor & Francis Web site at
http://www.taylorandfrancis.com

and the CRC Press Web site at
http://www.crcpress.com

*To My Parents, for their Love
and Support*

Preface

The eukaryotic cilium is a fascinating organelle, whose discovery stems from more than 200 years ago. Currently unofficially called "the cell's antenna", the eukaryotic cilium bears great similarity with the prokaryotic flagella, which prokaryotes use for their efficient movement in the environment. In eukaryotes, however, cilia present a more complex story, since not all eukaryotic cilia are motile. Some organs, such as the brain and trachea contain motile cilia, whereas others like the kidney and heart contain immotile, sensory cilia (also known as primary). More than 15 years of research using modern molecular biology and microscopy techniques and live cell imaging has demonstrated that eukaryotic cilia have related, but disparate structures and functions in the various organs.

It is now known that the eukaryotic cilium is crucial for the transduction of signalling pathways and that almost every mammalian cell will carry at least one cilium. Cilia are surrounded by the plasma membrane, but they do not contain any subcellular organelles. As a result, they do not synthesize proteins but they have an important role in regulating signalling. This modulation is achieved by the process of **intraflagellar** transport (IFT), a specialised translocation complex which also involves motor proteins. IFT particles consisting of IFT-A and IFT-B protein complexes are transported to the tip of the cilium by heterotrimeric kinesin-2, in a process commonly referred to as anterograde transport. Transport of the proteins back to the base of the cilium is facilitated by cytoplasmic dynein-2; this movement is called retrograde transport. The hedgehog (Hh) signalling pathway has been directly linked to the primary cilium, with Patched 1 (Ptch1), the Hh receptor, being present in the cilium in the absence of a ligand. Ptch1 moves out of the cilium in response to Hh ligand and it is replaced by the Smoothened (Smo) protein, which acts downstream of Ptch1. Smo localisation to the cilium is necessary for the activation of downstream Hh signalling. Further, enrichment of the Gli2 and Gli3 transcription factors has been observed at the ciliary tip, in such a way that their levels at the tip of the cilium increase in response to Hh ligand.

Besides Hh signalling, other pathways have also been linked to the eukaryotic cilium; these involve Wnt, Platelet Derived Growth Factor (PDGF), Fibroblast Growth Factor (FGF), Notch and Hippo signalling pathways. Work from a number of excellent labs across the globe has shown that many human diseases could arise due to ciliary loss or dysfunction. These diseases have since been renamed Ciliopathies and they have been the focus of significant research. The aim of this book is to shed light on these diseases, highlight their importance in development and disease and allow for a better understanding of the eukaryotic cilium. Our vision is that research developments in the area of Ciliopathies will assist in eliminating the burden that these diseases pose on the patients and their families.

Contents

CHAPTER 1

Cilia in Brain Development and Disease

Gilbert Lauter, Peter Swoboda* and Isabel Tapia-Páez

INTRODUCTION

Brain Development

The notion that the human brain is the most complex object in the universe may be a narcissistic statement. However, the roughly 1.3 kg of human brain does grant its bearer rather astonishing capabilities, ranging from the subconscious control of bodily functions and complex motor skills to consciousness, emotions, memory, language and reasoning. Still, the complex structure of the adult human brain develops gradually from a simple field of neural cells. Insights into embryonic neural development might thus contribute to a better understanding of how neuronal interaction can result in all the executive and cognitive functions carried out by the human brain. Vertebrate brain development follows an intrinsically orchestrated sequence of events that are highly conserved between species. During neurulation, the ectoderm-derived planar neural plate bends, folds and finally closes to form the neural tube. Morphologically distinguishable brain vesicles start to appear as the neural tube becomes progressively subdivided along its rostro-caudal axis. The well-known three-vesicle stage with the prominent bulges of the fore-(prosencephalon), mid-(mesencephalon) and hindbrain (rhombencephalon) thus forms. Subsequently it is replaced by the five-vesicle stage, which is derived through further subdivision of the pros- and rhombencephalon. Interestingly, this sequence of neuromeric bulges occurs in all vertebrates at a certain developmental stage and results from zones of

Karolinska Institute, Department of Biosciences and Nutrition, S-141 83 Huddinge, Sweden.
* Corresponding author: peter.swoboda@ki.se

high mitotic activity located at the center of each vesicle (Bergquist, 1952; Bergquist and Källén, 1954; Källén, 1952). Taking into account the actual curvature of the developing brain, Bergquist and Källén demonstrated that mitotic zones are arranged transversely. This transverse arrangement of so-called neuromeres along the brain axis argues in favor of a segmental organization of the brain, thereby questioning the back then predominating columnar model, which stated that more or less the entire brain is organized into longitudinal, functional columns similar to those found in the spinal cord (Herrick, 1910). During subsequent development, transverse and later-forming longitudinal proliferation zones intersect to partition the brain into a patchwork of mitotic areas or so-called migration areas, where each area constitutes the future site for differentiation and migration (Bergquist and Källén, 1954). Surprisingly, the topography of migration areas is highly conserved between species and therefore suitable for establishing homology relationships on the basis of shared origin instead of similar function, a concept summarized by the term *field homology* (Nieuwenhuys, 2009; Puelles and Medina, 2002). The conserved patchwork of migration areas can be seen as an example of a common vertebrate *bauplan* ("construction plan") for brain development. With the advent of modern molecular techniques, even more support for segmental brain organization and insights into the framework behind the common vertebrate *bauplan,* were gained.

During development, regional specification of the brain arises progressively through repeated subdivisions of larger areas into smaller ones. The underlying patterning mechanisms are similar to those employed for the development of other body parts. In general, transient cues of positional information are conveyed via signaling molecules and instruct small cell populations to act as self-organizing centers. In turn, these so-called local organizers or signaling centers control the release of morphogens to which the surrounding tissues respond. The initial, transient positional information is tightly coupled to early events in axis formation during which morphogen gradients of Fibroblast Growth Factor (FGF), WNT and retinoic acid are set up across the anterior-posterior aspect of the developing brain. Subsequently, the dose-dependent response of the receiving tissue results in the expression of transcription factors (TFs) that mark broad regional identities (Sanes et al., 2006). Organizers typically arise at the boundary between regional markers (Macdonald et al., 1994). One prominent example of an early organizer is the mid- to hindbrain boundary (MHB), located at the interphase between *Orthodenticle Homeobox 2 (OTX2)* and *Gastrulation Brain Homeobox 2 (GBX2)* expression. The MHB arises through mutual repression between *OTX2* and *GBX2*, which eventually leads to a sharp boundary and maintained expression of *FGF8*. In turn, positional information is mediated via the secretion of FGF8 to adjacent areas of the future mid- and hindbrain (reviewed in Raible and Brand, 2004). Thereby surrounding cells acquire different cell fates, depending on the concentration of a received morphogen and on their competence. Similar organizing centers, albeit with different molecular players at work, are set up at various positions along the neuraxis (reviewed in Cavodeassi and Houart, 2012). For example, the *zona limitans intrathalamica (ZLI)* is marked by the prominent expression of Sonic Hedghog (SHH) between the future rostral and caudal thalamus (reviewed in Scholpp and Lumsden, 2010). The anterior neural ridge found at the anterior-most end of the developing brain, is essential for induction and development of the telencephalon via regulation of the WNT signaling pathway (Houart et al., 1998; Shimamura and Rubenstein, 1997). By contrast to these centers controlling cell fate specification along the anterior-posterior axis, the floor plate is a prime example for a local organizer acting primarily along the dorsal-ventral axis. This part of the ventral midline constitutes a continuous source for secreted SHH, which eventually leads to the establishment of different neural progenitor zones by differential SHH responsiveness (Briscoe et al., 2000; reviewed in Caspary and Anderson, 2003).

In summary, pan-neural patterning events of axis determination set up the major brain areas by broad expression of regional TFs. Subsequently, the concerted action of organizing centers subdivide larger regional identities into smaller histogenetic fields marked by the expression of a unique transcription factor specification code. In vertebrates, these specification codes have been largely conserved across species and constitute the genetic framework behind a common vertebrate *bauplan* for brain development (Ferran et al., 2007; Hauptmann and Gerster, 2000; Lauter et al., 2013; Puelles and Rubenstein, 2003). Eventually, progenitor domains form and secondary histogenetic processes like proliferation, migration, differentiation, synaptogenesis and apoptosis follow an exact spatio-temporal sequence to shape the complex structure of the brain with its various specialized brain functions.

Brain Anatomy

Vertebrates inhabit very diverse environmental niches. It is therefore not surprising that their brains have evolved various specializations. Still, mature brains are the result of a sequence of highly conserved developmental steps and therefore, they share major brain subdivisions (see above). Specializations can thus be seen as deviations from a common, albeit hypothetical, archetype by further evolution and elaboration of only certain parts. For example, a dorsal pallium can be found in all vertebrates but in primates it tends to enlarge substantially, culminating in the human cortex (Striedter, 2005). Similarly, a cerebellum can be distinguished in all vertebrates, but the cerebellum of animals that move in a three-dimensional space, like birds and fish, often displays greater complexity. A simplified archetype proposes shared forebrain regions of the telencephalon, like the pallium and subpallium in addition to the olfactory bulbs and the hypothalamus. The diencephalon of the forebrain is built by the epithalamus together with rostral and caudal thalami, followed by pretectal areas. The midbrain consists of the tectum and tegmentum, whereas the cerebellum, pons and medulla represent the major parts of the hindbrain. It should be noted, however, that such a simplified archetype is of a hypothetical nature and it represents a principle that integrates conserved, universal features. It primarily functions as a template to facilitate cross-species comparisons (Striedter, 2005).

Inevitably, these universalities extend beyond the regional morphological structures and they can also be found at the cellular and molecular level. All brains are constructed with the same cellular building blocks, derivatives of neurons and glia together with a vascular supply. Furthermore, cells group together and they form defined aggregates of similar organization as seen in laminae, brain nuclei or loose, reticular formations. At the subcellular level, a key universal feature of the brain is the presence of specialized contact points for communication, referred to as the synapses. At the synaptic cleft the propagation of electrochemical signals is regulated in a way that allows neurons to transmit, receive and integrate information. The computational power of the synapses relies on the exact spatial and temporal organization of molecules like scaffolding proteins, neurotransmitters and neuromodulators. This molecular machinery is highly conserved. The same set of neurotransmitters enables neurons to perform their typical functions not only in all vertebrates, but also in invertebrates. Despite the wealth and detail of knowledge of many of the components of the brain, some uncertainty prevails about how the coordinated (inter-)action of all its parts guarantees brain function.

Brain Physiology

From a reductionist point of view, the brain represents a central hub, which coordinates interactions of the individual with the environment to ensure survival-promoting behavior. In order to decide upon an adequate behavioral response, information about the state of the environment has to be gathered.

Sensory perception of the environment starts when a specialized sensory cell picks up an adequate stimulus via the corresponding sensory molecule or receptor. During subsequent steps this signal is turned into an electrochemical (neuronal) signal, which is propagated to the respective brain areas. Interestingly, receptors are often concentrated and exposed to the environment in specialized subcellular structures that rely on cilia (see below). In most vertebrates the sensory system commonly includes the perception of light, sound, taste, odor, touch, heat, posture and pain. Further, depending on the environment and the concomitant selective pressure, animals have evolved astonishing specializations resulting in extreme sensitivity to the adequate stimulus, sometimes even reaching physical limits (Frings, 2012). Following receptor activation, a complex interplay between second messenger and ion channel dependent mechanisms results in signal amplification and the production of electrical charges. The generation of action potentials follows an all or nothing principle in the sense that sensory cells will either not respond at all or generate full-fledged potentials subsequent to stimulation. From any given stimulus the sensory system tries to acquire information not only about the type of stimulus received, but also about its duration, intensity and localization. The wealth of information sensed simultaneously by all the senses is truly enormous, keeping in mind that furthermore, the brain is constantly fed with information about the current body status through other systems. In addition, information can be compared to past, memorized events and the consequences of future actions can be assessed. But even though various information sources are of totally different nature, the readable output transmitted in the brain almost always consists of propagating action potentials, tiny localized bursts of electrical charges. The uniformity of action potentials implicates that information must be encoded by the timing and frequency of discharges. Frequency modulation seems to be the general *modus operandi* for encoding information in the brain (Frings, 2012). Albeit the format of exchangeable information is uniform and while we have good knowledge about the basic functional units, we still lack a fundamental understanding of the steps leading from cellular chemistry to cognition.

However, some typical features are associated with the appearance of cognitive functions. Less evolved brains tend to have fewer areas connected between primary sensory and executive structures. With the evolution of higher cognitive functions, more and more cortical areas are added into pathways. Thus, the addition of cortical areas that are highly inter-connected but only indirectly linked to sensory and executive entities seems to be essential. At the same time the connectivity of participating areas does not follow a linear, hierarchal logic. Instead of a single command center where everything converges, information seems to be processed in parallel and reciprocally in different brain areas by varying cell assemblies (Singer, 2012). It seems that convergence in time, not space, is essential for letting us enjoy such amazing brain functions as for example reading.

Cilia in the Brain

Cilia are hair-like cell protrusions found on many different cell types within the eukaryotic kingdom (Piasecki et al., 2010). The cilium consists of a microtubule based core structure, the axoneme, surrounded by a specialized ciliary membrane (Figure 1A). The axoneme originates from the basal body positioned close to the cell membrane as a ring of nine microtubular doublets (Figure 1B). Primary and sensory cilia typically display this 9 + 0 configuration, whereas motile cilia contain an extra pair of microtubules in the middle of the ring and therefore adopt a 9 + 2 configuration (Figure 1B). Auxiliary structures like outer and inner dynein arms together with radial spokes are required to ensure motility. Active, motor-protein driven intraflagellar transport (IFT) is essential for the formation and maintenance as well as for the function of the ciliary compartment. Despite their wide

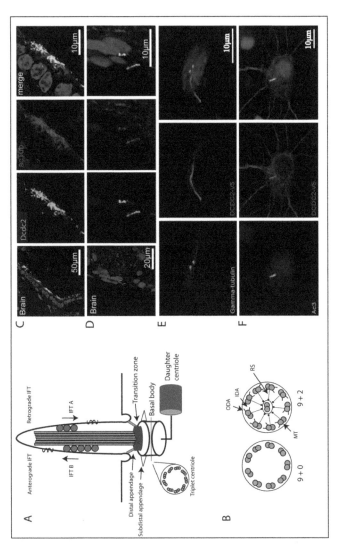

Figure 1. Cilia in the brain. (A) Schematic drawing of the structure of cilia showing the daughter centriole, the basal body, the transition zone, the axoneme with anterograde and retrograde intraflagellar transport (IFT) and a cross section of the basal body depicting the nine triplets of microtubules. **(B)** Cross section of the ciliary axoneme showing the structure of 9 + 0 primary cilia and 9 + 2 motile cilia, respectively. Outer dynein arms (ODA), inner dynein arms (IDA), microtubule doublets (MT) and radial spokes (RS) are depicted. **(C)** and **(D)** The ciliary marker acetylated tubulin (acTub) and the dyslexia candidate protein Dcdc2 stain the ciliary axoneme in mouse ependymal cells of the third brain ventricle (C) and cranial pia mater (D). Original pictures were kindly provided by Andrea Bieder and reproduced with permission (from Schueler et al., 2015). **(E)** and **(F)** Confocal images of Rat hippocampal neuron primary cells that were transfected with DCDC2-V5 and labeled with either the centriolar marker gamma-tubulin (E) or the neuronal ciliary marker adenylate cyclase 3 (Ac3) (F). The merged images to the right clearly show colocalization between the ciliary markers and DCDC2-V5 (Massinen et al., 2011). Images in (E) and (F) were reproduced with permission (from Massinen et al., 2011). In all images nuclei were stained with DAPI (blue).

distribution and functional importance, cilia have long been overlooked. This cellular organelle is present on most polarized cell types in the human body and is especially abundant in the brain (Sarkisian and Guadiana, 2015), where motile cilia exist on the ependymal lining of the ventricular walls (Spassky, 2013) and primary cilia are present on most neuronal cell types ranging from progenitor cells to differentiated neurons and astrocytes (Arellano et al., 2012; Bishop et al., 2007; Cohen, 1987; Dubreuil et al., 2007; Fawcett, 1954; Fuchs and Schwark, 2004; Tong et al., 2014). Examples for functional contributions of cilia can be found for most, if not all, morphogenetic processes active during brain development (Guo et al., 2015). The high number of brain conditions associated with human ciliopathies underscores the importance of proper cilia function for normal brain development.

Here we focus on cilia in connection to the vertebrate brain. We do not discuss highly relevant (and intensely studied) non-vertebrate organisms, like the sea squirt *Ciona intestinalis*, the fruit fly *Drosophila melanogaster* or the round worm *Caenorhabditis elegans*. All three develop highly structured brains or at least brain-like structures, central and peripheral nervous systems (CNS and PNS). *Ciona* belongs to a member group of the chordate animals, which also include vertebrates. Being a "sister" to vertebrates *Ciona* is considered evolutionarily significant because its development reveals features of the early evolution of the vertebrate body plan, including nervous system development. The *Ciona* nervous system includes ciliated neurons, for example ependymal cells with motile cilia as part of the neural tube of the CNS and epidermal sensory neurons of the PNS (Hozumi et al., 2015; Imai and Meinertzhagen, 2007a,b; Konno et al., 2010). For both the fruit fly *Drosophila* and the worm *C. elegans*, within the entire nervous system, respectively, cilia are found only on certain types of sensory neurons (*Drosophila*: Jana et al., 2016; Vieillard et al., 2015) (*C. elegans*: Doroquez et al., 2014; Inglis et al., 2007; Perkins et al., 1986; White et al., 1986). Both the *Drosophila* and *C. elegans* cilia research communities have contributed enormously to uncovering overall (neuronal) cilia structure and function, in particular to how cilia assemble and function as molecular machines serving motility and sensory tasks (Doroquez et al., 2014; Jana et al., 2016).

Primary Cilia in Early Brain Development

Primary cilia and SHH signaling

Primary cilia are often referred to as the cell's antenna, a phrase that pictures the unique structural features that make cilia ideal compartments for cell signaling. The notion of primary cilia as sensory hubs that coordinate multiple signal transduction pathways has gained traction with the increasing number of identified cilia-associated pathways. Primary cilia have been implicated in the regulation of a number of signal transduction pathways such as the ones involving SHH, WNT, planar cell polarity (PCP), specific G-protein coupled receptors (GPCRs), receptor tyrosine kinases (RTKs), Notch receptors as well as ion channels and transporters (Christensen et al., 2012; Goetz and Anderson, 2010; Guemez-Gamboa et al., 2014). Cilia-associated signaling pathways govern various aspects of cellular physiology including patterning, organogenesis, homeostasis and regeneration of the brain. We will focus our discussion on the role cilia play in the regulation of the SHH pathway, for which a substantial wealth of information exists (for information on other pathways please consult the respective references).

In vertebrates SHH signaling is tightly linked to primary cilia, as all major pathway components cluster in and around the cilium (reviewed in Goetz and Anderson, 2010). In the absence of the ligand SHH, the receptor Patched1 (PTCH1) localizes to the ciliary base and the axoneme preventing entry of the transmembrane protein Smoothened (SMO) into the ciliary compartment. In the presence of ligand, PTCH1 exits from its ciliary location whereupon SMO accumulates within the cilium, thereby initiating pathway activation (Corbit et al., 2005; Rohatgi, 2007). SHH signaling mediates changes in target gene expression through members of the Glioma (GLI) transcription factor family. Under normal conditions, the full-length GLI-activator forms (GLIAs) are proteolytically processed into shorter GLI-repressor forms (GLIRs). Upon exposure to SHH, processing is blocked and GLIAs out-balance the repressor versions thus enabling target gene expression. Concurrently with the translocation of SMO, activator versions of GLI2 and GLI3 start to accumulate at the ciliary tip (Haycraft et al., 2005; Kim et al., 2009). The ciliary localization of PTCH1, SMO and GLI factors together with other pathway modulators like SUFU and GPR161 underscore the direct, molecular link between cilia and SHH signaling (Haycraft et al., 2005; Mukhopadhyay et al., 2013).

As exemplified by SHH pathway components, the ciliary organelle acts as a hub for the accumulation of unique sets of receptors and transduction proteins. In part, this is achieved by a septin diffusion barrier between the contiguous plasma and ciliary membrane. At the same time, the transition zone acts as a selective filter at the ciliary base to allow exclusive entry only for cargoes targeted to the cilium (Reiter et al., 2012). Being a single, long cell surface projection with a small and confined volume, the cilium combines unique structural features that allow contemplating possible mechanisms for the assembly and amplification of signaling cascades. In the signaling compartment model the cilium simply acts as a type of a "reaction tube", where signaling components become locally concentrated to facilitate physical interaction (Nachury, 2014). The scaffold model relies on active trafficking and proposes that the IFT machinery functions as a scaffold for the assembly of signaling modules. Regardless of whether these two models are mutually exclusive or co-exist, IFT plays an important role, as it is essential for the formation and maintenance of the organelle and might even be an integral part of signaling cascades (Goetz and Anderson, 2010; Nachury, 2014).

IFT describes the molecular machinery that powers the movement of protein cargo along the ciliary axoneme. Two types of multi-protein aggregates, termed IFTA and IFTB particles, transport cargo in linear arrays in the space between the microtubular axoneme and the ciliary membrane, similar to freight trains on rails (reviewed in Ishikawa and Marshall, 2011). Anterograde movement, from the base to the tip, is driven by motor proteins of the KINESIN-2 family and requires the linear assembly of IFTB-anchored trains. In contrast, retrograde movement, from the tip to the base, is powered by cytoplasmic DYNEIN-2 motor proteins and requires trains of IFTA-anchored particles that can be somewhat shorter. Both complexes, IFTB (comprising of at least twenty known components) and IFTA (comprising of at least six known components), travel together along the axoneme (Ou et al., 2005; Qin et al., 2005). After the delivery of ciliary cargo mediated by IFTB, rearrangements occur at the tip and IFTA takes over to return proteins to the cytoplasm of the cell. This functional division of bi-directional transport between two separate systems gives rise to distinct mutant phenotypes associated with mutations affecting either IFTB or IFTA components.

IFTB is essential for cilia assembly and maintenance. Therefore mutations in IFTB components (*IFT52, -57, -88, -172*) tend to have a profound impact on ciliogenesis, culminating in the absence of cilia and entailing the shutting down of SHH signaling altogether (Huangfu et al., 2003). On the other hand, some IFTA components (*IFT122, -139, -144*) do not seem to be essential for cilia assembly. Hypomorphic alleles in the corresponding genes are associated with the accumulation of ciliary cargo at the tip, resulting in bulbous or abnormally long cilia (Liem et al., 2012; Qin, 2011; Tran et al., 2008). Surprisingly, these mutations cause increased SHH pathway activation and consequently a ventralization of the caudal brain is observed. The common assumption is that IFTB is essential for the buildup of activated GLIA at the tip and subsequent pathway activation, whereas IFTA-mediated trafficking differentially modulates this process. Therefore mutations in IFTA components can differ in phenotypes from IFTB mutations, where the absence of cilia precludes any conclusion on the influence of IFTB-mediated trafficking (Goetz and Anderson, 2010; Tran et al., 2008).

Studies on the trafficking effects of SHH pathway components have recently tried to shed light on the underlying molecular mechanisms. Systematic analyses of PTCH1 protein domains show that ciliary localization and binding of SHH is required for SMO regulation, although conveyed by different PTCH1 protein domains. It turns out that the actual clearance of PTCH1 from the ciliary compartment after SHH stimulation is not required for efficient signaling. The pathway is also activated when PTCH1 is artificially retained within the cilium after stimulation (Kim, 2015).

Whatever effect specific trafficking events have on the total pathway readout, the function of particular IFT components might not be restricted solely to transport in one or the other direction. Some IFT components might participate in both transport activities and carry out additional functions, like loading cargo proteins onto IFT trains. Further, multifunctional IFT components might add new levels of complexity to pathway regulation. For example, in hypomorphic *Ift122* mouse mutants GLI2, GLI3 and the IFTB particles IFT57, IFT88 are enriched at the ciliary tip. Secondly, also the ratio of GLI activator over GLI repressor forms is increased and SHH-signaling is over-activated (Qin, 2011). Interestingly, IFT122 acts downstream of SMO, which shows normal localization and trafficking in these mutants. Two important conclusions can be drawn from this: SMO trafficking does not depend on IFT122 and IFT122 differentially controls further pathway modulators (Qin, 2011). One can thus envisage the loss of a negative SHH modulator, which is IFT122 dependent. The tubby-like protein 3 (TULP3) functions as a SHH repressor, localizes to the ciliary tip in an IFT122 dependent manner and is responsible for the transport of G-protein coupled receptors (GPCRs) (Mukhopadhyay et al., 2010; Qin, 2011). The actual SHH-modulator itself seems to be the GPCR GPR161, whose interaction with the IFTA core complex is mediated by TULP3. In the wild-type situation, GPR161 promotes the processing of GLIAs into GLIRs and could thus function as a general negative modulator that suppresses signaling in the absence of SHH. When SHH-signaling is initiated, GPR161 is rapidly internalized, allowing switching to "power-on" (Mukhopadhyay et al., 2013). Cilia thus provide a regulatory platform to integrate the function of different pathway components and modulators that are active during signaling.

The role of primary cilia in patterning

Signal transduction pathways describe the actual molecular mechanisms by which transient information is transduced across the plasma membrane into changes in gene expression within the nucleus. As cells acquire different cell-type identities depending on their relative spatial locations, pattern formation is perceived as the net outcome of signaling events at the tissue and organ level. Examples of cilia-mediated patterning defects originate primarily from the inextricable link between cilia and SHH signaling.

The first evidence that cilia are involved in brain patterning in connection to the SHH signaling pathway came from genetic screens for embryonic patterning defects in mice. Mutant alleles of *Ift172 (wimple)*, *Ift88 (polaris/fxo)* and the motor-protein *Kif3a* perturb cilia development and phenocopy *Shh* mutants in mice (Huangfu et al., 2003). In mutant embryos ventral markers for the floor plate (*SHH*) and for the motor neuron progenitor domain (*HB9*) are lost, whereas dorsal markers (*PAX6*) expand ventrally (Huangfu et al., 2003). Generally, the spinal cord is the prime example of how different progenitor domains are specified by the combinatorial expression of two classes of SHH-responsive TFs: (a) dorsalizing factors repressed by SHH signaling and (b) ventralizing factors activated by SHH signaling (Briscoe et al., 2000; reviewed in Caspary and Anderson, 2003). Genetic analysis of double mutants further shows that *IFT172*, *IFT88* and *KIF3A* act downstream of *PTCH1* and *SMO*, most likely by affecting the proteolytic processing of GLI3 (Huangfu et al., 2003).

Interestingly, the *Ift88* hypomorphic allele *cobblestone* (*cbs*) has been instrumental for studying the influence of cilia in forebrain development. Brains of *cbs* mouse mutants display a severe disorganization of the dorsal telencephalon including medial structures such as the hippocampus, cortical hem and choroid plexus. Heterotopias, ectopic clusters of dividing cells, are observed as subpial bulges, giving the mutant its name. Patterning is also affected, as the dorsal telencephalon has severe regionalization defects, manifested by the disruption of the pallial-subpallial as well as the telencephalic-diencephalic boundary (Willaredt et al., 2008). Surprisingly, despite the loss of about 75% of IFT88 protein, primary cilia of normal length and ultrastructure still project into the ventricle. One possible explanation is that the remaining low level of IFT88 protein is still sufficient to make structurally normal, albeit not fully functional cilia. This functional impairment affects the outcome of cilia-dependent signaling processes. However, the *cbs* mutant phenotypes strongly resemble those observed earlier in a *Gli3* deletion mutant in the mouse (Fotaki et al., 2006; Theil et al., 1999; Tole et al., 2000; Willaredt et al., 2008). Indeed, although *GLI3* mRNA expression levels are unaltered in *cbs* mutants, proteolytic processing is affected and the balance of GLI3A versus GLI3R tilts in favor of the GLI3 activator form (Willaredt et al., 2008).

The *selective Lim binding protein (slb)* mouse mutant carries a hypomorphic allele for *Ift172* and embryonic cilia of all brain regions studied have an aberrant morphology with the microtubule axoneme missing (Gorivodsky et al., 2009). Morphological phenotypes like truncation of the forebrain, holoprosencephaly and failure in neural tube closure are indicative of disrupted anterior-posterior patterning. Reduced *FGF8* expression in secondary organizers like the anterior neural ridge and at the midbrain-hindbrain boundary further support this notion. In the ventral fore- and midbrain, ciliary defects correlate well with a strong reduction in the expression of *SHH* and its direct target *GLI1*. Interestingly, *slb* mutants also show disrupted Nodal signaling, which leads to the abnormal development of axial mesoderm. This points to a function of IFT172 in earlier events upstream of SHH during the development of tissues like the axial mesoderm, which in turn are important for the induction of the neural tube.

Another clear link between brain patterning and cilia stems from studies of Regulatory Factor X (RFX) transcription factors. RFX factors are evolutionarily conserved key regulators of the transcriptional networks governing ciliogenesis (reviewed in Choksi et al., 2014). For example in the nematode *Caenorhabditis elegans*, loss of *daf-19* (the sole *RFX* representative in this model) leads to the complete loss of all sensory cilia, thus emphasizing its central role for cilia formation and maintenance (Swoboda et al., 2000). In mammals, eight RFX factors exist, whereby several are expressed in the brain and have specific functions during brain development (Choksi et al., 2014).

Interestingly, *RFX4* shows specific expression in the CNS and homozygous mouse mutants lose medial and paramedial dorsal structures. In the telencephalon, this affects structures like the cortical hem and the choroid plexus, which fail to develop. Changes in expression of regional markers like *MSX2*, *WNT3A*, *WNT7B* and *BMP4* suggest that *RFX4* is required to establish dorsal signaling centers in the telencephalon (Blackshear et al., 2003). Examination of the telencephalon of *Rfx4* mutants reveals a partial ventralization, as seen by reduced *LHX2* and *WNT8B* expression and loss of the cortical hem and the hippocampal *anlagen* (Ashique et al., 2009). The dorso-ventral patterning effect bears strong resemblance to the one observed in *Gli3* mutants and is consistent with the notion that RFX4 influences Shh signaling by modulating GLI3 processing. Accordingly, SHH signaling expands dorsally and the level of GLI3A is increased in *Rfx4* mutants. The patterning effect is most likely secondary to changes in ciliary structure given the known role of RFX factors in cilia development and the finding that in *Rfx4* mutants *IFT172* expression is down-regulated and cilia structure is impaired (Ashique et al., 2009).

During early brain development, *RFX3* is present in ciliated ependymal cells of the ventricular lining and loss of *RFX3* function is often accompanied by severe hydrocephalus (Baas et al., 2006). *Rfx3* deficient mice exhibit defects in the development of the corpus callosum (CC), which normally connects the two brain hemispheres (Benadiba et al., 2012). This malformation traces back to expanded Fgf8 signaling in the cortical septal boundary in consequence of changes in the modulation of GLI3 during early development. The resulting patterning defect leads to misplacement of so-called guidepost neurons, which are required for the correct routing of callosal axons (Benadiba et al., 2012).

Strikingly, many of the phenotypes that are caused by disruption of well-known ciliary components can be connected with changes in GLI3 proteolytic processing (Ashique et al., 2009; Besse et al., 2011; Gorivodsky et al., 2009; Huangfu et al., 2003; Mukhopadhyay et al., 2013; Willaredt et al., 2008). Primary cilia seem to be required to maintain the right balance between GLI3 activators and repressors. Different mutant alleles of the same gene can shift the balance in either direction resulting in a net gain of either GLI3A or GLI3R, as illustrated by the opposing patterning defects observed with the hypomorphic alleles *fxo* and *cbs* for *Ift88* (Huangfu et al., 2003; Willaredt et al., 2008). Yet, how cilia actually impact GLI processing is less clear. One possible molecular mechanism for the regulation of GLI3R production might depend on the ciliary localization of GPR161 and its influence on the activity of protein kinase A (PKA). In the absence of ligand, GPR161 resides within the cilium and causes an elevation of cAMP levels sufficient to stimulate PKA activity, which in turn targets full-length GLI3 for degradation (Mukhopadhyay et al., 2013; Tuson et al., 2011; Wang et al., 2000). In summary, the fact that strong data converge on the processing of

SHH effectors can be seen as a consequence of the tight connection between primary cilia and SHH signaling in vertebrates.

Proliferation, migration and wiring

During early development the neural tube consists of a single layer of neuroepithelial cells with a uniform proliferative activity along the ventricular lining. The brain wall appears pseudo-stratified as cell nuclei show periodic movement along the entire apical-basal extension with mitosis always occurring at the ventricular lining (interkinetic nuclear migration—INM) (Sauer, 1935). Later, under the influence of patterning events, neural tube proliferation becomes progressively restricted to only certain areas at distinct spatial locations along the major brain axes. Primary cilia extend from mitotic progenitor cells into the ventricle and are essential for coordinated growth within each domain (Dahl, 1963; Wilson et al., 2012). Mutants of ciliary components like *IFT88*, *ODF1*, *RPGRIP1L*, *RFX4* and *KIF3A* often display structural defects in cilia. At the same time they share strikingly similar morphological phenotypes in the forebrain and show disturbed GLI3 processing (Ashique et al., 2009; Besse et al., 2011; D'Angelo et al., 2012; Stottmann et al., 2009; Willaredt et al., 2008; Wilson et al., 2012). Among the typical brain abnormalities observed in these mutants are exencephaly, dorsoventral patterning defects combined with loss of dorsal-midline structures and an overall increase in size. Conditional knockdown of *Kif3a* in mice shows that impairment of primary cilia function leads to substantial brain enlargement (Wilson et al., 2012). The increase in brain size is not due to reduced apoptosis (cell death is actually upregulated in mutant brains). Rather, it is better explained by the faster progression of progenitors through the cell cycle. Interestingly, cell cycle length directly correlates with GLI3A to GLI3R ratios, suggesting a model where cilia dependent SHH signaling controls cell cycle progression via GLI3 processing. High ratios correlate with high proliferative activity as seen during the early phases of development, while the subsequent slowdown in cell cycle progression during later developmental phases correlates with low GLI3A to GLI3R ratios. As primary cilia reach into the ventricular cavity, they are simultaneously exposed to any mitogens present in the cerebrospinal fluid (CSF), for example SHH (Huang et al., 2010; Lehtinen et al., 2011). It is therefore tempting to speculate that cilia regulate the differential growth of progenitor domains via SHH distributed in the CSF (Wilson et al., 2012). Cilia-dependent regulation of mitotic activity mediated through SHH effectors is not a phenomenon unique to forebrain development, but it has also been observed in mid- and hindbrain development (Blaess et al., 2008; Chizhikov et al., 2007; Spassky et al., 2008). Inactivation of *GLI3* at different developmental time points uncovers differential functions of GLI3 during development (Blaess et al., 2008). Under the control of SHH signaling, GLI3 mediates patterning events during early development and later on coordinates progression through the cell cycle

(Blaess et al., 2008; Wilson et al., 2012). However, one has to keep in mind that the ciliary axoneme germinates from the basal body, which is derived from the mother centriole. Therefore, there is a tight interdependence of cilia and cell cycle progression, as the basal body has to resume centriolar functions during mitosis. Since many ciliary proteins interact with structures of the basal body/centriole, it is not always easy to clearly distinguish between ciliary and centrosomal functions during cell proliferation (Guemez-Gamboa et al., 2014).

During radial glial cell transition, undifferentiated neuroepithelial cells adopt radial glia cell characteristics with a clear apical-basal polarization. Radial glia cells span the entire neural wall and extend a primary cilium at their apical side into the ventricle (Schmechel and Rakic, 1979). In the opposite direction, long radial processes reach all the way to the pial surface, where they form basal endfeet. Radial glia cells divide asymmetrically to generate newborn neurons that use the glial processes as a scaffold for directed migration to the periphery. As post-mitotic neurons migrate away from the proliferation site at the ventricular lining to the periphery, the formerly pseudo-stratified wall starts to adopt a more layered organization. Therefore, the development of a layered brain wall organization strongly depends on polarized radial glia cells to act as coordinators of neurogenesis and migration.

The ciliary component ARL13B is essential for the initial establishment of the correct apical-basal polarity of radial glia cells. ARL13B belongs to the superfamily of small ARF/ARL GTPases and localizes specifically to cilia, including those of radial glia, where it regulates microtubule dynamics of the axoneme (Caspary et al., 2007; Higginbotham et al., 2013). Neural wall organization is reversed in mouse embryos carrying an *Arl13b* null allele. The cell somata of radial glia localize close to the pial surface instead of the ventricular side and the characteristic glial endfeet attach to the ventricular lining. ARL13B function is required during an early developmental window important for the initial formation and apical-basal organization of radial glia cells, as revealed by the conditional inactivation of *ARL13B* at different developmental time points (Higginbotham et al., 2013). Strikingly, in *ARL13B* mutants cilia have reduced length and reduced plasticity. Diminished ciliary dynamics such as remodeling and re-orientation events of the cilium are indicative of impaired ciliary signaling (Kim et al., 2010; Mukhopadhyay et al., 2008). Taken together these aspects suggest that cilia function is needed in early developing glia to perceive and respond to critical extracellular cues. Failure to do so results in aberrant polarity and development of the glial scaffold, which in turn is essential for the morphogenetic construction of the brain wall.

Recently, important roles for cilia in asymmetric inheritance and neurogenesis have gained much attention (reviewed by Sarkisian and Guadiana, 2015). Interestingly, dividing progenitors display asymmetric re-localization of cilia on the resulting daughter cells, whereby the cell destined to delaminate re-grows a cilium at the basolateral side, while the cell remaining in the adherence junction belt does so at the apical side. Thereby, basolateral cilia are excluded from signals present in the CSF and face an entirely different signaling environment than their apical counterparts (Wilsch-Brauninger et al., 2012). Furthermore, at the onset of mitosis remnants of the ciliary membrane are endocytosed along with the mother centriole and become asymmetrically distributed (Paridaen et al., 2013). The cell that remains proliferative always inherits the mother centriole with the ciliary remnant and re-grows a fully functional cilium faster than its sibling cell destined to differentiate. Due to spatial and temporal differences, the cilia of any two daughter cells are thus exposed to discriminative signaling environments that presumably promote cell fate changes (Taverna et al., 2014; Wilsch-Brauninger et al., 2012).

During further development, isolated proliferation zones partition the embryonic brain into a network of so-called migration areas. Within each area a unique developmental program directs future differentiation and migration streams into other brain regions (Bergquist and Källén, 1954). A large subpopulation of interneurons generated in the medial and caudal ganglionic eminences (prominently marked by the expression of *DLX5* and *DLX6*) migrate tangentially to their destinations within the dorsal cerebral cortex. As the formation of functional brain circuits requires accurate and correct placement of all participating neurons, coordinated migration is thus of utmost importance. Inhibitory interneurons (marked by the expression of gamma-aminobutyric acid [GABA] stimulated receptors) possess a primary cilium that helps them to read the right road signs during their journey to dorsal areas (Baudoin et al., 2012; Higginbotham et al., 2012). Conditional inactivation of cilia function using floxed *Arl13b*$^{flx/flx}$ crossed with *Dlx5/6-Cre* mice reveals a migration defect, where significantly fewer GABA-ergic interneurons reach the dorsal cortex (Higginbotham et al., 2012). Inhibitory interneurons collectively seem to get stuck at the subpallial-pallial boundary and typical migration streams fail to form. Mutant cells show atypical cell morphology with increased branching. Although mutant interneurons show all phases of normal migratory behavior, pauses in between translocation events are substantially longer and the overall migration rate is reduced. Interestingly, during these pauses normal cilia become rather dynamic and start to elongate, retract and rotate as if probing their environment. Mutant cells lack this type of dynamic cilia and are also unable to navigate through gradients of guidance cues. Rescue experiments show that the migration phenotype depends on the ciliary localization of ARL13B, further corroborating the notion that primary cilia are specifically required during the migration of interneurons to sense guidance cues (Higginbotham et al., 2012).

Obviously SHH is a prime candidate for such a guidance cue. It was shown that interneurons assemble a short SHH-responsive cilium during migration (Baudoin et al., 2012). Interestingly, translocation of the nucleus during the migration cycles is linked to the subcellular positioning of the centrosome and the associated primary cilium. The cilium is presented to the external environment at the leading edge and bundles of microtubules extend from the basal body/mother centriole. These observations suggest a mechanism whereby pulling forces administered along MT bundles cause the distantly located nucleus to move closer to the centrosome/cilium. After nuclear translocation the centrosome adopts a perinuclear location within the cytoplasm and the cilium disappears or is internalized along with a ciliary vesicle. The centrosome/cilium will again dock to the plasma membrane at the leading edge as a new migration cycle is initiated. Conditional inactivation of *KIF3A* and *IFT88* cause abnormal migratory behavior, whereby interneurons fail to colonize the cortical plate. In this context, SHH seems to function as an exit signal that tells migratory cells when to leave the deep tangential migratory stream and start to invade the cortical plate (Baudoin et al., 2012).

Cilia During Late Development and Adulthood

The functions of primary cilia discussed above are all associated with a specific developmental process such as patterning, proliferation or migration occurring during a restricted developmental time window. Whether early primary cilia are therefore only transient structures or persist and adopt new functions during later development has not been completely resolved.

Specialized cilia of sensory neurons play an important role in the formation and function of sensory organelles, which are an integral part of fundamental physiological processes like olfaction, hearing and vision throughout adulthood. For example, the Organ of Corti in the inner ear houses mechanosensory hair cells, which transduce sheering of the overlying tectorial membrane by distortion of apical hair bundles. Interconnected stereovilli of ascending length organize into a V-shaped bundle, whereby the longest one has a typical ciliary microtubular configuration. This so-called kinocilium is essential for the formation of the entire hair cell bundle. Olfactory sensory neurons (OSN) on the other hand typically possess 10 to 30 non-motile cilia per cell with varying microtubular configurations along the axoneme. The cilia of OSN are of remarkable length (about 50 μm on average in mammals) and are distinguished by the expression of only a single type of odorant GPCR per cell (Falk et al., 2015). The rod and cone photoreceptors of the vertebrate retina possess a connecting cilium, which bridges the light sensitive outer-segment to the energy-producing inner-segment. The short connecting cilium resembles a primary cilium in its microtubular organization and is vital to manage the logistics of the massive transport occurring between inner- and outer-segments. Sensory cilia show varying degrees of specialization regarding structure, composition, signal-transduction and transport, which make them well adapted for their specific tasks. As is apparent, we only touch upon the field of sensory cilia to complete the picture of ciliary occurrence in the brain. For more extensive discussions please consult book Chapters 7 and 8, about cilia in the inner ear and eye.

Early primary cilia of progenitor cells and migrating interneurons seem to be distinct from the neuronal cilia found for example on cortical neurons during late development and adulthood. Early primary cilia are rapidly assembled structures of rather short length (ranging between 1–2 µm) (Ashique et al., 2009; Besse et al., 2011; Gorivodsky et al., 2009; Willaredt et al., 2008), whereas ciliogenesis in maturing cortical neurons is a protracted process that starts during later neuronal development, after migration has been completed. During the time course of several weeks, the cilium continually grows from a rudimentary procilium into a cilium with an average length of about 5 µm (Arellano et al., 2012).

On the other hand, motile cilia start to appear on mature ependymal cells only after the regression of the ventricular zone (Del Bigio, 2010). During development, ependymal cells first grow a single motile cilium and consecutively generate further cilia, whereby the exact number and distribution follows a species dependent pattern. Generally, motile cilia share a common orientation relative to the central canal, which is believed to be important for the cilia-driven caudally directed movement of the CSF (reviewed in Hoyer-Fender, 2013).

Mature neuronal primary cilia

Despite the omnipresence of neuronal primary cilia in the maturing and adult brain, studies on the function of this organelle have only recently begun to accumulate. Post-mitotic interneurons of the isocortex and hippocampus initiate the formation of slow-growing primary cilia once migration has been completed and the cells have reached their final destinations (Arellano et al., 2012; Higginbotham et al., 2012; Kumamoto et al., 2012; Sarkisian and Guadiana, 2015). This might explain why in the mouse, disruption of ciliary components like Arl13b and Stumpy, do not cause lamination changes in the cortex (Arellano et al., 2012; Higginbotham et al., 2012).

The hippocampus is one of the regions where new neurons are generated continuously during adulthood. Newborn hippocampal dentate gyrus cells (DGCs) have to integrate into functional circuits of the granule cell layer, by growing appropriate dendritic arbors and synapses. Strikingly, ciliogenesis in DGCs coincides with the time window when newborn cells actively develop dendritic arbors and form glutamatergic synapses with projections from the entorhinal cortex. In order to investigate whether the cilia of newborn DGCs might be required for dendritic refinement and/or synapse formation, Kumamoto et al. (2012) specifically suppressed ciliogenesis in these cells using an inducible, retroviral expression system for a dominant-negative form of KIF3A. This kind of cell autonomous ablation of cilia leads to reduced dendritic refinement and to fewer glutamatergic synapses being formed. Transduced cells show an elevated level of Wnt/β-catenin signaling. Furthermore, suppression of Wnt signaling can rescue—while over-activation can induce—arboration changes. These results corroborate the notion that the correct synaptic integration of DGCs depends upon cilia-mediated Wnt regulation (Kumamoto et al., 2012).

Granule neurons of the dentate gyrus receive unidirectional input foremost from the entorhinal cortex and are known to play an essential role in learning and memory formation. Interestingly, conditional inactivation of *Ift88* in cortical and hippocampal neurons is associated with deficits in aversive memory and object recognition at the behavioral level. At the tissue and cellular level, loss of IFT88 causes altered electrophysiological properties, supporting the finding that cilia have an influence on synaptic properties (Berbari et al., 2014).

Given the predominant role of cilia in the transduction of SHH signaling during early development, it makes sense to ask whether primary cilia in the adult brain serve a similar function. Strikingly, in the adult brain, pathway components like SHH, SMO and GLI1 localize to sites harboring neuronal stem cell niches, namely the aforementioned region in the dentate gyrus, the subgranular zone (SGZ), and the subventricular zone (SVZ) of the lateral ventricle (reviewed in Álvarez-Buylla and Ihrie, 2014). Conditional inactivation of several SHH components during late development causes a marked decrease in neurogenesis and a size reduction of the SGZ and SVZ, suggesting that *SHH* has a role in the establishment of neural stem cell niches (Álvarez-Buylla and Ihrie, 2014; Balordi and Fishell, 2007a,b; Machold et al., 2003).

There is an intriguing similarity of postnatal phenotypes between *SHH*-type and ciliary mutants. Conditional, genetic ablation of *Kif3a*, *Ift20*, *Ift88* as well as *Stumpy* causes size reduction and disorganization of the mature SGZ and SVZ (Amador-Arjona et al., 2011; Breunig et al., 2008; Han et al., 2008). For the dentate gyrus, the defects can be traced back to disturbances in the proliferation of maturing astrocyte-like neuronal precursors, which build up a pool of neuronal stem cells during adult neurogenesis. The loss of primary cilia function is associated with the down-regulation of *SHH* target genes and subsequent defective SHH signaling. In addition, the proliferation phenotype cannot be rescued by ectopic activation of the SHH pathway in ciliary mutants (Breunig et al., 2008; Han et al., 2008). Therefore, there is strong evidence that cilia also take part in the regulation of SHH signaling during postnatal development. Disruption of cilia function perturbs SHH-dependent proliferation and development of neuronal stem cell precursors, which has an impact on subsequent neurogenesis. It is noteworthy though that this function is mediated by cilia of short length during late development, during a time interval before migration has been completed and the characteristically long primary cilia have completely formed.

The widespread distribution of adult neuronal cilia has been foremost characterized by the expression of adenylate cyclase III (ACIII) and somatostatin receptor 3 (SSTR3), which both are part of GPCR mediated signaling processes (Arellano et al., 2012; Bishop et al., 2007; Händel et al., 1999; Stanic et al., 2009). In general, GPCRs constitute a vast and diverse superfamily of transmembrane receptors that mediate cellular responses to diverse stimuli such as hormones, neurotransmitters, lipids, light, odors and morphogens. Ciliary GPCRs in the brain include among others well-known mediators of SHH signaling like SMO, PTCH1 and GPR161, various dopamine receptor isoforms, serotonin receptor 6 (HTR6), melanin concentrating hormone receptor 1 (MCHR1) and kisspetin receptor 1 (KISS1R) (Schou et al., 2015). Given this heterogeneity it can be assumed that neuronal, ciliary GPCRs influence various signaling networks with effects on a plethora of cellular functions, including the physiological processes controlled by the brain. For example, many of the hormone receptors mentioned above are typically present on specific hypothalamic neurons that are involved in the regulation of the energy status via the hypothalamic-pituitary-adrenal axis. Conditional interference of cilia function in subsets of hypothalamic neurons is associated with obesity and memory deficits (Einstein et al., 2010; Green et al., 2012; Omori et al., 2015). An additional level of complexity is conferred by the finding that certain ciliary GPCRs also colocalize. The colocalization and interaction of MCHR1 and SSTR3 within cilia of distinct hippocampal neurons presumably has functional consequences on GPCR signaling (Green et al., 2012). Generally, neuronal primary cilia that express specific sets of GPCRs are believed to bind neuromodulators and act as non-synaptic sensory and signaling organelles (Green and Mykytyn, 2014). Disturbance of ciliary signaling, including GPCRs, might thus explain some of the cognitive deficits seen in many ciliopathies.

Motile cilia in the mature brain

Motile cilia with a typical 9 + 2 microtubule configuration are characteristic in mature brain morphology and protrude from multiciliated ependymal cells that cover the ventricular linings (Del Bigio, 2010; Hoyer-Fender, 2013). Ciliated ependymal cells are found throughout the entire ventricular system, whereby the degree of ciliation shows regional and species dependent variation. For example, in various mammals the lining of the third ventricle is highly ciliated in the upper two thirds, where tufts of multiple cilia extend from each ependymal cell. In contrast, the remaining ventricular surface is increasingly sparsely ciliated (Spassky, 2013). Motile cilia of the ventricular lining beat in a coordinated manner to generate a constant laminar flow of the CSF.

Although the presence of multiciliated ependymal cells is a feature of the mature brain, no proliferation of ependymal cells has been observed in adult mice using different labeling techniques. Instead, during embryonic development the ependymal cell lineage originates (around E15 in mice) from a subpopulation of polarized radial glia cells that express both glial and ependymal markers (e.g., GLAST and S100β, respectively) and extend a primary cilium from their apical surface (Spassky et al., 2005). Around birth, deuterosomes, electron-dense masses that function as nucleation centers for basal body formation during multiciliogenesis, can be found in the cytoplasm of ependymal progenitor cells (Hoyer-Fender, 2013; Spassky et al., 2005). Shortly thereafter, a multitude of basal bodies detach from the deuterosome and dock to the apical membrane, where ciliary shafts with the typical 9 + 2 configuration start to grow and reach maturity within the second postnatal week (Spassky, 2013). Multiciliogenesis of ependymal cells includes several aspects of planar cell polarity. Firstly, in a process described as translational polarity, the initially evenly distributed basal bodies start to accumulate within the anterior aspect of the apical membrane (Hirota et al., 2010). Surprisingly, this event of planar cell polarity depends on the original polarization of the founder radial glia cell, which was set up by its primary cilium (Mirzadeh et al., 2010). Furthermore, when establishing rotational polarity the initially randomly oriented basal bodies start to reorient and align their basal feet in the direction of the effective cilia stroke. This process seems to depend in part on the hydrodynamic forces exerted on cilia by the flow of the CSF (Guirao et al., 2010). Through the events of translational and rotational polarity, ciliary beating becomes coordinated both on individual as well as neighboring cells. The result is a metachronal wave travelling along the ventricular lining that facilitates constant laminar flow of the CSF (Spassky, 2013). Genetic disruption of cilia function is often associated with the occurrence of hydrocephalus frequently caused by clogging of the cerebral aqueduct, which exemplifies the importance of ciliary movement for proper brain function. Motile cilia might have additional functions associated with the propelling of the CSF. For example, ependymal cilia are required for the formation of concentration gradients of guidance cues that help adult-born neuroblast to migrate from the subventricular zone to their final destination in the olfactory bulb (Sawamoto et al., 2006).

Ciliopathies with Brain Phenotypes

Ciliopathies are a group of genetic diseases caused by the disruption of ciliary structure and function. So far, mutations in more than a hundred ciliary genes leading to disease have been described. In recent years, these numbers have steadily increased with the availability and accessibility of next-generation sequencing in both clinical and research laboratories (Wheway et al., 2015). Many ciliopathies feature broad and pleiotropic phenotypes, which often include several organs that are affected. The spectrum of these symptoms vary from mild to very severe and may include retinal degeneration, *situs inversus*, infertility, cardiac defects, polydactyly, kidney cysts, hepatic fibrosis, early fetal lethality, obesity, laterality defects, skeletal defects, but also neurodevelopmental conditions and cognitive impairments.

Interestingly, a number of genes mutated in ciliopathy patients have functions not only in the cilium itself, but also in other cellular processes.

Particularly for ciliopathies with brain or cognitive phenotypes, whether the phenotype is caused by the ciliary or by non-ciliary function of the implicated genes and their protein products remains to be elucidated. Some of these proteins are implicated in non-ciliary processes, including mitotic spindle orientation and chromosome segregation, cell cycle and cytokinesis, DNA damage, immunological synapse formation and endosome trafficking (Vertii et al., 2015).

Most cells in the vertebrate brain have primary cilia, but there are also cells that have motile cilia, such as the ependymal cells lining the ventricles. The ventricles are filled with CSF secreted by the choroid plexus. Motile cilia in the ventricles play an important role during brain development, by coordinating the flow of CSF through constant beating. In the adult human brain, the CSF is replaced approximately three times per day: it removes metabolites and provides nutrients to the brain (Spassky, 2013). Malfunction of motile cilia of the ependymal cells may lead to hydrocephaly due to CSF accumulation.

While some ciliopathies have a brain-related or cognitive deficit as a hallmark, others do not display such a phenotype. In this section we briefly describe those syndromes and disorders that include a prominent brain or cognitive phenotype caused by the malfunction of cilia; these include Meckel-Gruber syndrome (MKS), Joubert syndrome (JBS), Nephronophthisis (NPHP) and Bardet-Biedl syndrome (BBS). We further expand the discussion and include dyslexia. Although dyslexia is not classified as a ciliopathy, several of the dyslexia candidate genes have demonstrated functions at the cilium. Children with dyslexia have specific difficulties in learning to read properly despite normal IQ, good learning opportunities as well as normal social conditions. Genetically, dyslexia is a complex disorder with about a dozen of genes linked to it.

Meckel-Gruber syndrome (MKS)

MKS is a lethal recessive neurodevelopmental condition displaying phenotypic variability and high heterogeneity. Several of the genes responsible for MKS show extensive allelism, as mutations in the respective genes are also associated with JBS, NPHP and BBS (Garcia-Gonzalo et al., 2011) (Table 1). The clinical phenotypes of MKS include cognitive impairment, optic nerve hypoplasia, occipital encephalocele (Figure 2A), but also bilateral polycystic kidneys and post axial polydactyly (Parelkar et al., 2013). The prevalence of MKS in the population varies between 1:13,000 and 1:140,000 live births with the highest incidence in Gujarati Indians, Kuwait Bedouins and Finns (Parelkar et al., 2013). Interestingly, several of the proteins encoded by genes mutated in MKS and JBS localize to a special compartment at the ciliary base known as the transition zone (TZ) (Figure 1A). These proteins form complexes known as the "MKS-JBS module" and are thought to regulate ciliogenesis and the ciliary membrane composition (Garcia-Gonzalo et al.,

Table 1. Ciliopathies with brain and cognitive phenotypes.[1]

Disease/disorder with ciliary phenotype	Brain phenotype/clinical features	Genes involved	References
Meckel-Gruber syndrome (MKS)	Posterior fossa abnormalities, corpus callosum defects, encephalocele, optic nerve hypoplasia, cognitive impairment	**MKS1, NPHP3, TCTN2,** B9D1, B9D2, **CEP290, TMEM67, TMEM216, TMEM231, CSPP1, RPGRIP1L,** CC2D2, **BBS2, BBS4, BBS6**	(Szymanska et al., 2014; Waters and Beales, 2011)
Joubert syndrome (JBS)	Cerebellar vermis hypoplasia or aplasia "molar tooth sign", developmental delay or mental retardation, ataxia, retinal dystrophy, oculomotor apraxia, cognitive impairment	INPP5E, TMEM237, TMEM138, **TMEM216, TMEM67, TMEM231,** AHIL, **NPHP1,** CEP41, ARL13B, OFD1, TTC21B, KIF7, TCTN1, **TCTN2,** TCTN3, **ZNF423,** C5orf42, **CSPP1,** PDE6D, **CEP290,** CC2D2A, **RPGRIP1L,** CXORF5, KIAA0556	(Sanders et al., 2015; Szymanska et al., 2014; Waters and Beales, 2011)
Nephronophthisis and related ciliopathies (NPHP-RC)	Learning disability, cerebellar hypoplasia, hydrocephalus	NPHP1, NPHP2/INV, **NPHP3,** NPHP4, NPHP5, **CEP290/NPHP6,** NPHP7/GLIS2, NPHP8, NEK8/NPHP9, NPHP10, **TMEM67/ NPHP11,** NPHP12, NPHP13/WDR19, ZNF423/NPHP14, CEP164/NPHP15, ANKS6/ NPHP16, **IFT172/NPHP17, RPGRIP1L/ NPHP18,** NPHP1L/XPNPEP3, NPHP2L, IFT20, DCDC2	(Waters and Beales, 2011; Wolf, 2015)
Bardet-Biedl syndrome (BBS)	Intellectual disability, cognitive impairment, cerebellar hypoplasia, retinal dystrophy	BBS1, **BBS2,** BBS3, **BBS4,** BBS5, **BBS6,** BBS7, BBS8, BBS9, BBS10, BBS11, BBS12, BBS15, **IFT172/BBS20,** MGC1203, CCDC28B, **TMEM67, MKS1, CEP290**	(Schaefer et al., 2016; Waters and Beales, 2011)
Dyslexia[2]	Unexpected difficulties in learning to read despite normal IQ; small anatomical defects reported in a few dyslexic human individuals; neuronal migration impairment during development in rats	DYX1C1, **DCDC2,** KIAA0319, ROBO1, C2Orf3, MRPL19, CYP19A1, PCNT, DIP2A, S100B, PRMT2, MC5R, DYM, NEDD4L, KIAA0319L, DGKI, CEP63	(Einarsdottir et al., 2015; Kere, 2014; Scerri and Schulte-Korne, 2010)

[1]The genes marked in bold are mutated in several of the ciliopathies presented in this Table.
[2]Dyslexia is not classified as a ciliopathy, but several of the genes associated with the disorder function at or are localized to cilia.

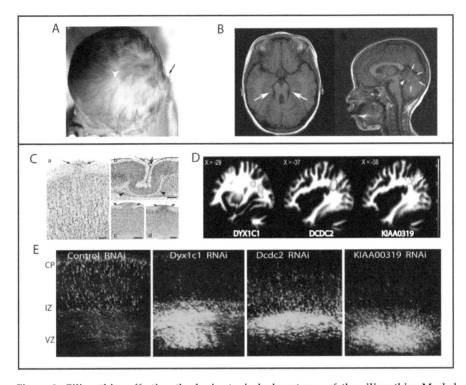

Figure 2. Ciliopathies affecting the brain: typical phenotypes of the ciliopathies Meckel Gruber syndrome (MKS) and Joubert syndrome (JBS), and of dyslexia. (A) Encephalocele in an MKS child, the black arrrow shows the protrusion of the occiput of the patient skull. (B) MRI images of the brain from a JBS patient with the typical "molar tooth sign". The white arrows in the axial image show the elongated superior cerebellar peduncles and the white arrowhead the deepened interpeduncular fossa. In the sagittal MRI the small white arrows show the hypoplastic cerebellar vermis. Panels (A) and (B) were reproduced with permission (from Parisi, 2009). In panels (C) to (E) phenotypes associated with dyslexia and its candidate genes are shown. (C) **a.** Ectopia in the post mortem brain of a dyslexic individual. **b.** Microgyria in the brain of a rat. **c.** Ectopia in the brain of an immune-defective mouse. **d.** Ectopia in a rat brain in which the dyslexia candidate gene *dyx1c1* was silenced by *in utero* electroporation of siRNA. Images were reproduced with permission (from Galaburda et al., 2006). (D) MRI images showing white matter clusters of dyslexia associated single nucleotide polymorphisms (SNPs) from the dyslexia candidate genes *DYX1C1, DCDC2* and *KIAA0319*. Images were reproduced with permission (Darki et al., 2012). (E) Neuronal migration impairment in the brains of rat embryos caused by RNAi knockdown of the dyslexia candidate genes *Dyx1c1, Dcdc2* and *Kiaa0319*. Cortical plate (CP), intermediate zone (IZ) and ventricular zone (VZ) are indicated. Images were reproduced with permission (from Gabel et al., 2010).

2011). Similarly, further protein studies have shown that NPHP proteins also form complexes at the TZ (Garcia-Gonzalo et al., 2011; Sang et al., 2011). The TZ plays the role of a gatekeeper for proteins that enter and exit the cilium and several of the proteins mutated in different ciliopathies are located at the TZ (Szymanska et al., 2014).

Joubert syndrome (JBS)

JBS is a developmental brain disorder with recessive inheritance characterized by heterotopias in several brain regions and compromised brain tissue architecture (Juric-Sekhar et al., 2012). The hallmark of JBS is the "molar tooth sign" detected by axial brain magnetic resonance imaging (MRI), which depicts the hypoplasia of the cerebellar vermis with deepened interpeduncular fossa and elongated superior cerebellar peduncles (Parisi, 2009) (Figure 2B). JBS is often associated with intellectual disability and autism spectrum disorders. Other clinical manifestations are polydactyly, cleft lip, seizures and retinal degeneration (Yuan and Sun, 2013). The estimated incidence of JBS in the United States is 1:100,000 (Parisi, 2009). JBS displays a very large phenotypic heterogeneity even inside affected families, probably due to environmental modifier effects (Lee and Gleeson, 2010). To date, twenty-five genes have been associated with JBS (Table 1) and as mentioned above, there is strong allelism between JBS and MKS with several genes mutated in both conditions, including TCTN2, CEP290, TMEM67, TMEM216, TMEM231, CSPP1, RPGRIP1L and CC2D2A. Some of these genes are also mutated in NPHP such as CEP290, NPHP1, TMEM67, TMEM216, RPGRIP1L and CC2D2A (Table 1). The proteins encoded by the genes mutated in JBS function in a variety of aspects of ciliary structure and function.

Nephronophthisis (NPHP)

NPHP is a rare autosomal, recessive disorder characterized by kidney tubular cysts, polyuria, polydipsia, secondary enuresis and anemia (Wolf, 2015). Brain-related phenotypes include learning disabilities, cerebellar hypoplasia and hydrocephalus. The incidence varies in different populations and it is estimated to be 1:50,000 in Canada, 1 > 60,000 in Finland and $1 \leq 1$ million in the United States (Simms et al., 2011). Although rare, NPHP represent the most frequent genetic cause of end-stage kidney disease in the first three decades of life (Hildebrandt and Zhou, 2007). Approximately 15–20% of the NPHP cases have other extrarenal phenotypes, the most common being ophthalmological defects such as retinitis pigmentosa (also known as Senior-Løken syndrome) or oculomotor apraxia type Cogan; other extrarenal phenotypes are liver fibrosis, mental retardation and ataxia. This group of disorders is known as the NPHP-related ciliopathies (NPHP-RC) (Wolf, 2015). To date, 22 genes causing NPHP have been identified (Table 1) and all together are responsible for only one third of all the NPHP cases, showing the high heterogeneity of the disorder. Interestingly, almost all the proteins encoded by these genes localize to the basal body and the centrosome. An exception is NPHP1L/XPNPEP3, which localizes to mitochondria (O'Toole et al., 2010).

Bardet-Biedl syndrome (BBS)

BBS is an autosomal, recessive disorder characterized by obesity, retinitis pigmentosa, polydactyly, hypogonadism, renal failure and cognitive and learning impairment. BBS is a genetically heterogeneous disorder with more than a dozen of genes identified that upon mutation are linked to the disease. The incidence varies among populations with an incidence of 1:160,000 in Northern Europe and 1:13,500 in Kuwait (Forsythe and Beales, 2013). Interestingly, several of the proteins encoded by these genes interact with each other forming what is known as the BBSome protein complex. The complex includes BBS1, BBS2, BBS4, BBS5, BBS7, BBS8, BBS9 and BBIP10; all these proteins were identified as BBS4-associated proteins by tandem affinity purification followed by mass spectrometry (Loktev et al., 2008; Nachury et al., 2007). The BBSome protein complex localizes to a specific ciliary compartment, the basal body; it also plays an important role in the assembly and maintenance of the IFT transport system (Figure 1A). Remarkably, BBSome proteins are evolutionarily conserved among all ciliated organisms, but are absent from non-ciliated organisms such as plants, yeast and amoebae (Jin and Nachury, 2009), underscoring the importance of the BBSome for cilia structure and function. Later studies have reproduced human mutations in rodent models and those displayed BBS-like phenotypes, including thinning of the cerebral cortex, enlargement of the lateral and third ventricles, and structural changes of cilia. BBS as a ciliopathy is thus of particular interest in the light of recent studies that have implicated primary neuronal cilia in neuronal signal transduction (Agassandian et al., 2014).

Cilia and Dyslexia

Interestingly, some of the genes that have previously been linked to dyslexia encode proteins functioning in cilia, such as *DYX1C1*, *DCDC2* and *KIAA0319* (Kere, 2014). These genes were originally identified as dyslexia candidate genes through classical positional cloning (Cope et al., 2005; Schumacher et al., 2006; Taipale et al., 2003). Later, RNAi-mediated gene silencing in rat embryos revealed that the proteins are involved in neuronal migration (Gabel et al., 2010) (Figure 2E); in the case of *DYX1C1* one could even observe brain ectopia (Galaburda et al., 2006) (Figure 2C).

The first observations that connect dyslexia candidate genes and cilia came when Massinen et al. (2011) reported that the gene *DCDC2* affects the length and signaling of cilia in neurons. This information has opened new and unexpected insights into the etiology of dyslexia. Further, gene expression studies from two independent groups pointed to the same direction. In the first study, an *in silico* analysis of gene expression differences between multi-ciliated and less ciliated tissues was performed and three of the known dyslexia candidate genes *DYX1C1*, *DCDC2* and *KIAA0319* were found to be significantly upregulated (Ivliev et al., 2012). In a second study, Hoh et al. (2012) performed a very interesting experiment in which the transcriptome of mouse tracheal epithelial cells was studied during the process of ciliogenesis. As a result, several dyslexia candidate genes were differentially expressed including *DYX1C1, DCDC2, KIAA0319, ROBO1, S100B, ACOT13, PRMT2, TTRAP* and *PCNT*. The link between dyslexia candidate genes and cilia function has been further strengthened by recent work in cell, tissue and animal models. It was demonstrated that in different human cell models the genes *DYX1C1* and *DCDC2* are regulated by ciliogenic RFX factors (Tammimies et al., 2016), and that the encoded proteins localize to cilia (Massinen et al., 2011; Schueler et al., 2015; Tammimies et al., 2016) (Figure 1C–F). In another example, the morpholino knockdown of *Dyx1c1* and *Dcdc2* in zebrafish results in typical ciliopathy-related phenotypes, characterized by a ventrally curved body axis, kinky tails, hydrocephalus, *situs inversus* and kidney cysts (Chandrasekar et al., 2013; Schueler et al., 2015). In mouse, knockout of *Dyx1c1* leads to a strong phenotype that includes *situs inversus totalis* and hydrocephalus, whereby this phenotype resembles some of the characteristics of primary ciliary dyskinesia (PCD) (Tarkar et al., 2013). Recent studies in humans show that patients with different ciliopathies such as primary cilia dyskinesia (PCD) (Casey et al., 2015; Tarkar et al., 2013), nephronopthisis-related ciliopathies (NPHP-RC) (Schueler et al., 2015) and deafness also bear mutations in *DYX1C1* and *DCDC2* (Grati et al., 2015). In yet another interesting example, MRI studies of school-age children from Sweden, demonstrates that genetic variants in *DYX1C1, DCDC2* and *KIAA0319* have an effect on white matter volume in temporo-parietal regions of the brain, whereby the affected regions overlap (Darki et al., 2012) (Figure 2D). Interestingly, the regions affected have been shown to be less active in dyslexic individuals (Paulesu et al., 2001).

Another type of study investigated candidate genes for more severe neuropsychiatric disorders such as schizophrenia, autism, bipolar disorder, but also for intellectual disability (ID). The possible cellular function of the proteins encoded by these genes was looked at by RNAi in the ciliated cell line NIH3T3. It was found that 23 out of 41 genes converge on cilia, 20 genes reduced ciliation and 3 increased cilia length (Marley and von Zastrow, 2012). Yet another example is the gene *PQBP1* linked to ID. PQBP1 protein localizes to neuronal cilia and knockdown of the gene by RNAi leads to a decrease in ciliated neurons in cells of the hippocampus (Ikeuchi et al., 2013). These results show a clear link between ciliary function and brain phenotypes. Moreover, primary cilia may act as a cellular "node" on to which neuropsychiatric protein function converges.

Conclusion

In the last decade the number of diseases categorized as ciliopathies has strongly increased and based on the data presented above, it is very appealing to categorize dyslexia as a new ciliopathy. However, there are still some difficulties into making such a statement. While the experimental connection between dyslexia candidate genes and cilia is strong, the connection between cilia and dyslexia is still tenuous. Most of the ciliopathies are rare diseases with high pleiotropy and heterogeneity, while dyslexia is a relatively common, yet genetically complex disorder. Additionally, loss-of-function mutations in the genes *DYX1C1* and *DCDC2* found in ciliopathy patients without any evidence of dyslexia suggest that the mechanisms involved in the etiology of dyslexia are complex and work at different levels. Thus, further functional and genetic studies are needed to elucidate the underlying cellular mechanisms of dyslexia and the role that cilia play in the etiology of this human brain disorder.

Acknowledgements

We thank Andrea Bieder for critically reviewing this text and for help with the figures. Work in the laboratory of P.S. was supported by the Swedish Research Council, the Torsten Söderberg and Åhlén Foundations and by the Karolinska Institute Strategic Neurosciences Program.

References

Agassandian, K., Patel, M., Agassandian, M., Steren, K. E., Rahmouni, K., Sheffield, V. C. and Card, J. P. 2014. Ciliopathy is differentially distributed in the brain of a Bardet-Biedl syndrome mouse model. PLoS One 9: e93484.

Álvarez-Buylla, A. and Ihrie, R. A. 2014. Sonic hedgehog signaling in the postnatal brain. Seminars in Cell & Developmental Biology 33: 105–111.

Amador-Arjona, A., Elliott, J., Miller, A., Ginbey, A., Pazour, G. J., Enikolopov, G., Roberts, A. J. and Terskikh, A. V. 2011. Primary cilia regulate proliferation of amplifying progenitors in adult hippocampus: implications for learning and memory. J. Neurosci. 31: 9933–9944.

Arellano, J. I., Guadiana, S. M., Breunig, J. J., Rakic, P. and Sarkisian, M. R. 2012. Development and distribution of neuronal cilia in mouse neocortex. J. Comp. Neurol. 520: 848–873.

Ashique, A. M., Choe, Y., Karlen, M., May, S. R., Phamluong, K., Solloway, M. J., Ericson, J. and Peterson, A. S. 2009. The Rfx4 transcription factor modulates Shh signaling by regional control of ciliogenesis. Sci. Signal 2, ra70.

Baas, D., Meiniel, A., Benadiba, C., Bonnafe, E., Meiniel, O., Reith, W. and Durand, B. 2006. A deficiency in RFX3 causes hydrocephalus associated with abnormal differentiation of ependymal cells. Eur. J. Neurosci. 24: 1020–1030.

Balordi, F. and Fishell, G. 2007a. Hedgehog signaling in the subventricular zone is required for both the maintenance of stem cells and the migration of newborn neurons. J. Neurosci. 27: 5936–5947.

Balordi, F. and Fishell, G. 2007b. Mosaic removal of hedgehog signaling in the adult SVZ reveals that the residual wild-type stem cells have a limited capacity for self-renewal. J. Neurosci. 27: 14248–14259.

Baudoin, J. P., Viou, L., Launay, P. S., Luccardini, C., Espeso Gil, S., Kiyasova, V., Irinopoulou, T., Alvarez, C., Rio, J. P., Boudier, T. et al. 2012. Tangentially migrating neurons assemble a primary cilium that promotes their reorientation to the cortical plate. Neuron. 76: 1108–1122.

Benadiba, C., Magnani, D., Niquille, M., Morle, L., Valloton, D., Nawabi, H., Ait-Lounis, A., Otsmane, B., Reith, W., Theil, T. et al. 2012. The ciliogenic transcription factor RFX3 regulates early midline distribution of guidepost neurons required for corpus callosum development. PLoS Genet 8: e1002606.

Berbari, N. F., Malarkey, E. B., Yazdi, S. M., McNair, A. D., Kippe, J. M., Croyle, M. J., Kraft, T. W. and Yoder, B. K. 2014. Hippocampal and cortical primary cilia are required for aversive memory in mice. PLoS One 9: e106576.

Bergquist, H. 1952. Studies on the cerebral tube in vertebrates. The neuromeres. Acta Zoologica 33: 117–187.

Bergquist, H. and Källén, B. 1954. Notes on the early histogenesis and morphogenesis of the central nervous system in vertebrates. J. Comp. Neurol. 100: 627–659.

Besse, L., Neti, M., Anselme, I., Gerhardt, C., Ruther, U., Laclef, C. and Schneider-Maunoury, S. 2011. Primary cilia control telencephalic patterning and morphogenesis via Gli3 proteolytic processing. Development 138: 2079–2088.

Bishop, G. A., Berbari, N. F., Lewis, J. and Mykytyn, K. 2007. Type III adenylyl cyclase localizes to primary cilia throughout the adult mouse brain. J. Comp. Neurol. 505: 562–571.

Blackshear, P. J., Graves, J. P., Stumpo, D. J., Cobos, I., Rubenstein, J. L. and Zeldin, D. C. 2003. Graded phenotypic response to partial and complete deficiency of a brain-specific transcript variant of the winged helix transcription factor RFX4. Development 130: 4539–4552.

Blaess, S., Stephen, D. and Joyner, A. L. 2008. Gli3 coordinates three-dimensional patterning and growth of the tectum and cerebellum by integrating Shh and Fgf8 signaling. Development 135: 2093–2103.

Breunig, J. J., Sarkisian, M. R., Arellano, J. I., Morozov, Y. M., Ayoub, A. E., Sojitra, S., Wang, B., Flavell, R. A., Rakic, P. and Town, T. 2008. Primary cilia regulate hippocampal neurogenesis by mediating sonic hedgehog signaling. Proc. Natl. Acad. Sci. USA 105: 13127–13132.

Briscoe, J., Pierani, A., Jessel, T. M. and Ericson, J. 2000. A homeodomain protein code specifies progenitor cell identity and neuronal fate in the ventral neural tube. Cell 101: 435–445.

Casey, J. P., McGettigan, P. A., Healy, F., Hogg, C., Reynolds, A., Kennedy, B. N., Ennis, S., Slattery, D. and Lynch, S. A. 2015. Unexpected genetic heterogeneity for primary ciliary dyskinesia in the Irish Traveller population. Eur. J. Hum. Genet. 23: 210–217.

Caspary, T. and Anderson, K. V. 2003. Patterning cell types in the dorsal spinal cord: what the mouse mutants say. Nat. Rev. Neurosci. 4: 289–297.

Caspary, T., Larkins, C. E. and Anderson, K. V. 2007. The graded response to Sonic Hedgehog depends on cilia architecture. Dev. Cell 12: 767–778.

Cavodeassi, F. and Houart, C. 2012. Brain regionalization: of signaling centers and boundaries. Dev. Neurobiol. 72: 218–233.

Chandrasekar, G., Vesterlund, L., Hultenby, K., Tapia-Paez, I. and Kere, J. 2013. The zebrafish orthologue of the dyslexia candidate gene DYX1C1 is essential for cilia growth and function. PLoS One 8: e63123.

Chizhikov, V. V., Davenport, J., Zhang, Q., Shih, E. K., Cabello, O. A., Fuchs, J. L., Yoder, B. K. and Millen, K. J. 2007. Cilia proteins control cerebellar morphogenesis by promoting expansion of the granule progenitor pool. J. Neurosci. 27: 9780–9789.

Choksi, S. P., Lauter, G., Swoboda, P. and Roy, S. 2014. Switching on cilia: transcriptional networks regulating ciliogenesis. Development 141: 1427–1441.

Christensen, S. T., Clement, C. A., Satir, P. and Pedersen, L. B. 2012. Primary cilia and coordination of receptor tyrosine kinase (RTK) signalling. J. Pathol. 226: 172–184.

Cohen, E. and Meininger, V. 1987. Ultrastructural analysis of primary cilium in the embryonic nervous tissue of mouse. International Journal of Developmental Neuroscience 5: 43–51.

Cope, N., Harold, D., Hill, G., Moskvina, V., Stevenson, J., Holmans, P., Owen, M. J., O'Donovan, M. C. and Williams, J. 2005. Strong evidence that KIAA0319 on chromosome 6p is a susceptibility gene for developmental dyslexia. Am. J. Hum. Genet. 76: 581–591.

Corbit, K. C., Aanstad, P., Singla, V., Norman, A. R., Stainier, D. Y. and Reiter, J. F. 2005. Vertebrate Smoothened functions at the primary cilium. Nature 437: 1018–1021.

D'Angelo, A., De Angelis, A., Avallone, B., Piscopo, I., Tammaro, R., Studer, M. and Franco, B. 2012. Ofd1 controls dorso-ventral patterning and axoneme elongation during embryonic brain development. PLoS One 7: e52937.

Dahl, H. A. 1963. Fine structure of cilia in rat cerebral cortex. Zeitschrift für Zellforschung 60: 369–386.

Darki, F., Peyrard-Janvid, M., Matsson, H., Kere, J. and Klingberg, T. 2012. Three dyslexia susceptibility genes, DYX1C1, DCDC2, and KIAA0319, affect temporo-parietal white matter structure. Biol. Psychiatry 72: 671–676.

Del Bigio, M. R. 2010. Ependymal cells: biology and pathology. Acta Neuropathol. 119: 55–73.

Doroquez, D. B., Berciu, C., Anderson, J. R., Sengupta, P. and Nicastro, D. 2014. A high-resolution morphological and ultrastructural map of anterior sensory cilia and glia in Caenorhabditis elegans. eLife 3: e01948.

Dubreuil, V., Marzesco, A. M., Corbeil, D., Huttner, W. B. and Wilsch-Brauninger, M. 2007. Midbody and primary cilium of neural progenitors release extracellular membrane particles enriched in the stem cell marker prominin-1. J. Cell Biol. 176: 483–495.

Einarsdottir, E., Svensson, I., Darki, F., Peyrard-Janvid, M., Lindvall, J. M., Ameur, A., Jacobsson, C., Klingberg, T., Kere, J. and Matsson, H. 2015. Mutation in CEP63 co-segregating with developmental dyslexia in a Swedish family. Hum. Genet. 134: 1239–1248.

Einstein, E. B., Patterson, C. A., Hon, B. J., Regan, K. A., Reddi, J., Melnikoff, D. E., Mateer, M. J., Schulz, S., Johnson, B. N. and Tallent, M. K. 2010. Somatostatin signaling in neuronal cilia is critical for object recognition memory. J. Neurosci. 30: 4306–4314.

Falk, N., Losl, M., Schroder, N. and Giessl, A. 2015. Specialized cilia in mammalian sensory systems. Cells 4: 500–519.

Fawcett, D. W. and Porter, K. R. 1954. A study on the fine structure of ciliated epithelia. Journal of Morphology 94: 221–281.

Ferran, J. L., Sanchez-Arrones, L., Sandoval, J. E. and Puelles, L. 2007. A model of early molecular regionalization in the chicken embryonic pretectum. J. Comp. Neurol. 505: 379–403.

Forsythe, E. and Beales, P. L. 2013. Bardet-Biedl syndrome. Eur. J. Hum. Genet. 21: 8–13.

Fotaki, V., Yu, T., Zaki, P. A., Mason, J. O. and Price, D. J. 2006. Abnormal positioning of diencephalic cell types in neocortical tissue in the dorsal telencephalon of mice lacking functional Gli3. J. Neurosci. 26: 9282–9292.

Frings, S. 2012. Sensory cells and sensory organs. pp. 5–22. In: Barth, F. G., Giampieri-Deutsch, P. and Klain, H.-D. (eds). Sensory Perception: Mind and Matter (Vienna, Austria: Springer-Verlag).

Fuchs, J. L. and Schwark, H. D. 2004. Neuronal primary cilia: a review. Cell Biol. Int. 28: 111–118.

Gabel, L. A., Gibson, C. J., Gruen, J. R. and LoTurco, J. J. 2010. Progress towards a cellular neurobiology of reading disability. Neurobiol. Dis. 38: 173–180.

Galaburda, A. M., LoTurco, J., Ramus, F., Fitch, R. H. and Rosen, G. D. 2006. From genes to behavior in developmental dyslexia. Nat. Neurosci. 9: 1213–1217.

Garcia-Gonzalo, F. R., Corbit, K. C., Sirerol-Piquer, M. S., Ramaswami, G., Otto, E. A., Noriega, T. R., Seol, A. D., Robinson, J. F., Bennett, C. L., Josifova, D. J. et al. 2011. A transition zone complex regulates mammalian ciliogenesis and ciliary membrane composition. Nature Genetics 43: 776–784.

Goetz, S. C. and Anderson, K. V. 2010. The primary cilium: a signalling centre during vertebrate development. Nat. Rev. Genet. 11: 331–344.

Gorivodsky, M., Mukhopadhyay, M., Wilsch-Braeuninger, M., Phillips, M., Teufel, A., Kim, C., Malik, N., Huttner, W. and Westphal, H. 2009. Intraflagellar transport protein 172 is essential for primary cilia formation and plays a vital role in patterning the mammalian brain. Dev. Biol. 325: 24–32.

Grati, M., Chakchouk, I., Ma, Q., Bensaid, M., Desmidt, A., Turki, N., Yan, D., Baanannou, A., Mittal, R., Driss, N. et al. 2015. A missense mutation in DCDC2 causes human recessive deafness DFNB66, likely by interfering with sensory hair cell and supporting cell cilia length regulation. Hum. Mol. Genet. 24: 2482–2491.

Green, J. A., Gu, C. and Mykytyn, K. 2012. Heteromerization of ciliary G protein-coupled receptors in the mouse brain. PLoS One 7: e46304.

Green, J. A. and Mykytyn, K. 2014. Neuronal primary cilia: an underappreciated signaling and sensory organelle in the brain. Neuropsychopharmacology 39: 244–245.

Guemez-Gamboa, A., Coufal, N. G. and Gleeson, J. G. 2014. Primary cilia in the developing and mature brain. Neuron. 82: 511–521.

Guirao, B., Meunier, A., Mortaud, S., Aguilar, A., Corsi, J. M., Strehl, L., Hirota, Y., Desoeuvre, A., Boutin, C., Han, Y. G. et al. 2010. Coupling between hydrodynamic forces and planar cell polarity orients mammalian motile cilia. Nat. Cell Biol. 12: 341–350.

Guo, J., Higginbotham, H., Li, J., Nichols, J., Hirt, J., Ghukasyan, V. and Anton, E. S. 2015. Developmental disruptions underlying brain abnormalities in ciliopathies. Nat. Commun. 6: 7857.

Han, Y. G., Spassky, N., Romaguera-Ros, M., Garcia-Verdugo, J. M., Aguilar, A., Schneider-Maunoury, S. and Alvarez-Buylla, A. 2008. Hedgehog signaling and primary cilia are required for the formation of adult neural stem cells. Nat. Neurosci. 11: 277–284.

Händel, M., Schulz, S., Stanarius, A., Schreff, M., Erdmann-Vouliotis, M., Schmidt, H., Wolf, G. and Höllt, V. 1999. Selective targeting of somatostatin receptor 3 to neuronal cilia. Neuroscience 89: 909–926.

Hauptmann, G. and Gerster, T. 2000. Regulatory gene expression patterns reveal transverse and longitudinal subdivisions of the embryonic zebrafish forebrain. Mech. Dev. 91: 105–118.

Haycraft, C. J., Banizs, B., Aydin-Son, Y., Zhang, Q., Michaud, E. J. and Yoder, B. K. 2005. Gli2 and Gli3 localize to cilia and require the intraflagellar transport protein polaris for processing and function. PLoS Genet 1: e53.

Herrick, C. 1910. The morphology of the forebrain in amphibia and reptilia. J. Comp. Neurol. 20: 413–547.

Higginbotham, H., Eom, T. Y., Mariani, L. E., Bachleda, A., Hirt, J., Gukassyan, V., Cusack, C. L., Lai, C., Caspary, T. and Anton, E. S. 2012. Arl13b in primary cilia regulates the migration and placement of interneurons in the developing cerebral cortex. Dev. Cell 23: 925–938.

Higginbotham, H., Guo, J., Yokota, Y., Umberger, N. L., Su, C. Y., Li, J., Verma, N., Hirt, J., Ghukasyan, V., Caspary, T. et al. 2013. Arl13b-regulated cilia activities are essential for polarized radial glial scaffold formation. Nat. Neurosci. 16: 1000–1007.

Hildebrandt, F. and Zhou, W. 2007. Nephronophthisis-associated ciliopathies. J. Am. Soc. Nephrol. 18: 1855–1871.

Hirota, Y., Meunier, A., Huang, S., Shimozawa, T., Yamada, O., Kida, Y. S., Inoue, M., Ito, T., Kato, H., Sakaguchi, M. et al. 2010. Planar polarity of multiciliated ependymal cells involves the anterior migration of basal bodies regulated by non-muscle myosin II. Development 137: 3037–3046.

Hoh, R. A., Stowe, T. R., Turk, E. and Stearns, T. 2012. Transcriptional program of ciliated epithelial cells reveals new cilium and centrosome components and links to human disease. PLoS One 7: e52166.

Houart, C., Westerfield, M. and Wilson, S. 1998. A small population of anterior cells patterns the forebrian during zebrafish gastrulation. Nature 391: 788–792.

Hoyer-Fender, S. 2013. Primary and motile cilia: Their ultrastructure and ciliogenesis. pp. 1–53. In: Caspary, T. (ed.). Cilia and Nervous System Development and Function (Dordrecht: Springer).

Hozumi, A., Horie, T. and Sasakura, Y. 2015. Neuronal map reveals the highly regionalized pattern of the juvenile central nervous system of the ascidian Ciona intestinalis. Dev. Dyn. 244: 1375–1393.

Huang, X., Liu, J., Ketova, T., Fleming, J. T., Grover, V. K., Cooper, M. K., Litingtung, Y. and Chiang, C. 2010. Transventricular delivery of Sonic hedgehog is essential to cerebellar ventricular zone development. Proc. Natl. Acad. Sci. USA 107: 8422–8427.

Huangfu, D., Liu, A., Rakeman, A., Murcia, N. S., Niswander, L. and Anderson, K. V. 2003. Hedgehog signalling in the mouse requires intraflagellar transport proteins. Nature 426: 83–87.

Ikeuchi, Y., de la Torre-Ubieta, L., Matsuda, T., Steen, H., Okazawa, H. and Bonni, A. 2013. The XLID protein PQBP1 and the GTPase Dynamin 2 define a signaling link that orchestrates ciliary morphogenesis in postmitotic neurons. Cell Rep. 4: 879–889.

Imai, J. H. and Meinertzhagen, I. A. 2007a. Neurons of the ascidian larval nervous system in Ciona intestinalis: I. Central nervous system. J. Comp. Neurol. 501: 316–334.

Imai, J. H. and Meinertzhagen, I. A. 2007b. Neurons of the ascidian larval nervous system in Ciona intestinalis: II. Peripheral nervous system. J. Comp. Neurol. 501: 335–352.

Inglis, P. N., Ou, G., Leroux, M. R. and Scholey, J. M. 2007. The Sensory Cilia of Caenorhabditis Elegans. WormBook, 1–22.

Ishikawa, H. and Marshall, W. F. 2011. Ciliogenesis: building the cell's antenna. Nat. Rev. Mol. Cell Biol. 12: 222–234.

Ivliev, A. E., t Hoen, P. A., van Roon-Mom, W. M., Peters, D. J. and Sergeeva, M. G. 2012. Exploring the transcriptome of ciliated cells using *in silico* dissection of human tissues. PLoS One 7: e35618.

Jana, S. C., Bettencourt-Dias, M., Durand, B. and Megraw, T. L. 2016. Drosophila melanogaster as a model for basal body research. Cilia 5: 22.

Jin, H. and Nachury, M. V. 2009. The BBSome. Curr. Biol. 19: R472–473.

Juric-Sekhar, G., Adkins, J., Doherty, D. and Hevner, R. F. 2012. Joubert syndrome: brain and spinal cord malformations in genotyped cases and implications for neurodevelopmental functions of primary cilia. Acta Neuropathol. 123: 695–709.

Källén, B. 1952. Notes on the proliferation processes in the neuromeres in vertebrate embryos. Acta Societatis Medicorum Upsaliensis 57: 111–118.

Kere, J. 2014. The molecular genetics and neurobiology of developmental dyslexia as model of a complex phenotype. Biochem. Biophys. Res. Commun. 452: 236–243.

Kim, J., Kato, M. and Beachy, P. A. 2009. Gli2 trafficking links Hedgehog-dependent activation of Smoothened in the primary cilium to transcriptional activation in the nucleus. Proc. Natl. Acad. Sci. USA 106: 21666–21671.

Kim, J., Lee, J. E., Heynen-Genel, S., Suyama, E., Ono, K., Lee, K., Ideker, T., Aza-Blanc, P. and Gleeson, J. G. 2010. Functional genomic screen for modulators of ciliogenesis and cilium length. Nature 464: 1048–1051.

Kim, J., Hsia, E. Y. C., Brigui, A., Plessis, A., Beachy, P. A. and Zheng, X. 2015. The role of ciliary trafficking in Hedgehog receptor signaling. Sci. Signal 8: ra55.

Konno, A., Kaizu, M., Hotta, K., Horie, T., Sasakura, Y., Ikeo, K. and Inaba, K. 2010. Distribution and structural diversity of cilia in tadpole larvae of the ascidian Ciona intestinalis. Dev. Biol. 337: 42–62.

Kumamoto, N., Gu, Y., Wang, J., Janoschka, S., Takemaru, K., Levine, J. and Ge, S. 2012. A role for primary cilia in glutamatergic synaptic integration of adult-born neurons. Nat. Neurosci. 15: 399–405, S391.

Lauter, G., Söll, I. and Hauptmann, G. 2013. Molecular characterization of prosomeric and intraprosomeric subdivisions of the embryonic zebrafish diencephalon. J. Comp. Neurol. 521: 1093–1118.

Lee, J. H. and Gleeson, J. G. 2010. The role of primary cilia in neuronal function. Neurobiol. Dis. 38: 167–172.

Lehtinen, M. K., Zappaterra, M. W., Chen, X., Yang, Y. J., Hill, A. D., Lun, M., Maynard, T., Gonzalez, D., Kim, S., Ye, P. et al. 2011. The cerebrospinal fluid provides a proliferative niche for neural progenitor cells. Neuron. 69: 893–905.

Liem, K. F., Jr., Ashe, A., He, M., Satir, P., Moran, J., Beier, D., Wicking, C. and Anderson, K. V. 2012. The IFT-A complex regulates Shh signaling through cilia structure and membrane protein trafficking. J. Cell Biol. 197: 789–800.

Loktev, A. V., Zhang, Q., Beck, J. S., Searby, C. C., Scheetz, T. E., Bazan, J. F., Slusarski, D. C., Sheffield, V. C., Jackson, P. K. and Nachury, M. V. 2008. A BBSome subunit links ciliogenesis, microtubule stability and acetylation. Dev. Cell 15: 854–865.

Macdonald, R., Xu, Q., Barth, K. A., Mikkola, I., Holder, N., Fjose, A., Krauss, S. and Wilson, S. W. 1994. Regulatory gene expression boundaries demarcate sites of neuronal differentiation in the embryonic zebrafish forebrain. Neuron. 13: 1039–1053.

Machold, R., Hayashi, S., Rutlin, M., Muzumdar, D., Nery, S., Corbin, J. G., Gritli-Linde, A., Dellovade, T., Porter, J. A., Rubin, L. L. et al. 2003. Sonic hedgehog is required for progenitor cell maintenance in telencephalic stem cell niches. Neuron. 39: 937–950.

Marley, A. and von Zastrow, M. 2012. A simple cell-based assay reveals that diverse neuropsychiatric risk genes converge on primary cilia. PLoS One 7: e46647.

Massinen, S., Hokkanen, M. E., Matsson, H., Tammimies, K., Tapia-Paez, I., Dahlstrom-Heuser, V., Kuja-Panula, J., Burghoorn, J., Jeppsson, K. E., Swoboda, P. et al. 2011. Increased expression of the dyslexia candidate gene DCDC2 affects length and signaling of primary cilia in neurons. PLoS One 6: e20580.

Mirzadeh, Z., Han, Y. G., Soriano-Navarro, M., Garcia-Verdugo, J. M. and Alvarez-Buylla, A. 2010. Cilia organize ependymal planar polarity. J. Neurosci. 30: 2600–2610.

Mukhopadhyay, S., Lu, Y., Shaham, S. and Sengupta, P. 2008. Sensory signaling-dependent remodeling of olfactory cilia architecture in C. elegans. Dev. Cell 14: 762–774.

Mukhopadhyay, S., Wen, X., Chih, B., Nelson, C. D., Lane, W. S., Scales, S. J. and Jackson, P. K. 2010. TULP3 bridges the IFT-A complex and membrane phosphoinositides to promote trafficking of G protein-coupled receptors into primary cilia. Genes Dev. 24: 2180–2193.

Mukhopadhyay, S., Wen, X., Ratti, N., Loktev, A., Rangell, L., Scales, S. J. and Jackson, P. K. 2013. The ciliary G-protein-coupled receptor Gpr161 negatively regulates the Sonic hedgehog pathway via cAMP signaling. Cell 152: 210–223.

Nachury, M. V., Loktev, A. V., Zhang, Q., Westlake, C. J., Peranen, J., Merdes, A., Slusarski, D.C., Scheller, R. H., Bazan, J. F., Sheffield, V. C. et al. 2007. A core complex of BBS proteins cooperates with the GTPase Rab8 to promote ciliary membrane biogenesis. Cell 129: 1201–1213.

Nachury, M. V. 2014. How do cilia organize signalling cascades? Philos. Trans. R. Soc. Lond. B. Biol. Sci. 369.

Nieuwenhuys, R. 2009. The structural organization of the forebrain: a commentary on the papers presented at the 20th Annual Karger Workshop 'Forebrain evolution in fishes'. Brain Behav. Evol. 74: 77–85.

O'Toole, J. F., Liu, Y., Davis, E. E., Westlake, C. J., Attanasio, M., Otto, E. A., Seelow, D., Nurnberg, G., Becker, C., Nuutinen, M. et al. 2010. Individuals with mutations in XPNPEP3, which encodes a mitochondrial protein, develop a nephronophthisis-like nephropathy. J. Clin. Invest. 120: 791–802.

Omori, Y., Chaya, T., Yoshida, S., Irie, S., Tsujii, T. and Furukawa, T. 2015. Identification of G Protein-Coupled Receptors (GPCRs) in primary cilia and their possible involvement in body weight control. PLoS One 10: e0128422.

Ou, G., Blacque, O. E., Snow, J. J., Leroux, M. R. and Scholey, J. M. 2005. Functional coordination of intraflagellar transport motors. Nature 436: 583–587.

Parelkar, S. V., Kapadnis, S. P., Sanghvi, B. V., Joshi, P. B., Mundada, D. and Oak, S. N. 2013. Meckel-Gruber syndrome: A rare and lethal anomaly with review of literature. J. Pediatr. Neurosci. 8: 154–157.

Paridaen, J. T., Wilsch-Brauninger, M. and Huttner, W. B. 2013. Asymmetric inheritance of centrosome-associated primary cilium membrane directs ciliogenesis after cell division. Cell 155: 333–344.

Parisi, M. A. 2009. Clinical and molecular features of Joubert syndrome and related disorders. Am. J. Med. Genet. C. Semin. Med. Genet. 151C: 326–340.

Paulesu, E., Démonet, J. -F., Fazio, F., McCrory, E., Chanoine, V., Brunswick, N., Cappa, S. F., Cossu, G., Habib, M., Frith, C. D. et al. 2001. Dyslexia cultural diversity and biological unity. Science 291: 2165–2167.

Perkins, L. A., Hedgecock, E. M., Thomson, J. N. and Culotti, J. G. 1986. Mutant sensory cilia in the nematode Caenorhabditis elegans. Dev. Biol. 117: 456–487.

Piasecki, B., Burghoorn, J. and Swoboda, P. 2010. Regulatory Factor X (RFX)-mediated transcriptional rewiring of ciliary genes in animals. Proc. Natl. Acad. Sci. USA 107: 12969–12974.

Puelles, L. and Medina, L. 2002. Field homology as a way to reconcile genetic and developmental variability with adult homology. Brain Research Bulletin 57: 243–255.

Puelles, L. and Rubenstein, J. L. 2003. Forebrain gene expression domains and the evolving prosomeric model. Trends Neurosci. 26: 469–476.

Qin, H., Burnette, D. T., Bae, Y. K., Forscher, P., Barr, M. M. and Rosenbaum, J. L. 2005. Intraflagellar transport is required for the vectorial movement of TRPV channels in the ciliary membrane. Curr. Biol. 15: 1695–1699.

Qin, J., Lin, Y., Norman, R. X., Ko, H. W. and Eggenschwiler, J. T. 2011. Intraflagellar transport protein 122 antagonizes Sonic Hedgehog signaling and controls ciliary localization of pathway components. Proc. Natl. Acad. Sci. USA 108: 1456–1461.

Raible, F. and Brand, M. 2004. Divide et Impera—the midbrain-hindbrain boundary and its organizer. Trends Neurosci. 27: 727–734.

Reiter, J. F., Blacque, O. E. and Leroux, M. R. 2012. The base of the cilium: roles for transition fibres and the transition zone in ciliary formation, maintenance and compartmentalization. EMBO Rep. 13: 608–618.

Rohatgi, R., Milenkovic, L. and Scott, M. P. 2007. Patched1 regulates hedgehog signaling at the primary cilium. Science 317: 372–376.

Sanders, A. A., de Vrieze, E., Alazami, A. M., Alzahrani, F., Malarkey, E. B., Sorusch, N., Tebbe, L., Kuhns, S., van Dam, T. J., Alhashem, A. et al. 2015. KIAA0556 is a novel ciliary basal body component mutated in Joubert syndrome. Genome Biol. 16: 293.

Sanes, D., Reh, T. and Harris, W. 2006. Development of the nervous system, 2nd edn. (Elsevier Academic Press).

Sang, L., Miller, J. J., Corbit, K. C., Giles, R. H., Brauer, M. J., Otto, E. A., Baye, L. M., Wen, X., Scales, S. J., Kwong, M. et al. 2011. Mapping the NPHP-JBTS-MKS protein network reveals ciliopathy disease genes and pathways. Cell 145: 513–528.

Sarkisian, M. R. and Guadiana, S. M. 2015. Influences of primary cilia on cortical morphogenesis and neuronal subtype maturation. Neuroscientist 21: 136–151.

Sauer, F. C. 1935. Mitosis in the neural tube. Journal of Comparative Neurology 62: 37–405.

Sawamoto, K., Wichterle, H., Gonzales-Perez, O., Cholfin, J. A., Yamada, O., Spassky, N., Murcia, N. S., Garcia-Verdugo, J. M., Marin, O., Rubenstein, J. L. et al. 2006. New neurons follow the flow of cerebrospinal fluid in the adult brain. Science 311: 629–632.

Scerri, T. S. and Schulte-Korne, G. 2010. Genetics of developmental dyslexia. Eur. Child Adolesc. Psychiatry 19: 179–197.

Schaefer, E., Stoetzel, C., Scheidecker, S., Geoffroy, V., Prasad, M. K., Redin, C., Missotte, I., Lacombe, D., Mandel, J. L., Muller, J. et al. 2016. Identification of a novel mutation confirms the implication of IFT172 (BBS20) in Bardet-Biedl syndrome. J. Hum. Genet. 61: 447–450.

Schmechel, D. E. and Rakic, P. 1979. A golgi study of radial glia cells in developing monkey telencephalon: Morphogenesis and transformation into astrocytes. Anat. Embryol. 156: 115–152.

Scholpp, S. and Lumsden, A. 2010. Building a bridal chamber: development of the thalamus. Trends Neurosci. 33: 373–380.

Schou, K. B., Pedersen, L. B. and Christensen, S. T. 2015. Ins and outs of GPCR signaling in primary cilia. EMBO Rep. 16: 1099–1113.

Schueler, M., Braun, D. A., Chandrasekar, G., Gee, H. Y., Klasson, T. D., Halbritter, J., Bieder, A., Porath, J. D., Airik, R., Zhou, W. et al. 2015. DCDC2 mutations cause a renal-hepatic ciliopathy by disrupting Wnt signaling. Am. J. Hum. Genet. 96: 81–92.

Schumacher, J., Anthoni, H., Dahdouh, F., Konig, I. R., Hillmer, A. M., Kluck, N., Manthey, M., Plume, E., Warnke, A., Remschmidt, H. et al. 2006. Strong genetic evidence of DCDC2 as a susceptibility gene for dyslexia. Am. J. Hum. Genet. 78: 52–62.

Shimamura, K. and Rubenstein, J. L. R. 1997. Inductive interactions direct early regionalization of the mouse forebrain. Development 124: 2709–2718.

Simms, R. J., Hynes, A. M., Eley, L. and Sayer, J. A. 2011. Nephronophthisis: a genetically diverse ciliopathy. Int. J. Nephrol. 2011: 527137.

Singer, W. 2012. What binds it all togehter? Synchronized oscillatory activity in normal and pathological conditions. pp. 57–70. *In*: Barth, F. G., Giampieri-Deutsch, P. and Klain, H. -D. (eds.). Sensory Perception: Mind and Matter. Springer-Verlag, Vienna, Austria.

Spassky, N., Merkle, F. T., Flames, N., Tramontin, A. D., Garcia-Verdugo, J. M. and Alvarez-Buylla, A. 2005. Adult ependymal cells are postmitotic and are derived from radial glial cells during embryogenesis. J. Neurosci. 25: 10–18.

Spassky, N., Han, Y. G., Aguilar, A., Strehl, L., Besse, L., Laclef, C., Ros, M. R., Garcia-Verdugo, J. M. and Alvarez-Buylla, A. 2008. Primary cilia are required for cerebellar development and Shh-dependent expansion of progenitor pool. Dev. Biol. 317: 246–259.

Spassky, N. 2013. Motile cilia and brain function: Ependymal motile cilia development, organization, function and their associated pathologies. pp. 193–207. *In*: Tucker, K. L. and Caspary, T. (eds.). Cilia and Nervous System Development and Function. Springer, Dordrecht.

Stanic, D., Malmgren, H., He, H., Scott, L., Aperia, A. and Hokfelt, T. 2009. Developmental changes in frequency of the ciliary somatostatin receptor 3 protein. Brain Res. 1249: 101–112.

Stottmann, R. W., Tran, P. V., Turbe-Doan, A. and Beier, D. R. 2009. Ttc21b is required to restrict sonic hedgehog activity in the developing mouse forebrain. Dev. Biol. 335: 166–178.

Striedter, G. F. 2005. Principles of Brain Evolution (Sunderland, MA, USA: Sinauer Associates).

Swoboda, P., Adler, H. T. and Thomas, J. H. 2000. The RFX-type transcription factor DAF-19 regulates sensory neuron cilium formation in *C. elegans*. Mol. Cell 5: 411–421.

Szymanska, K., Hartill, V. L. and Johnson, C. A. 2014. Unraveling the genetics of Joubert and Meckel-Gruber syndromes. J. Pediatr. Genet. 3: 65–78.

Taipale, M., Kaminen, N., Nopola-Hemmi, J., Haltia, T., Myllyluoma, B., Lyytinen, H., Muller, K., Kaaranen, M., Lindsberg, P. J., Hannula-Jouppi, K. et al. 2003. A candidate gene for developmental dyslexia encodes a nuclear tetratricopeptide repeat domain protein dynamically regulated in brain. Proc. Natl. Acad. Sci. USA 100: 11553–11558.

Tammimies, K., Bieder, A., Lauter, G., Sugiaman-Trapman, D., Torchet, R., Hokkanen, M., Burghoorn, J., Castren, E., Kere, J., Tapia-Paez, I. et al. 2016. Ciliary dyslexia candidate genes DYX1C1 and DCDC2 are regulated by Regulatory Factor X (RFX) transcription factors through X-box promoter motifs. FASEB J. 10: 3578–3587.

Tarkar, A., Loges, N. T., Slagle, C. E., Francis, R., Dougherty, G. W., Tamayo, J. V., Shook, B., Cantino, M., Schwartz, D., Jahnke, C. et al. 2013. DYX1C1 is required for axonemal dynein assembly and ciliary motility. Nat. Genet. 45: 995–1003.

Taverna, E., Gotz, M. and Huttner, W.B. 2014. The cell biology of neurogenesis: toward an understanding of the development and evolution of the neocortex. Annu. Rev. Cell Dev. Biol. 30: 465–502.

Theil, T., Alvarez-Bolado, G., Walter, A. and Rüther, U. 1999. Gli3 is required for Emx gene expression during dorsal telencephalon development. Development 126: 3561–3571.

Tole, S., Ragsdale, C. W. and Grove, E. A. 2000. Dorsoventral patterning of the telencephalon is disrupted in the mouse mutant extra-toes. Dev. Biol. 217: 254–265.

Tong, C. K., Han, Y. G., Shah, J. K., Obernier, K., Guinto, C. D. and Alvarez-Buylla, A. 2014. Primary cilia are required in a unique subpopulation of neural progenitors. Proc. Natl. Acad. Sci. USA 111: 12438–12443.

Tran, P. V., Haycraft, C. J., Besschetnova, T. Y., Turbe-Doan, A., Stottmann, R. W., Herron, B. J., Chesebro, A. L., Qiu, H., Scherz, P. J., Shah, J. V. et al. 2008. THM1 negatively modulates mouse sonic hedgehog signal transduction and affects retrograde intraflagellar transport in cilia. Nat. Genet. 40: 403–410.

Tuson, M., He, M. and Anderson, K. V. 2011. Protein kinase A acts at the basal body of the primary cilium to prevent Gli2 activation and ventralization of the mouse neural tube. Development 138: 4921–4930.

Vertii, A., Bright, A., Delaval, B., Hehnly, H. and Doxsey, S. 2015. New frontiers: discovering cilia-independent functions of cilia proteins. EMBO Rep. 16: 1275–1287.

Vieillard, J., Duteyrat, J. L., Cortier, E. and Durand, B. 2015. Imaging cilia in *Drosophila melanogaster*. Methods Cell Biol. 127: 279–302.

Wang, B., Fallon, J. F. and Beachy, P. A. 2000. Hedgehog-regulated processing of Gli3 produces an anterior/posterior repressor gradient in the developing vertebrate limb. Cell 100: 423–434.

Waters, A. M. and Beales, P. L. 2011. Ciliopathies: an expanding disease spectrum. Pediatr. Nephrol. 26: 1039–1056.

Wheway, G., Schmidts, M., Mans, D. A., Szymanska, K., Nguyen, T. M., Racher, H., Phelps, I. G., Toedt, G., Kennedy, J., Wunderlich, K. A. et al. 2015. An siRNA-based functional genomics screen for the identification of regulators of ciliogenesis and ciliopathy genes. Nat. Cell Biol. 17: 1074–1087.

White, J. G., Southgate, E., Thomson, J. N. and Brenner, S. 1986. The structure of the nervous system of the nematode Caenorhabditis elegans. Philos. Trans. R. Soc. Lond. B. Biol. Sci. 314: 1–340.

Willaredt, M. A., Hasenpusch-Theil, K., Gardner, H. A., Kitanovic, I., Hirschfeld-Warneken, V. C., Gojak, C. P., Gorgas, K., Bradford, C. L., Spatz, J., Wolfl, S. et al. 2008. A crucial role for primary cilia in cortical morphogenesis. J. Neurosci. 28: 12887–12900.

Wilsch-Brauninger, M., Peters, J., Paridaen, J. T. and Huttner, W. B. 2012. Basolateral rather than apical primary cilia on neuroepithelial cells committed to delamination. Development 139: 95–105.

Wilson, S. L., Wilson, J. P., Wang, C., Wang, B. and McConnell, S. K. 2012. Primary cilia and Gli3 activity regulate cerebral cortical size. Dev. Neurobiol. 72: 1196–1212.

Wolf, M.T. 2015. Nephronophthisis and related syndromes. Curr. Opin. Pediatr. 27: 201–211.

Yuan, S. and Sun, Z. 2013. Expanding horizons: ciliary proteins reach beyond cilia. Annu. Rev. Genet. 47: 353–376.

CHAPTER 2

Cilia in Lung Development and Disease

Laura L. Yates,[1] Hannah M. Mitchison[2] and
Charlotte H. Dean[1,3],*

INTRODUCTION

The primary function of the lungs is as a site of gaseous exchange, since the lungs enable delivery of oxygen (O_2) to the blood and removal of carbon dioxide (CO_2). For this gaseous exchange to take place, almost the entire output of blood from the heart is directed through the lungs, so that it can be brought close enough to the alveoli where the exchange of gases can occur by diffusion. As a consequence of this primary function, the lungs are directly connected to the external environment through the nose and mouth via the trachea. This direct connection with the outside means that the lungs must have efficient mechanisms in place to help protect the body from inhaled insults, such as toxins or pathogens. One of the most important of these mechanisms is the action of motile cilia. In the lung airways, uniform orientation of these cilia beating together in a co-ordinated manner induces a directed flow of the protective mucous layer present on the surface of the airway epithelium called the "mucociliary escalator". This flow is used to direct inhaled foreign particles and microorganisms away from descending

[1] Inflammation Repair and Development, NHLI, Imperial College London, Exhibition Road, South Kensington, London SW7 2AZ, UK.
[2] Genetics and Genomic Medicine, UCL Institute of Child Health, 30 Guilford Street, London WC1N 1EK, UK.
[3] Mammalian Genetics Unit, MRC Harwell, Oxon OX11 0RD, UK.
* Corresponding author: c.dean@imperial.ac.uk

further down the network of airways into the gas-exchanging region of the lungs. Once trapped in the mucous, the particles will be actively transported back up the airways to the pharynx, via the mucociliary escalator, from where they can be swallowed or ejected (Tilley et al., 2015). Particles that are small enough to reach the alveolar region become stuck to the walls of the alveoli by tension and can then be engulfed by macrophages that either carry them to the mucociliary escalator or to the blood and lymph for removal.

The mature lungs are comprised of many different cell types including several sub-types of epithelial cells (see below), fibroblasts (also consisting of different sub-types), smooth muscle cells, macrophages, neuroendocrine cells, and endothelial cells. These cells are organized into the epithelial component (airways and alveoli) with its underlying basal lamina in addition to a vascular network, smooth muscle, nerves and cartilage.

As well as multiciliated epithelia, cells in the lungs including epithelia and fibroblasts also contain non-motile, primary, cilia but much less is known about the primary cilia in lung development and disease.

In this chapter we outline the structure and function of the lungs, discuss how the lungs develop and describe current evidence about the development of primary cilia and multicilia. We also consider how in different lung diseases respiratory function can be affected directly through impaired cilia structure and/or function, e.g., in primary ciliary dyskinesia (PCD), or through indirect impairment of cilia function, e.g., in asthma and cystic fibrosis, and describe how lung function is assessed in humans.

Overview of Lung Structure and Function

Below the upper respiratory tract, which consists of the nose, mouth, sinuses and inner ear, the most rostral region of the lungs is the trachea. This main airway is surrounded by cartilaginous rings, which help to support the tracheal tube. The trachea divides into two main bronchi, one of which extends into the left and the other into the right lung. In humans the overall lung structure below the trachea is arranged into two lobes on the left and three on the right. The lobes themselves are encased inside pleural membranes. Each lobe contains a ramified network of tubes (the airways) and the gas-exchanging portion of the lungs (alveoli and interstitium). The majority of the lung volume is occupied by alveoli surrounding airspaces and the volume of air within the lungs is substantial enough to enable them to float when placed in water. During mammalian development, the network of air conducting tubes (airways) is generated first. This begins when the primary lung buds grow out from the foregut endoderm and undergo 16 dichotomous rounds of branching to produce an ever-larger network of tubes; the airways. The larger airways (bronchi) become narrower as the respiratory tree extends, with narrower airways known as bronchioles. Distal to the bronchioles are the terminal bronchioles beyond which lies the major gas-exchanging region of the lungs consisting of alveolar ducts and finally alveolar sacs. The architecture of the adult lungs provides a huge surface area to enable efficient gas-exchange that can be packed into the relatively small space of the thoracic cavity.

Gaseous exchange

Gaseous exchange can be thought of in three parts: (1) transfer of oxygen into the alveoli from the air, (2) the pulmonary circulation which brings the gases (O_2 and CO_2) and blood together so that gas-exchange can take place and (3) the removal of CO_2 back out of the lungs. O_2 levels in the blood are maintained though inspiration of air into the alveoli and subsequent passive diffusion of oxygen through the alveolar wall cells and into the blood. To facilitate this process, the plasma membranes of the alveolar epithelial cells, pulmonary capillaries and associated basement membranes are extremely thin. For gaseous exchange to take place, blood from the heart is directed through the lungs where it can pass through the fine network of capillaries that are intricately laced around the alveolar epithelial cells. Here, the close proximity of the capillaries and alveolar epithelium allows diffusion of O_2 in and CO_2 out of the blood. CO_2 can diffuse back through the alveoli and then be removed upon expiration. Oxygenated blood flows from the lungs to the heart where it can then be pumped out and around the body through the systemic circulation (Davies, 2003).

At the cellular level, the airways and the alveoli are lined by mixed populations of epithelial cells that are adapted to perform specific functions for example, goblet cells and ciliated cells in the airways and the type I and type II alveolar epithelial cells (Figure 1). These specialized cell types are found in different locations within the lungs and the relative proportions of each cell type also varies along the rostral to caudal axis of the lungs (Figure 1). Mature lungs as mentioned above contain a number of different cell types but of these, the epithelia and fibroblast populations exhibit the greatest variety in form and function.

Epithelial cell types

AIRWAY

The epithelium lining the human airways is pseudostratified, consisting of a number of differentiated sub-types. The majority of these epithelial cells are columnar in structure with the exception of the basal cells that reside below the cells immediately adjacent to the lumen, and usually adopt a slightly flatter morphology.

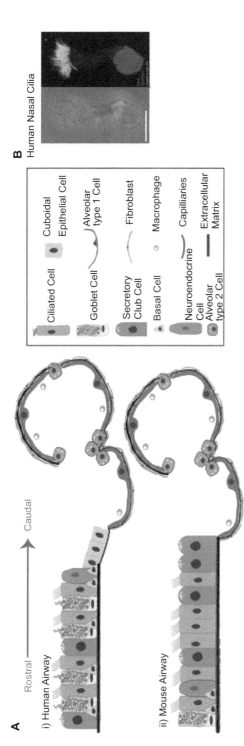

Figure 1. (*A*) *Organisation of Epithelial Cell Types in Human and Mouse Lungs.*

The composition of epithelial cells varies quite dramatically in the airways of humans (i) and mice (ii). The human airways (i) are lined with a pseudostratified epithelium consisting of roughly equal numbers of basal, ciliated and secretory cells. The secretory cells are comprised mostly of goblet cells with a few Club cells. When moving caudally down the airways and into the bronchioles, the epithelium changes from a pseudostratified epithelium to a simple cuboidal epithelium. In the mouse (ii) only the most rostral (tracheal) epithelia are pseudostratified, containing mostly ciliated cells, basal cells and a few neuroendocrine cells. The airways are lined with a simple columnar epithelium comprised mostly of ciliated cells, followed by secretory and neuroendocrine cells. Most of the secretory cells in the mouse are Club cells with few goblet cells and no basal cells in the caudal airways. Fewer ciliated cells are observed in the terminal bronchioles towards the alveoli duct compared to more rostral airways. The alveolar walls are predominantly (90%) lined by flat elongated AT1 cells that connect to a more numerous population of smaller, cuboidal ATII cells at the corners of the alveoli. Alveolar epithelial cells form close contacts with an extensive capillary network to allow for efficient gas exchange, as well as with the extracellular matrix (ECM) and mesenchymal cells including fibroblasts.

(*B*) *Human Bronchial Ciliated Cell*

Human bronchial epithelial cell obtained by brush biopsy, immunostained for gamma-tubulin and acetylated alpha-tubulin (green), markers for the cilia axoneme and basal bodies, and RSPH4A (red) a protein mutated in primary ciliary dyskinesia that localises to the axoneme. 4', 6-diamidino-2-phenylindole, DAPI (blue). A differential interference contrast (DIC) image is shown on the left. Image courtesy of Mitali Patel.

Club cells (secretory)

Club cells are a secretory cell population that are thought to act as resident stem cells and have been shown to give rise to ciliated cells (Rawlins and Hogan, 2006; Morrisey and Hogan, 2010).

Ciliated cells

This population bears multiple '9 + 2' structure motile cilia that protrude from their apical surface into the airways to participate in mucociliary clearance. On each cell, 200–300 motile cilia beat in a co-ordinated fashion using a planar forward power stroke and recovery stroke, to keep the airways clear and to direct invading particles out of the lungs (Chilvers and O'Callaghan, 2000).

Goblet cells

Goblet cells are interspersed between the ciliated and club cells. They are usually present in fewer numbers than club or ciliated cells, however, the number of these cells in the airways is highly plastic and can rapidly change in response to invading substances, such as pathogens or allergens. These secrete protective mucus, which lines the airways (McCauley and Guasch, 2015). By secreting mucus, their role is to increase viscosity and make the airways stickier to efficiently trap the invading particles, however, the number of goblet cells and the mucus consistency must be finely tuned to avoid blocking of the airways and mucus plug formation.

Basal cells

These cells lie below the other epithelial cell types, immediately adjacent to the basal lamina. They form tight contacts with the basal lamina through hemidesmosomes and they function as a stem cell population (Hogan et al., 2014). The shape of basal cells does not appear to be as static as the other airway epithelial cell populations; rather they can be flattened, pyramidal, cuboidal as well as occasionally columnar (Gaultier, 1999). Their shape probably depends on the state of the epithelium, e.g., healthy or diseased, as well as the number of basal cells present in a particular region of the airways.

ALVEOLAR

Type I cells (ATI)

The majority of the alveolar walls are comprised of ATI cells. These extremely flat, elongated cells occupy more than 90% of the alveolar surface area and are the major epithelial cell type involved in gaseous exchange. They are particularly susceptible to damage and have been shown to be selectively lost upon injury (Patel et al., 2013; Yang et al., 2016).

Type II cells (ATII)

These cuboidal cells are usually located in the corners of alveoli. They form connections with ATI cells. They secrete and are the major storage site for surfactant proteins (Mao et al., 2015). ATII cells appear to be more resistant to damage and upon injury, they can proliferate to replace damaged cells. Studies have also shown that the ATII cells are precursors for the ATI cells (Barkauskas et al., 2013).

Fibroblast cell types

Lying below the basal cells of the airways is the basal lamina and underneath this are blood vessels, smooth muscle, cartilage and nerves as well as fibroblasts (Hogan et al., 2014). Fibroblast sub-types play key roles in the alveoli both during development and in the adult lungs. This has been an increased area of research interest and it is now clear that as with the epithelial cells, there are several distinct populations of fibroblasts that play distinct roles in the lungs (McGowan and McCoy, 2014). Moreover, changes in the number or type of fibroblast cells can significantly contribute to lung diseases, such as idiopathic pulmonary fibrosis and asthma (Wilson and Wynn, 2009; Chambers and Mercer, 2015).

Myofibroblasts

These cells express some components typically associated with smooth muscle cells, such as alpha-smooth muscle actin. They are capable of secreting basal lamina components, including collagen and elastin. The function of these cells in the adult lungs is not well understood, however, in pulmonary fibrosis which is a component of a number of lung diseases, these cells proliferate leading to deposition of scar tissue, thickening of basement membrane and impaired lung function. Myofibroblasts are also thought to contribute to the thickening of the smooth muscle surrounding airways that occurs in asthma (Halayko et al., 2006; Schmidt and Mattoli, 2013).

Interstitial fibroblasts

This population is particularly important during alveolar development. There are two phenotypes of interstitial fibroblasts, lipid and non-lipid containing (Rehan et al., 2006). As yet, it is unclear whether these represent two distinct populations or whether their phenotype changes due to the environment (Rehan et al., 2006). It has been shown that the lipid-laden cells (lipofibroblasts) express distinct marker proteins such as CD90 (Thy1), whereas the non-lipid containing cells express CD166 (Hagood et al., 1999; McQualter et al., 2013). The lipid containing cells are thought to have a positive impact on alveolarization and tissue regeneration and may act as reservoirs for a number of factors important for lung development and repair. Lipofibroblasts also contribute to surfactant protein production (McGowan, 2014; Al Alam et al., 2015).

Pericytes

These mural cells line the capillaries in the lungs. They are specifically a population of vessel-associated mesenchymal cells, which are positive for the marker proteins desmin, NG2 and CD146. Pericytes provide both structural and biological support to capillaries and are likely to be important for tissue repair. It has been shown that pericytes can differentiate into myofibroblasts (Hung et al., 2013).

Differences between Human and Mouse Lungs

Much of our knowledge about lung development has come from studies in the mouse. Although this is an extremely useful model system, particularly for genetic manipulation, there are some important differences that are likely to be relevant to studies of development and disease. The first is that macroscopically, the human lungs are comprised of two lobes on the left and three on the right whilst the mouse lungs have one left lobe and four on the right.

Relevant to studies of cilia and ciliopathies is the respective composition of epithelial cell types found at different locations within the lungs (Figure 1). For example, human lungs have many more basal and mucin-secreting goblet cells than mouse lungs. There are relatively few mucus-secreting cells in normal adult mouse lung, however, large numbers can be seen in the airways within a few days of allergen exposure suggesting that they are rapidly expanded either from stem cells or by trans-differentiation from another adult population (Rawlins and Hogan, 2006). The relative paucity of goblet and other mucus-secreting cells in the mouse lungs may be particularly relevant in ciliopathies and diseases with secondary cilia defects, where mucocilliary clearance is one of the most debilitating features (Rock et al., 2010). Another difference between these species is the location of airways lined by pseudo-stratified epithelium. This is significant because it is only the pseudo-stratified epithelium that contains basal cells, which are thought to be an important stem cell population (Rock et al., 2010; Nakajima et al., 1998). In humans, this type of epithelium, which contains basal cells, extends all the way down from the main bronchi into the small airways and terminal bronchioles, whereas in mouse it is only the trachea that contains pseudo-stratified epithelium. The remainder of the mouse airways, from the main stem bronchi downwards, are lined by a simple columnar epithelium. These differences may have some bearing on the response of the human and mouse lungs to disease or environmental insults.

The Phases of Lung Development

Despite the notable differences in human and mouse lung structure and physiology described above, both mouse and human lung development can be divided into four key phases (Table 1).

1. Pseudoglandular phase

In the mouse, *NK2 homeobox 1 (Nkx2-1)* also known as *thyroid transcription factor 1 (Titf1)* expression at embryonic day (E) 9.0–9.5 (Minoo et al., 1995) in the ventral wall of the anterior foregut initiates the outgrowth of two epithelial buds that will give rise to the respiratory tree. The Wingless (Wnt) ligands Wnt2 and Wnt2b are expressed in the ventral mesoderm surrounding the anterior foregut endoderm and activate canonical Wnt signalling. This in turn localises *Nkx2-1* expression to the anterior foregut and restricts SRY (sex determining region Y)-box 2, also known as Sox2, to the distal foregut where the oesophagus will develop (Goss et al., 2009; Swarr and Morrisey, 2015). Together with bone morphogenetic protein 4 (Bmp4), its agonist Noggin and fibroblast growth factors (Fgfs) in the mesoderm, these reciprocal signalling networks ensure the correct placement of the lung along the proximal-distal axis of the foregut (Shu et al., 2005; Domyan et al., 2011). In response to localised *Nkx2-1* expression the two primary asymmetrical epithelial buds begin to expand into the surrounding mesoderm. Simultaneously, the foregut tube just anterior of the lung buds begins to separate in two, to form the dorsal oesophagus that leads to the stomach and the ventral trachea that remains connected to the lung buds.

Branching morphogenesis

In the mouse, the primary lung buds extend into the surrounding mesenchyme in response to Fibroblast growth factor 10 (FGF10) (Min et al., 1998) which signals to Fibroblast growth factor receptor 2 (FGFR2) (De Moerlooze et al., 2000) in the epithelium, initiating branching morphogenesis.

Table 1. Stages of lung development in mice and humans.

Phases of lung development	Mouse lung	Human lung
Pseudoglandular Phase *(Branching Morphogenesis)*	E9.5–E16.5	4–17 weeks
Canalicular phase *(Narrowing of terminal buds, epithelial differentiation)*	E16.5–E17.5	17–26 weeks
Saccular phase *(Pre-cursers of alveoli develop, lung continues to increase in size)*	E18.5–P5	26–36 weeks
Alveologenesis/Alveolarization *(Formation of alveoli to promote efficient gas exchange)*	P5–P21	~ 36 weeks to maturity *(most active 6 months after birth)*

Reciprocal signalling between the epithelium and mesenchyme is essential for the regulation of normal lung branching and involves a number of key signalling molecules including; Retinoic acid receptors (RARs) (Mendelsohn et al., 1994), Sonic Hedgehog (Shh), Wnt2/7b and BMP4 (Pepicelli et al., 1998) that work together to control downstream effector molecules such as N-myc (Myc, v-myc avian myelocytomatosis viral oncogene homolog), Ets Variant gene 4 (Etv4/56), Extracellular signal-related protein kinases 1 and 2 (ERK1/2 or Mapk3 and Mapk1) and p38 mitogen activated kinase (Mapk14) (Shu et al., 2005; Liu et al., 2008; Goss et al., 2009). Recent studies have also begun to highlight an important role for small non-coding microRNAs (miRNAs) such as the MiR-17 family, during pseudoglandular lung development (Ventura et al., 2008; Carraro et al., 2009).

Mouse lung undergoes sixteen rounds of stereotypical branching which involves four key steps; (1) bud elongation, (2) cessation of outgrowth, (3) expansion of the tip and (4) bifurcation (Metzger et al., 2008). In 2008, Metzger and colleagues fixed and immunostained embryonic mouse lungs aged E11–E15 in order to visualise the pattern of lung branching during development. From these studies they proposed that lung branching follows a stereotypical program consisting of three modes of branching; domain branching, orthogonal branching and planar branching. Domain branching is defined as the sprouting of new buds at specific distances along the length of a single stalk. Domain branching is thought to be required for the establishment of the main structure of the respiratory tree (Metzger et al., 2008). Orthogonal and planar branching occurs at the tips as the buds bifurcate. Planar branching occurs when a tip splits into two buds within a single plane and orthogonal branching describes when a new bud forms orthogonal (90°) to the parent stalk (Metzger et al., 2008). However, more recent studies have used 3D optical projection tomography (OPT) (Sharpe et al., 2002) to visualise mouse lungs *in situ*, in combination with a specifically designed Tree Surveyor Program to computationally analyse the branching pattern of these organs (Short et al., 2013). Data generated from these studies in general agree with the original findings from Metzger et al., in that the overall pattern and direction of branching, especially the initial major branches, matched what was previously proposed (Short et al., 2013). However, Short et al. (2013) described a greater degree of variability in the number of subsequent dorsal, ventral, anterior and posterior branches (Short et al., 2013), suggesting these later bifurcation events are not as stereotypical and do not fit into one specific mode of branching (planar or orthogonal) as described previously (Metzger et al., 2008).

Non-canonical WNT/Planar cell polarity (PCP) signalling is known to be a key regulator of lung branching morphogenesis with mouse mutants of PCP genes displaying severe lung branching defects (Paudyal et al., 2010; Yates et al., 2010; Yates et al., 2013). Moreover, Kadzik et al., have shown that the Wnt receptor Frizzled 2 (Fzd2) is required for domain branch formation and that Wnt/Fzd2 signalling is important in regulating epithelial cell shape and tube morphology (Kadzik et al., 2014). Cytoskeletal dynamics essential for changes in cell shape and migration as well as cell—extra cellular matrix (ECM) interactions are also key processes that are required for the correct establishment of the respiratory tree (Sakai et al., 2003; Liu et al., 2004; Yates et al., 2010; Chen and Krasnow, 2012; Kadzik et al., 2014). The PCP pathway has also been shown to be important for orienting multicilia in lung airway epithelial cells so that directional motility can take place (Vlader et al. 2012). Further, PCP mutant lung epithelia have misaligned cilia (Vlader et al. 2016).

2. *Canalicular phase*

During the canalicular phase, the terminal lung buds narrow, the mesenchyme thins, capillaries begin to form and the currently undifferentiated cells that line the airways begin to specify (Chung and Andrew, 2008; Chao et al., 2015). Branching morphogenesis and alveolar differentiation appear to be antagonistic, therefore it is only upon cessation of branching that epithelial differentiation can proceed (Chang et al., 2013).

Epithelial cell differentiation

By adulthood the lungs comprise a large number of diverse cell types (see above). The mechanisms determining the complex patterning of differentiated cell types lining the airways is still poorly understood, but it is known to be influenced by autocrine-paracrine signalling that controls cell differentiation (Morrisey and Hogan, 2010). Thyroid transcription factor 1 (TTF-1), Nuclear Factor 1 beta (NF-1β), GATA binding protein 6 (GATA-6) along with other transcription factors including Retinoblastoma (RB), E26 transformation-specific (ETS), SOX and Forkhead box (FOX) family members play a role in cell specific differentiation and gene expression in the conducting airways (Maeda et al., 2007; Rock et al., 2011; Alanis et al., 2014). There are significant differences between the cellular composition and organisation of the airway epithelium of mouse and human lungs. In the mouse lung, the epithelium is either cuboidal or columnar and a pseudostratified layer is found only in the trachea and main bronchi. In humans, the lung epithelium is mostly pseudostratified with only the most distal tubes lined by simple epithelium. The pseudostratified epithelium is composed of ciliated and secretory cells and basal stem cells. Ciliated cells are the most dominant cell type in the lung, existing in a ratio of 7–8 to 1 secretory cell. The correct formation and maintenance of the pseudostratified epithelium is critical for the process of mucociliary clearance. During normal senescence or upon lung injury as a result of exposure to environmental toxins or pathogens, the basal cells generate intermediate undifferentiated cells that are then stimulated to become ciliated or secretory cells that restore the epithelium (Crystal et al., 2008). Notch signalling has been identified as a key regulator of lung epithelial differentiation (Tsao et al., 2008; Tsao et al., 2009). Sustained Notch activation in the adult trachea has been shown to inhibit the differentiation of basal cells into ciliated cells and promote the formation of secretory cells (Rock et al., 2011). Furthermore, Notch signalling has been shown to inhibit ciliogenesis in the developing mouse lung and in human airway epithelium (Tsao et al., 2009; Morimoto et al., 2010; Marcet et al., 2011; Morimoto et al., 2012). Multicilin (*Mcidas*), a transcriptional coregulator that acts downstream of Notch has been shown to control centriole biogenesis and the assembly of cilia via transcription factors Myeloblastosis (Myb) and forkhead box protein J1 (Foxj1) (Stubbs et al., 2012). Activation of Foxj1 induces basal-cell derived progenitors to fully differentiate into ciliated cells (Rock et al., 2011). MicroRNAs (miRNAs) of the miR-34/449 family are also known to promote ciliogenesis by suppressing a number of genes including *Notch1*, delta-like1 (*Dll1*) and a centriolar protein that is known to inhibit cilia formation, *Ccp110* (Marcet et al., 2011). More recently, Tadokoro et al., have shown that Interleukin 6 (IL-6) and downstream Signal transducer and activator of transcription 3 (STAT3) pathway components are positive regulators of multiciliogenesis in the lung, through their inhibition of Notch signalling and *Mcidas* (Tadokoro et al., 2014). The development, transcriptional control and maintenance of airway cilia remains poorly understood, however, it is clear that FOXJ1 and inhibition of Notch signalling are essential for airway epithelial ciliogenesis.

3. Saccular phase

By the saccular stage, small sacs, the precursors of alveoli start to develop at the bud tips and surfactant production begins. Multipotent progenitor cells begin to differentiate into alveolar type 1 and type 2 cells that together comprise the differentiated alveolar epithelium (Treutlein et al., 2014). The blood vessels that have been developing in parallel with the airways now become closely associated with type 1 alveolar epithelial cells to enable gas exchange that is essential for efficient lung function after birth. Alveolar type 2 cells produce surfactant proteins and lipids required to reduce surface tension in the airways, in addition to proteins involved in innate immunity. The lungs continue to increase in size after birth as a result of increases in the length and diameter of the airways that have formed *in utero* and the subdivision of the immature alveoli into smaller subunits.

4. Alveolar phase

The timing of alveolar development varies between species, in mice it occurs between postnatal day 5–30. In humans, while some alveoli form before birth, alveolarization is most active in the first 6 months after birth (Schittny et al., 2008). However, evidence suggests that this stage continues well into childhood though its precise end point remains unclear (Narayanan et al.,

2012). It is during the saccular-alveolar period when the foetus is born that the lungs move from a fluid-filled to an air-filled system. At birth, pulmonary vascular resistance falls, pulmonary blood flow increases, lung fluid is reabsorbed and pulmonary surfactant is secreted into the peripheral saccules of the lung, thus reducing surface tension and preventing alveolar collapse once the lung is filled with air. Lack of surfactant results in respiratory distress syndrome in preterm infants, an important cause of morbidity and mortality in newborns (Whitsett and Weaver, 2015). During this phase the alveolar surface area increases massively at the expense of the mesenchyme through subdividing of the alveolar sacs that form into the mature alveoli (alveolarization/alveologenesis). This process requires the deposition of elastin in primary septae within the wall of the alveolar sacs. The septae elongate across the airspace in a process known as septation, to subdivide and form new mature alveoli (Swarr and Morrisey, 2015; Whitsett and Weaver, 2015). The primary septae contain a double layer of capilliaries that thin to a single layer allowing more efficient gas exchange (Chao et al., 2015).

Non-Motile (Primary) Cilia in Lung Development and Disease

In contrast to other organs such as the kidney, there has been very little research into lung primary cilia. This may largely be because the lungs contain multiciliated cells and therefore most research has focused on this more visible population of cilia. However, studies of mouse mutants with defects in primary (non-motile) cilia have often reported developmental lung defects (Goggolidou et al., 2014) and pulmonary hypoplasia has been reported in some ciliopathy patients, for example those with Joubert and short-rib polydactyly syndromes (Goggolidou et al., 2014). *OFD1* mutations are a cause of the primary ciliopathy disorder orofacial digital syndrome and a subtype of disease with respiratory ciliary dysfunction has been reported (Budny et al., 2006). In some cases the lung defects are likely to be secondary to other phenotypes that result from the cilia dysfunction, for example mouse mutants of Jeune syndrome and short-rib polydactyly syndrome mice have a restricted thoracic space which can itself significantly impair lung development. However, there are other mice with primary cilia defects such as *Kerouac (Krc)*, a mouse model of Meckel syndrome, that do show significant developmental lung abnormalities, and these mice do not have a notably reduced thorax. As is common with mutations affecting primary cilia, *Krc* mice exhibit abnormalities in a number of structures including limbs, neural tube, bone and kidneys in addition to pulmonary hypoplasia. Examination of these mice revealed a defect in primary cilia formation (Weatherbee et al., 2009). At E12.5, the lung airway epithelial cells were monociliated and fewer monocilia were present in *Krc* homozygous mutant lungs compared to wildtype. However, by E18.5 multiple motile cilia were present on airway epithelial cells and these appeared grossly normal in *Krc* homozygotes. In fact a temporal relationship has been proposed between primary and motile cilia in the lungs (Jain et al., 2010). By examining expression of genes unique to either primary or motile cilia, it has been shown that in early stages of lung development (approx. E12.5 in mice) single, primary cilia can be observed on airway epithelial cells but later on (approx. E16.5–18.5) multicilia are present (Francis et al., 2009; Jain et al., 2010) and the primary cilia are no longer observed. Flow itself appears to be established in early post-natal life once the lungs are fully functional (Francis et al., 2009).

Much recent work has focused on the importance of primary cilia for normal Hedgehog signaling (Berbari et al., 2009) and since Sonic Hedgehog (Shh) is critical for lung development (Kugler et al., 2015), it is currently difficult to determine whether a pulmonary hyperplasia phenotype in primary cilia mouse mutants results directly from structural cilia defects or is due to resulting signaling deficiencies.

Published data specifically investigating primary lung cilia is limited but some reports are beginning to emerge. A recent study of polycystic kidney disease proteins in lung related cell lines identified primary cilia on both A549 (ATII type human alveolar basal epithelial cells) and Calu-3 cells (sub-mucosal cell line) (Hu et al., 2014). The authors proposed that the primary cilia might be involved in regulating mucociliary sensing and transport in the airways and primary cilia in airway smooth muscle cells have been reported to play a role in airway remodelling (Trempus et al., 2017). In another study the *Talpid3* chicken mutant was found to have hypoplastic lungs. *Talpid3* encodes a centrosomal protein, KIAA0586 and loss of this protein in the *Talpid3* chicken leads to loss of both motile and non-motile cilia (Stephen et al., 2013). The phenotype of this mutant resembles that of the short-rib polydactyly syndromes, with abnormal Shh signaling and reduced thoracic space. A specific loss of primary cilia from the lungs of *Talpid3* mutants was also shown (Davey et al., 2014). A key question for future research is to determine whether lungs from these primary cilia mutants still exhibit pulmonary hypoplasia even when cultured *ex-vivo*, in the absence of any space restriction.

Lung Disorders and Primary Defects in Multicilia

Two major disease associations are known to arise from inherited gene mutations affecting the motile cilia of the respiratory airway epithelia: primary ciliary dyskinesia (PCD; OMIM#244400) (Lucas et al., 2014) and an overlapping condition that has been termed mucociliary clearance disorder with reduced generation of multiple motile cilia (RGMC; OMIM#)(Wallmeier et al., 2014). PCD and RGMC are rare conditions collectively affecting an estimated 10,000–15,000 individuals per live births (Barbato et al., 2009; Lucas et al., 2014). PCD is caused by mutations in genes affecting the structure and thereby the motility, rather than number, of the respiratory cilia functioning in multiciliated epithelia. In contrast, RGMC is caused by mutations in genes involved in the control of human multiciliogenesis, and multiciliogenesis in other vertebrates. RGMC is characterized by reduced numbers of cilia on muliticiliated cells of the respiratory epithelia, but these cilia may or may not appear to have disrupted motility (Boon et al., 2014; Wallmeier et al., 2014; Funk et al., 2015). Thus, although individuals affected by these motile ciliopathy conditions are clinically almost indistinguishable, with markedly overlapping clinical symptoms requiring similar specialized diagnosis and management by respiratory physicians, the molecular basis for each is distinct. As described below, in PCD the cilia of patients can be completely immotile or they can retain even quite extensive motility depending on the underlying genetic defect, but in either case an ineffective waveform results that is deficient for normal mucociliary clearance, whilst in RGMC the cilia numbers are insufficient for effective mucociliary flow (Figure 2).

The shared clinical features of PCD and RGMC reflect the restricted distribution of specialized motile cilia-bearing cells in the body. Due to deficient mucociliary clearance by multicilia in the airways, affected individuals have chronic respiratory symptoms greatly affecting lifelong morbidity and quality of life. These usually manifest early in the newborn period with respiratory distress syndrome (Lucas et al., 2014). Through life, deficient mucociliary clearance results in recurrent airway bacterial infections, pneumonia, chronic cough, rhinosinusitis, nasal polyps, congestion and infection, as well as infection and blockage of the middle ear associated with hearing loss and a need for corrective grommet surgery. The diffuse airway obstructions, mucus plugging and recurrent infections ultimately progress to permanent lung destruction (bronchiectasis) that are untreatable. Regular use of antibiotics, a healthy lifestyle and a rigorous regime of physiotherapy are used to reduce the speed of progression. In severe cases, a heart-lung transplant may be performed but this carries a significant survival risk in individuals with poor lung function. Additional features of motile cilia diseases are female infertility due to defective fallopian tube multicilia movement of the female gametes and an increased incidence of hydrocephalus arising from defective motility of ependymal motile multicilia lining the brain ventricles to facilitate cerebrospinal fluid movement (Ibanez-Tallon et al., 2004). In PCD, defects of left-right body axis development arise due to defects in early embryonic cilia, and since sperm flagella are structurally related to cilia there is a high incidence of male infertility arising from defective sperm tail motility.

Defects of laterality occur only in PCD and not in RGMC. Normal situs composition of the organs is seen in all RGMC affected individuals since RGMC genes play no role in left-right axis determination (Boon et al., 2014; Wallmeier et al., 2014). In PCD only, around half of patients have laterality defects (Kartagener syndrome). These primarily manifest as harmless mirror-image situs inversus, but in about 6% of these cases more complex left-right isomerism related heterotaxies occur that affect heart development and give rise to congenital heart defects. Placement of the visceral organs is

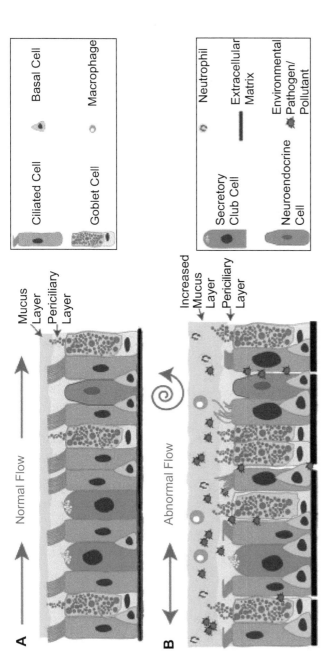

Figure 2. *Mucociliary Clearance in Health and Disease.*

(A) In healthy lungs goblet cells secrete mucins that form a mucus layer and barrier between the lung epithelia and the external environment. This mucus layer traps external particles and transports them up and out of the lungs via ciliary beating. The periciliary layer forms between the cilia and mucus layer and controls the distribution of water allowing the cilia to beat efficiently. Effective cilia beating occurs by the initial extension of cilia into the mucus layer, followed by a power forward stroke that propels the mucus forward. The cilia then withdraw back into the periciliary layer before extending back out into the mucus layer ready to propel forward again. Normal cilia structure, synchronisation and frequency of beating are all crucial for effective mucociliary clearance.

(B) Primary Ciliary Dyskinesia (PCD) and Reduced Generation of Multiple Motile Cilia (RGMC) are genetic diseases that result in structural and therefore functional defects in cilia or result in a reduced number of cilia. Both of which result in impaired mucociliary clearance leading to airway obstruction, mucus plugging and recurrent infections. Exposure to environmental pathogens and toxins such as cigarette smoke can damage the airway epithelium, alter cilia structure and function that also results in impaired mucociliary clearance. The airways of smokers often display basal cell hyperplasia, mucus overproduction, damage to ciliated cells, such that some cells lack the correct number of cilia and/or display atypical nuclei. These defects are more severe in smokers with an associated respiratory disease such as Chronic Obstructive Pulmonary Disease (COPD), where giant cilia, compound cilia and/or structural ciliary defects have been observed.

also affected, e.g., causing abnormal spleen number (Shapiro et al., 2014). Laterality defects arise from deficient motility of single monocilia present on the surface of epithelial cells within the embryonic node of developing embryos that beat in a circular motion to direct the correct flow of key laterality-determining morphogens in the earliest stages of development. The connection between cilia motility and laterality specification was first made in studies of mouse models with mutations affecting ciliary nodal flow (Nonaka et al., 1998), followed by confirmation using mechanical disruption of mouse nodal flow (Nonaka et al., 1998; Nonaka et al., 2002). There is a notable exception in the subtype of PCD arising from mutations that cause a loss of the multicilia central microtubular pairs (*HYDIN, RSPH1, RSPH3, RSPH4A, RSPH9* mutations). None of the affected individuals have laterality defects and this is thought to be because typically nodal cilia lack the central pair, having a 9 + 0 arrangement of microtubules, and therefore central pair loss does not affect their function (Castleman et al., 2009; Onoufriadis et al., 2014b). The process of ciliogenesis of the motile node monocilia differs from that of multicilated cells, however at least one gene mutated in RGMC, *CCNO*, is specifically expressed at the node and its exact role there will require further investigation (Funk et al., 2015). RSPH9 mutations cause PCD with central pair loss and no laterality defects, but this gene is also expressed at the node where its exact function remains unclear (Castleman et al., 2009).

Another difference is a higher incidence of hydrocephalus detected in RGMC compared to PCD linked to ciliogenesis defects (Boon et al., 2014; Amirav et al., 2016). It could be that cilia loss is more detrimental than cilia dysmotility to fluid movement in the brain, but this has not been explored and other cell biological consequences may also influence this variability. Lastly, whilst in PCD there is a high rate of male infertility due to reduced sperm counts and dysmotility of sperm flagella (tails) that are structurally related to cilia with a similar 9 + 2 axonemal arrangement, no male infertility has been associated with RGMC since sperm flagella development is not subject to the multiciliogenesis specification program.

Intensive human genetic studies in the last decade have defined biallelic mutations in over 30 different genes that can cause recessive PCD and RGMC, with autosomal and X-linked syndromic forms. Many cases remain undiagnosed and additional genes are likely to be identified through similar studies and large scale next-generation sequencing projects such as the UK's Genomics England 100,000 Genomes Project (Marx, 2015). The causal genes are summarized in Table 2. There is one syndromic form of PCD which is caused by mutations in the X-linked *RPGR* gene, with affected individuals also displaying retitinis pigmentosa likely linked to the function of this gene in protein transport across the connecting cilium of retinal photoreceptors (Moore et al., 2006; Bukowy-Bieryllo et al., 2013). The classical forms of PCD are all caused by mutations in genes encoding proteins involved in motile cilia structure and assembly. A loss of the outer dynein arm motors (ODA) that power ciliary beating and contain heavy, intermediate and light dynein

Table 2. Genes mutated in motile ciliopathies.

Disorder	Gene	Protein	Proposed function	Electron microscopy defect	Reference
PCD	ARMC4	Armadillo repeat containing 4	Outer dynein arm docking/targeting protein	ODA absent	(1, 2)
PCD	CCDC114	Coiled-coil domain containing 114	Outer dynein arm docking complex protein	ODA absent	(3, 4)
PCD	TTC25	Tetratricopeptide repeat domain-containing protein 25	Outer dynein arm docking complex protein	ODA absent	(5)
PCD	CCDC151	Coiled-coil domain containing 151	Outer dynein arm docking complex-related protein	ODA absent	(6)
PCD	DNAH5	Dynein, axonemal, heavy chain 5	Outer dynein arm heavy chain	ODA absent	(7)
PCD	DNAI1	Dynein, axonemal, intermediate chain 1	Outer dynein arm intermediate chain	ODA absent	(8)
PCD	DNAI2	Dynein, axonemal, intermediate chain 2	Outer dynein arm intermediate chain	ODA absent	(9)
PCD	DNAL1	Dynein, axonemal, light chain 1	Outer dynein arm light chain	ODA absent	(10)
PCD	NME8	Thioredoxin domain-containing protein 3	Outer dynein arm light chain	ODA absent	(11)
PCD	DNAH11	Dynein, axonemal, heavy chain 11	Outer dynein arm heavy chain	Normal	(12, 13)
PCD	CCDC103	Coiled-coil domain containing 103	Dynein arm microtubule attachment	ODA and IDA absent	(14)
PCD	C21ORF59	Chromosome 21 open reading frame 59	Cytoplasmic dynein arm assembly factor	ODA and IDA absent	(15)
PCD	DNAAF1 (LRRC50)	Dynein, axonemal, assembly factor 1	Cytoplasmic dynein arm assembly factor	ODA and IDA absent	(16, 17)

PCD	DNAAF2 (KTU)	Dynein, axonemal, assembly factor 2	Cytoplasmic dynein arm assembly factor	ODA and IDA absent	(18)
PCD	DNAAF3	Dynein, axonemal, assembly factor 3	Cytoplasmic dynein arm assembly factor	ODA and IDA absent	(19)
PCD	DNAAF4 (DYX1C1)	Dynein, axonemal, assembly factor 4	Cytoplasmic dynein arm assembly factor	ODA and IDA absent	(20)
PCD	DNAAF5 (HEATR2)	Dynein, axonemal, assembly factor 5	Cytoplasmic dynein arm assembly factor	ODA and IDA absent	(21)
PCD	LRRC6	Leucine rich repeat containing 6	Cytoplasmic dynein arm assembly factor	ODA and IDA absent	(22)
PCD	SPAG1	Sperm associated antigen 1	Cytoplasmic dynein arm assembly factor	ODA and IDA absent	(23)
PCD	PIH1D3	PIH1 domain-containing protein	Cytoplasmic dynein arm assembly factor	ODA and IDA absent	(24, 25)
PCD	ZMYND10	Zinc finger, MYND-type containing 10	Cytoplasmic dynein arm assembly factor	ODA and IDA absent	(26, 27)
PCD	HYDIN	Axonemal central pair apparatus protein	Subunit of the central microtubular pair	Normal/central pair absent	(28)
PCD	RSPH1	Radial spoke head 1 homolog	Radial spoke head protein	CP absent (transposition defect)	(29, 30)
PCD	RSPH4A	Radial spoke head 4A homolog	Radial spoke head protein	CP absent (transposition defect)	(31)
PCD	RSPH9	Radial spoke head 9 homolog	Radial spoke head protein	CP absent (transposition defect)	(31)
PCD	RSPH3	Radial spoke head 3 homolog	Radial spoke stalk protein	CP absent (transposition defect) + radial spokes absent (MT disorganised)	(32)

Table 2 contd. ....

...*Table 2 contd.*

Disorder	Gene	Protein	Proposed function	Electron microscopy defect	Reference
PCD	DNAJB13	DNAJ/HSP40 homolog, subfamily b, member 13	Radial spoke stalk protein	CP absent (transposition defect)	(33)
PCD	STK36	Serine/threonine protein kinase 36	Radial spoke-central pair protein	CP absent	(34)
PCD	CCDC39	Coiled-coil domain containing 39	"Molecular ruler" protein, nexin and inner arm stability	Nexin-dynein regulatory complex (MT disorganised) + IDA deficient	(35)
PCD	CCDC40	Coiled-coil domain containing 40	"Molecular ruler" protein, nexin and inner arm stability	Nexin-dynein regulatory complex (MT disorganised) + IDA deficient	(36)
PCD	CCDC65	Coiled-coil domain containing 65	Nexin-dynein regulatory complex subunit	Nexin-dynein regulatory complex (MT disorganised)	(15, 37)
PCD	DRC1 (CCDC164)	Dynein regulatory complex subunit 1	Nexin-dynein regulatory complex subunit	Nexin-dynein regulatory complex (MT disorganised)	(38)
PCD	GAS8	Growth arrest specific 8	Nexin-dynein regulatory complex subunit	Normal/misaligned outer MT doublets	(39)
RGMC	MCIDAS	Multicilin	Master regulator of multiciliated cell differentiation/multiciliogenesis	Reduced cilia number (few per cell)	(40)
RGMC	CCNO	Cyclin O	Mother centriole amplification and maturation, for multiciliogenesis	Reduced cilia number (few per cell)	(41)
PCD + RP	RPGR	Retinitis Pigmentosa GTPase Regulator	Photoreceptor connecting cilium protein transport	Mixed—normal and microtubular disorganisation	(42)

1. Hjeij, R., Lindstrand, A., Francis, R., Zariwala, M. A., Liu, X., Li, Y., Damerla, R., Dougherty, G. W., Abouhamed, M., Olbrich, H. et al. 2013. ARMC4 mutations cause primary ciliary dyskinesia with randomization of left/right body asymmetry. American Journal of Human Genetics 93: 357–367.

2. Onoufriadis, A., Shoemark, A., Munye, M. M., James, C. T., Schmidts, M., Patel, M., Rosser, E. M., Bacchelli, C., Beales, P. L., Scambler, P. J. et al. 2014. Combined exome and whole-genome sequencing identifies mutations in ARMC4 as a cause of primary ciliary dyskinesia with defects in the outer dynein arm. Journal of Medical Genetics 51: 61–67.

3. Onoufriadis, A., Paff, T., Antony, D., Shoemark, A., Micha, D., Kuyt, B., Schmidts, M., Petridi, S., Dankert-Roelse, J. E., Haarman, E. G. et al. 2013. Splice-site mutations in the axonemal outer dynein arm docking complex gene CCDC114 cause primary ciliary dyskinesia. Am. J. Hum. Genet. 92: 88–98.

4. Knowles, M. R., Leigh, M. W., Ostrowski, L. E., Huang, L., Carson, J. L., Hazucha, M. J., Yin, W., Berg, J. S., Davis, S. D., Dell, S. D. et al. 2013. Genetic Disorders of Mucociliary Clearance C. Exome sequencing identifies mutations in CCDC114 as a cause of primary ciliary dyskinesia. American Journal of Human Genetics 92: 99–106.

5. Wallmeier, J., Shiratori, H., Dougherty, G. W., Edelbusch, C., Hjeij, R., Loges, N. T., Menchen, T., Olbrich, H., Pennekamp, P., Raidt, J. et al. 2016. TTC25 deficiency results in defects of the outer dynein arm docking machinery and primary ciliary dyskinesia with left-right body asymmetry randomization. Am. J. Hum. Genet. 99: 460–469.

6. Hjeij, R., Onoufriadis, A., Watson, C. M., Slagle, C. E., Klena, N. T., Dougherty, G. W., Kurkowiak, M., Loges, N. T., Diggle, C. P., Morante, N. F. et al. 2014. CCDC151 mutations cause primary ciliary dyskinesia by disruption of the outer dynein arm docking complex formation. Am. J. Hum. Genet. 95: 257–274.

7. Olbrich, H., Haffner, K., Kispert, A., Volkel, A., Volz, A., Sasmaz, G., Reinhardt, R., Hennig, S., Lehrach, H., Konietzko, N. et al. 2002. Mutations in DNAH5 cause primary ciliary dyskinesia and randomization of left-right asymmetry. Nature Genetics 30: 143–144.

8. Pennarun, G., Escudier, E., Chapelin, C., Bridoux, A. M., Cacheux, V., Roger, G., Clement, A., Goossens, M., Amselem, S. and Duriez B. 1999. Loss-of-function mutations in a human gene related to Chlamydomonas reinhardtii dynein IC78 result in primary ciliary dyskinesia. American Journal of Human Genetics 65: 1508–1519.

9. Loges, N. T., Olbrich, H., Fenske, L., Mussaffi, H., Horvath, J., Fliegauf, M., Kuhl, H., Baktai, G., Peterffy, E., Chodhari, R. et al. 2008. DNAI2 mutations cause primary ciliary dyskinesia with defects in the outer dynein arm. American Journal of Human Genetics 83: 547–558.

10. Mazor, M., Alkrinawi, S., Chalifa-Caspi, V., Manor, E., Sheffield, V. C., Aviram, M. and Parvari, R. 2011. Primary ciliary dyskinesia caused by homozygous mutation in DNAL1, encoding dynein light chain 1. American Journal of Human Genetics 88: 599–607.

11. Duriez, B., Duquesnoy, P., Escudier, E., Bridoux, A. M., Escalier, D., Rayet, I., Marcos, E., Vojtek, A. M., Bercher, J. F. and Amselem, S. 2007. A common variant in combination with a nonsense mutation in a member of the thioredoxin family causes primary ciliary dyskinesia. Proc. Natl. Acad. Sci. USA. 104: 3336–3341.

12. Bartoloni, L., Blouin, J. L., Pan, Y., Gehrig, C., Maiti, A. K., Scamuffa, N., Rossier, C., Jorissen, M., Armengot, M., Meeks, M. et al. 2002. Mutations in the DNAH11 (axonemal heavy chain dynein type 11) gene cause one form of situs inversus totalis and most likely primary ciliary dyskinesia. Proceedings of the National Academy of Sciences of the United States of America 99: 10282–10286.

Table 2 contd. ...

...Table 2 contd.

13. Schwabe, G. C., Hoffmann, K., Loges, N. T., Birker, D., Rossier, C., de Santi, M. M., Olbrich, H., Fliegauf, M., Failly, M., Liebers, U. et al. 2008. Primary ciliary dyskinesia associated with normal axoneme ultrastructure is caused by DNAH11 mutations. Human Mutation 29: 289–298.

14. Panizzi, J. R., Becker-Heck, A., Castleman, V. H., Al-Mutairi, D. A., Liu, Y., Loges, N. T., Pathak, N., Austin-Tse, C., Sheridan, E., Schmidts, M. et al. 2012. CCDC103 mutations cause primary ciliary dyskinesia by disrupting assembly of ciliary dynein arms. Nat. Genet. 44: 714–719.

15. Austin-Tse, C., Halbritter, J., Zariwala, M. A., Gilberti, R. M., Gee, H. Y., Hellman, N., Pathak, N., Liu, Y., Panizzi, J. R., Patel-King, R. S. et al. 2013. Zebrafish ciliopathy screen plus human mutational analysis identifies C21orf59 and CCDC65 defects as causing primary ciliary dyskinesia. American Journal of Human Genetics 93: 672–686.

16. Loges, N. T., Olbrich, H., Becker-Heck, A., Haffner, K., Heer, A., Reinhard, C., Schmidts, M., Kispert, A., Zariwala, M. A., Leigh, M. W. et al. 2009. Deletions and point mutations of LRRC50 cause primary ciliary dyskinesia due to dynein arm defects. American Journal of Human Genetics 85: 883–889.

17. Duquesnoy, P., Escudier, E., Vincensini, L., Freshour, J., Bridoux, A. M., Coste, A., Deschildre, A., de Blic, J., Legendre, M., Montantin, G. et al. 2009. Loss-of-function mutations in the human ortholog of Chlamydomonas reinhardtii ODA7 disrupt dynein arm assembly and cause primary ciliary dyskinesia. American Journal of Human Genetics 85: 890–896.

18. Omran, H., Kobayashi, D., Olbrich, H., Tsukahara, T., Loges, N. T., Hagiwara, H., Zhang, Q., Leblond, G., O'Toole, E., Hara et al. 2008. Ktu/PF13 is required for cytoplasmic pre-assembly of axonemal dyneins. Nature 456: 611–616.

19. Mitchison, H. M., Schmidts, M., Loges, N. T., Freshour, J., Dritsoula, A., Hirst, R. A., O'Callaghan, C., Blau, H., Al Dabbagh, M., Olbrich, H. et al. 2012. Mutations in axonemal dynein assembly factor DNAAF3 cause primary ciliary dyskinesia. Nature Genetics 44: 381–389, S381–382.

20. Tarkar, A., Loges, N. T., Slagle, C. E., Francis, R., Dougherty, G. W., Tamayo, J. V., Shook, B., Cantino, M., Schwartz, D., Jahnke, C. et al. 2013. DYX1C1 is required for axonemal dynein assembly and ciliary motility. Nature Genetics 45: 995–1003.

21. Horani, A., Druley, T. E., Zariwala, M. A., Patel, A. C., Levinson, B. T., Van Arendonk, L. G., Thornton, K. C., Giacalone, J. C., Albee, A. J., Wilson, K. S. et al. 2012. Whole-exome capture and sequencing identifies HEATR2 mutation as a cause of primary ciliary dyskinesia. American Journal of Human Genetics 91: 685–693.

22. Kott, E., Duquesnoy, P., Copin, B., Legendre, M., Dastot-Le Moal, F., Montantin, G., Jeanson, L., Tamalet, A., Papon, J. F., Siffroi, J. P. et al. 2012. Loss-of-function mutations in LRRC6, a gene essential for proper axonemal assembly of inner and outer dynein arms, cause primary ciliary dyskinesia. American Journal of Human Genetics 91: 958–964.

23. Knowles, M. R., Ostrowski, L. E., Loges, N. T., Hurd, T., Leigh, M. W., Huang, L., Wolf, W. E., Carson, J. L., Hazucha, M. J., Yin, W. et al. 2013. Mutations in SPAG1 cause primary ciliary dyskinesia associated with defective outer and inner dynein arms. American Journal of Human Genetics 93: 711–720.

24. Olcese, C., Patel, M. P., Shoemark, A., Kiviluoto, S., Legendre, M., Williams, H. J., Vaughan, C. K., Hayward, J., Goldenberg, A., Emes, R. D. et al. 2017. X-linked primary ciliary dyskinesia due to mutations in the cytoplasmic axonemal dynein assembly factor PIH1D3. Nat. Commun. 8: 14279.

25. Paff, T., Loges, N. T., Aprea, I., Wu, K., Bakey, Z., Haarman, E. G., Daniels, J. M., Sistermans, E. A., Bogunovic, N., Dougherty, G. W. et al. 2017. Mutations in PIH1D3 cause X-linked primary ciliary dyskinesia with outer and inner dynein arm defects. Am. J. Hum. Genet. 100: 160–168.

26. Moore, D. J., Onoufriadis, A., Shoemark, A., Simpson, M. A., zur Lage, P. I., de Castro, S. C., Bartoloni, L., Gallone, G., Petridi, S., Woollard, W. J. et al. 2013. Mutations in ZMYND10, a gene essential for proper axonemal assembly of inner and outer dynein arms in humans and flies, cause primary ciliary dyskinesia. American Journal of Human Genetics 93: 346–356.

27. Zariwala, M. A., Gee, H. Y., Kurkowiak, M., Al-Mutairi, D. A., Leigh, M. W., Hurd, T. W., Hjeij, R., Dell, S. D., Chaki, M., Dougherty, G. W. et al. 2013. ZMYND10 is mutated in primary ciliary dyskinesia and interacts with LRRC6. American Journal of Human Genetics 93: 336–345.

28. Olbrich, H., Schmidts, M., Werner, C., Onoufriadis, A., Loges, N. T., Raidt, J., Banki, N. F., Shoemark, A., Burgoyne, T., Al Turki, S. et al. 2012. Recessive HYDIN mutations cause primary ciliary dyskinesia without randomization of left-right body asymmetry. Am. J. Hum. Genet. 91: 672–684.

29. Kott, E., Legendre, M., Copin, B., Papon, J. F., Dastot-Le Moal, F., Montantin, G., Duquesnoy, P., Piterboth, W., Amram, D., Bassinet, L. et al. 2013. Loss-of-function mutations in RSPH1 cause primary ciliary dyskinesia with central-complex and radial-spoke defects. American Journal of Human Genetics 93: 561–570.

30. Onoufriadis, A., Shoemark, A., Schmidts, M., Patel, M., Jimenez, G., Liu, H., Thomas, B., Dixon, M., Hirst, R. A., Rutman, A. et al. 2014. Targeted NGS gene panel identifies mutations in RSPH1 causing primary ciliary dyskinesia and a common mechanism for ciliary central pair agenesis due to radial spoke defects. Hum. Mol. Genet. 23: 3362–3374.

31. Castleman, V. H., Romio, L., Chodhari, R., Hirst, R. A., de Castro, S. C., Parker, K. A., Ybot-Gonzalez, P., Emes, R. D., Wilson. S. W., Wallis, C. et al. 2009. Mutations in radial spoke head protein genes RSPH9 and RSPH4A cause primary ciliary dyskinesia with central-microtubular-pair abnormalities. American Journal of Human Genetics 84: 197–209.

32. Jeanson, L., Copin, B., Papon, J. F., Dastot-Le Moal, F., Duquesnoy, P., Montantin, G., Cadranel, J., Corvol, H., Coste, A., Desir, J. et al. 2015. RSPH3 mutations cause primary ciliary dyskinesia with central-complex defects and a near absence of radial spokes. American Journal of Human Genetics 97: 153–162.

33. El Khouri, E., Thomas, L., Jeanson, L., Bequignon, E., Vallette, B., Duquesnoy, P., Montantin, G., Copin, B., Dastot-Le Moal, F., Blanchon, S. et al. 2016. Mutations in DNAJB13, encoding an HSP40 family member, cause primary ciliary dyskinesia and male infertility. Am. J. Hum. Genet. 99: 489–500.

34. Edelbusch, C., Cindric, S., Dougherty, G. W., Loges, N. T., Olbrich, H., Rivlin, J., Wallmeier, J., Pennekamp, P., Amirav, I. and Omran, H. 2017. Mutation of serine/threonine protein kinase 36 (STK36) causes primary ciliary dyskinesia with a central pair defect. Hum. Mutat. 38: 964–969.

35. Merveille, A. C., Davis, E. E., Becker-Heck, A., Legendre, M., Amirav, I., Bataille, G., Belmont, J., Beydon, N., Billen, F., Clement, A. et al. 2011. CCDC39 is required for assembly of inner dynein arms and the dynein regulatory complex and for normal ciliary motility in humans and dogs. Nature Genetics 43: 72–78.

36. Becker-Heck, A., Zohn, I. E., Okabe, N., Pollock, A., Lenhart, K. B., Sullivan-Brown, J., McSheene, J., Loges, N. T., Olbrich, H., Haeffner, K. et al. 2011. The coiled-coil domain containing protein CCDC40 is essential for motile cilia function and left-right axis formation. Nature Genetics 43: 79–84.

37. Horani, A., Brody, S. L., Ferkol, T. W., Shoseyov, D., Wasserman, M. G., Ta-shma, A., Wilson, K. S., Bayly, P. V., Amirav, I., Cohen-Cymberknoh, M. et al. 2013. CCDC65 mutation causes primary ciliary dyskinesia with normal ultrastructure and hyperkinetic cilia. PloS One 8: e72299.

38. Wirschell, M., Olbrich, H., Werner, C., Tritschler, D., Bower, R., Sale, W. S., Loges, N. T., Pennekamp, P., Lindberg, S., Stenram, U. et al. 2013. The nexin-dynein regulatory complex subunit DRC1 is essential for motile cilia function in algae and humans. Nature Genetics 45: 262–268.

Table 2 contd. …

...Table 2 contd.

39. Olbrich, H., Cremers, C., Loges, N. T., Werner, C., Nielsen, K. G., Marthin, J. K., Philipsen, M., Wallmeier, J., Pennekamp, P., Menchen, T. et al. 2015. Loss-of-function GAS8 mutations cause primary ciliary dyskinesia and disrupt the nexin-dynein regulatory complex. American Journal of Human Genetics 97: 546–554.

40. Boon, M., Wallmeier, J., Ma, L., Loges, N. T., Jaspers, M., Olbrich, H., Dougherty, G. W., Raidt, J., Werner, C., Amirav, I. et al. 2014. MCIDAS mutations result in a mucociliary clearance disorder with reduced generation of multiple motile cilia. Nature Communications 5: 4418.

41. Wallmeier, J., Al-Mutairi, D. A., Chen, C. T., Loges, N. T., Pennekamp, P., Menchen, T., Ma, L., Shamseldin, H. E., Olbrich, H., Dougherty, G. W. et al. 2014. Mutations in CCNO result in congenital mucociliary clearance disorder with reduced generation of multiple motile cilia. Nature Genetics 46: 646–651.

42. Moore, A., Escudier, E., Roger, G., Tamalet, A., Pelosse, B., Marlin, S., Clement, A., Geremek, M., Delaisi, B., Bridoux, A. M. et al. 2006. RPGR is mutated in patients with a complex X linked phenotype combining primary ciliary dyskinesia and retinitis pigmentosa. Journal of Medical Genetics 43: 326–333.

chains, generally confers cilia immotility and this is the most common defect seen in PCD patients. ODA defects arise from mutations either in structural components of the outer dynein arms, the most commonly mutated being the dynein motor subunits *DNAH5*, *DNAH11* and *DNAI1*; or in components of the ODA-docking complex system required for attachment of ODAs to the axonemal microtubules (Hjeij et al., 2013; Lucas et al., 2014; Onoufriadis et al., 2014a) (Table 2). Notably, *DNAH11* mutations do not result in a structural defect that is visible by TEM, and the cilia have a specific stiff, hyperkinetic pattern rather than being completely static (Knowles et al., 2012).

Mutations in a set of genes cause PCD with combined loss of the inner and outer dynein arm motors. These are all thought to be involved in the cytoplasmically localized system for pre-assembly of dynein arm motors that occurs prior to their import into cilia (Omran et al., 2008; Mitchison et al., 2012), with the exception of CCDC103 which is was recently reported to be an integral axonemal protein involved in dynein arm microtubule attachment (King and Patel-King, 2015) (Table 2). Other structural axonemal components that govern ciliary beating and waveform are also mutated in PCD, comprising proteins of the central microtubular pair apparatus, radial spokes and nexin-dynein regulatory (N-DRC) complexes (Table 2). Since the outer dynein arms and often also the inner dynein arms are retained in the axoneme in these disease subtypes, the cilia can move but generally have defective waveforms which are all ineffective for mucociliary clearance. The different mutation classes in this patient subset confer specific related ultrastructural and motility defects (Table 2) that allow a further stratification of PCD disease and can assist with clinical management. For example, a loss of the central pairs confers a circular motility pattern to the motile cilia reminiscent of the movement of 9 + 0 node cilia that lack the central pair, and correspondingly these mutations do not confer laterality defects to the patients since the node cilia do not require central pairs. Loss of the non-motor radial spokes and N-DRC complex structures can confer more subtle defects in motility that are starting to be linked to milder disease symptoms (Knowles et al., 2014). However, disease may be more severe in cases of PCD caused by *CCDC39* and *CCDC40* mutations, where there is disruption to the N-DRC accompanied by loss of the inner dynein arms (Davis et al., 2015). CCDC39 and CCDC40 form a complex acting as a 96-nanometer long 'molecular ruler' that maintains the regular, repetitive distribution of axonemal motor components along the axoneme. This is key to ciliary motility and its loss disrupts the dispersal and maintenance of the axonemal components (Oda et al., 2014).

Only two genes are known to cause RGMC: *CCNO* and *MCIDAS* (Table 2) (Boon et al., 2014; Wallmeier et al., 2014). As mentioned above, microRNAs of the miR-34/449 family trigger multiciliated cell differentiation and ciliogenesis by repressing cell cycle genes and the Notch pathway. The *MCIDAS* and *CCNO* genes are proximal to each other within the miR-34/449 locus with their expression also negatively controlled by the Notch pathway activity, such that repression of Notch leads to increased *MCIDAS, CCNO* and *miR-449* expression. *MCIDAS* encoding multicilin is a master regulator directing the transcriptional activation of genes required for multicilia formation and its expression is sufficient to generate multiciliated cell differentiation in mouse airway epithelial cultures (Stubbs et al., 2012). *CCNO* encoding cyclin O works downstream in this pathway, with its expression driven by *MCIDAS*, playing a role in the deuterosome-mediated amplification of mother centrioles that mature into basal bodies which form the platform from which the multicilia can nucleate and grow (Funk et al., 2015). RGMC patients with *MCIDAS* and *CCNO* mutations can assemble only a few cilia per cell in the respiratory epithelia and *CCNO* mutants seem to assemble the normal motility apparatus as their ciliary beating appears normal (Boon et al., 2014; Wallmeier et al., 2014).

Progress is being made in understanding how the families affected by motile ciliopathy diseases can benefit from these advances in genetics, with such clinically useful correlations between the underlying genotype of an affected individual and their predicted disease course starting to emerge. With clinical variability and likely underdiagnosis of these poorly recognized conditions, improved genetics assists in more rapid and earlier diagnosis, with more relevant counselling, which is clearly linked to improved disease outcomes through the earlier management of symptoms (Lucas et al., 2014). In a different clinical association, there has been recent interest in understanding the role of motile cilia gene defects in idiopathic pulmonary fibrosis (IPF), since the finding that ciliary gene expression is markedly increased in lung samples from IPF patients and particularly associated with increased irreversible fibrotic cystic 'honeycombing' of the lungs (Yang et al., 2013). IPF is the commonest form of irreversible interstitial lung disease, being a progressive and often lethal condition with inherited familial forms caused by rare mutations in surfactant and telomerase complex genes that are likely to exist alongside a genetic component of more common gene variants causing disease (Kropski et al., 2015). Whether there is a causative role for cilia mutations in this disease remains to be fully investigated with further human next-generation sequencing efforts. In many adult lung diseases it is known that the cilia become affected as a consequence of the disease pathogenesis and their deterioration or functional impairment then contributes to the disease symptoms. These secondary cilia defects are discussed below.

Lung Disorders and Secondary Motile Cilia Defects

During normal breathing a range of different particles including bacteria, viruses, environmental and workplace pollutants can become deposited in the lungs. All of these have the potential to cause damage to the epithelial cells lining the airways, and if exposure is persistent or chronic, this damage can result in impaired lung function. In healthy lungs, airway mucus secreted by goblet cells and submucosal glands forms a barrier and aids in the protection of the lungs by trapping external particles, which are then cleared from the lungs by three mechanisms: mucociliary clearance, coughing and alveolar clearance. Mucociliary clearance via the mucociliary escalator clears the conducting airways of secreted mucus along with particles that might be trapped in it. Coughing aids this process especially if mucus clearance is impaired by structural or functional defects to motile cilia, or as a result of recurrent infections, or in other disease states whereby mucus secretion is increased or the composition of mucus is abnormal, such as in Cystic Fibrosis. Finally, alveolar clearance removes the insoluble particles deposited on the respiratory surface of the lungs.

Pathogens

Infection by a microorganism can alter airway cilia function leading to impaired mucociliary clearance (Amitani et al., 1991; Look et al., 2001). Microorganisms can affect cilia function in different ways; by targeting ciliated cells for adherence, reducing cilia beat frequency and/or disrupting cilia coordination and inducing ciliary dyskinesia (Look et al., 2001; Balder et al., 2009). Lung samples from patients infected with respiratory syncytial virus show that epithelial damage and loss of cilia are associated with decreased expression of the differentiation regulator forkhead box J1 (FOXJ1), with similar findings observed in mouse models (Look et al., 2001). The immune and **inflammatory** responses to infections can themselves result in impaired mucociliary clearance. High concentrations of human neutrophil elastase and reactive oxygen species generated by polymorphonuclear leukocytes (neutrophils, eosinophils and basophils) can reduce ciliary beat frequency (Amitani et al., 1991; Kantar et al., 1994).

Smoking

Smoking is a significant risk factor for the development of respiratory diseases, such as Chronic Obstructive Pulmonary Disease (COPD; discussed below) and has been associated with exacerbations of other lung diseases such as asthma (discussed below). Exposure to cigarette smoke can result in dramatic changes to the airway epithelium, including basal cell hyperplasia, mucus overproduction, and squamous metaplasia as well as structural and functional abnormalities of ciliated cells and an increase in airway barrier permeability (Sisson et al., 1994; Wistuba and Gazdar, 2006; Leopold et al., 2009; Shaykhiev et al., 2013).

Cigarette smoke induces oxidative stress within the lungs that can disrupt normal cell differentiation, repair and function. This oxidative stress has recently been linked to epidermal growth factor receptor (EGFR) activation in human lung cells *in vitro* (Takeyama et al., 2000; Casalino-Matsuda et al., 2006; Filosto et al., 2011; Shaykhiev et al., 2013). It has been proposed that in response to cigarette smoke, ciliated cells produce epidermal growth factors (EGFs) that stimulate basal cell differentiation towards a squamous phenotype and thereby suppress ciliated cell differentiation (Shaykhiev et al., 2013). Exposure to cigarette smoke, including passive exposure, can result in damage to ciliated cells, such that some cells lack the correct number of cilia and/or display atypical nuclei. Healthy smokers display shorter motile cilia in the large and small airways compared to cilia of nonsmokers (Leopold et al., 2009; Hessel et al., 2014). These defects are more severe in smokers with an associated respiratory disease such as COPD or chronic bronchitis, where giant cilia, compound cilia and/or structural ciliary defects have been observed (Yaghi et al., 2012; Tilley et al., 2015).

Mucociliary clearance is also impaired in smokers compared to non-smokers and cessation of smoking results in measurable improved nasal mucociliary clearance (Ramos et al., 2011). A number of studies have investigated whether smoking results in altered motile cilia beating. However, current data remains controversial with different studies reporting conflicting results as to whether cilia beat frequency is disrupted in healthy/diseased smokers versus healthy/diseased non-smokers (Tilley et al., 2015b).

COPD

COPD is an inflammatory lung disease that causes airway narrowing and breathlessness. It is currently the 3rd leading cause of death worldwide (Burney et al., 2015). Emphysema and chronic bronchitis are the two main components of COPD. Emphysema is the destruction of peripheral lung tissue (alveoli), such that lung function is impaired. Chronic bronchitis is defined by long-term inflammation of the bronchi and an increase in mucus production that together causes damage to the airways. COPD is predominantly caused by cigarette smoke; however continual exposure to other environmental pollutants and a predisposition to α1-antitrypsin deficiency are also known causal factors. In patients with pure emphysema induced by α1-antitrypsin deficiency, mucociliary clearance is normal, but it is impaired in smokers both with and without COPD, as discussed above.

The pathogenesis of COPD is complex and remains poorly defined. Current understanding suggests cigarette smoke and other environmental pollutants inhibit the normal function of airway epithelial cells by reducing motile cilia length, inducing airway epithelial cell death and increasing the goblet cell population, all of which subsequently increases mucus production (Bartalesi et al., 2005; Mercer et al., 2006; Haswell et al., 2010; Cloonan et al., 2014). Together these factors result in impaired mucociliary clearance such that pathogens and external particles cannot be efficiently removed from the lungs, resulting in chronic and/or recurrent infections that can exacerbate the disease (Barnes et al., 2003; Cloonan et al., 2014).

Autophagy is a process by which cellular components are degraded or recycled in order to promote cell survival under stress conditions, but excessive autophagy can **influence** apoptosis and other forms of cell death (Levine and Yuan, 2005; Chen et al., 2008). Through its multifunctional roles in regulating organelle homeostasis, **inflammation** and immune responses and protein turnover, autophagy may play a key role in lung disease progression (Choi et al., 2013). Autophagic protein expression is increased in COPD patients and recent studies have begun to attribute the shortened cilia seen in COPD patients to an autophagy-dependent mechanism mediated by Histone Deacetylase 6 (HDAC6), via a process newly referred to as ciliophagy (Lam et al., 2013).

Asthma

Electron microscopy of epithelial biopsies in both children and adults with asthma shows damage to ciliated cells which includes both loss of cilia and abnormal cilia structure. Indeed, autopsies of patients that suffered fatal asthma show loss of cilia, shedding of the bronchial epithelium and bronchial plugging (Kuyper et al., 2003). Consistent with these findings a number of functional studies have shown asthmatics to have reduced mucociliary clearance compared to healthy controls (Erle and Sheppard, 2014). Ciliary defects worsen as the severity of asthma increases; moderate and severe asthmatics have more dyskinetic and immotile cilia, with ciliary beat frequency appearing reduced compared to controls (Thomas et al., 2010; Erle and Sheppard, 2014). Severe asthma patients have more cilia disorientation, cilia depletion and microtubule defects than either controls or patients with mild asthma (Thomas et al., 2010). IL-13, a Th2-type cytokine, is overexpressed in the airways of asthma patients (Erle and Sheppard, 2014). IL-13 promotes goblet cell differentiation and a reduction in ciliated cells as well as the number of cilia per cell (Gomperts et al., 2007). In addition, via its interaction with ezrin, excess IL-13 can reduce the number of basal bodies (Gomperts et al., 2007; Erle and Sheppard, 2014). IL-13 has also been shown to slow and in some cases eliminate cilia beat frequency, by reducing FOXJ1 expression (Gomperts et al., 2007).

Cystic Fibrosis

Cystic Fibrosis (CF) is an autosomal recessive disease that affects the lungs, pancreas, liver, kidneys and intestine and is caused by mutations in the cystic fibrosis transmembrane conductance regulator (CFTR) (Ehre et al., 2014). Almost 2,000 mutations have been identified in the CFTR gene that encodes a cAMP-regulated Cl⁻ channel that modulates fluid absorption. The most severe manifestations of this disease occur in the airways, whereby dysfunctional Cl⁻ channels lead to dehydration of the epithelial surface, an increase in mucus viscosity and compromised mucociliary clearance (Ehre et al., 2014). Together these defects promote chronic infection and inflammation, ultimately resulting in respiratory failure (Davies, 2002; Rosenbluth et al., 2004). Defects in normal mucociliary clearance in CF patients arise due to the abnormal viscosity of airway mucus and not as a result of ciliopathy (Ehre et al., 2014). Electron microscopy of airway cilia from CF patients' lungs reveals an array of ciliary defects including regions of missing cilia, cilia with missing inner dynein arms, compound cilia and multiple cilia (Piorunek et al., 2008). As abnormal mucus production is the major cause of the CF phenotype, treatment strategies are currently focusing on reducing mucus viscosity in an attempt to restore normal mucociliary clearance (Ehre et al., 2014).

Cancer

The majority of cancers (85%) arise from epithelial tissues. Generally, cancer-initiating epithelial cells display primary cilia rather than motile cilia, therefore most studies have focused on the role of primary cilia in cancer (Hassounah et al., 2012). Cancer biologists first became interested in primary cilia due to their known function in cell cycle regulation. Primary cilia play an important role during embryonic development where they regulate Hedgehog signalling (Hh) that is required for normal cell growth and differentiation (Goetz et al., 2009; Hassounah et al., 2012; Shpak et al., 2014). More recently, primary cilia and Hh signalling have been implicated in tumourigenesis (Wong et al., 2009; Shpak et al., 2014). Primary cilia have been shown to both promote and inhibit tumourigenesis, depending on when early oncogenic events take place, i.e., upstream or downstream of Hh signaling (Wong et al., 2009). In breast and pancreatic cancer, epithelial cells appear to lose their primary cilia, whereas primary cilia remain in basal cell carcinoma and medulloblastoma (Han et al., 2009). Importantly, loss of primary cilia is often associated with aggressive disease and poorer prognosis (Hassounah et al., 2012). Loss of primary cilia could be attributed to an increase in cell proliferation, which is known to influence the presence of cilia. However, studies have shown that loss of primary cilia from cancerous cells was independent of Ki67 staining, a marker of proliferation, and therefore suggests that loss of cilia on cancerous cells may be a result of mutations in cilia-related genes and/or disruption of ciliogenesis (Hassounah et al., 2012; Shpak et al., 2014).

Lung cancer is the leading cause of cancer deaths worldwide (Cancer Research UK, 2015) and is initiated by changes in the airway epithelium caused by environmental insults, predominantly smoking and/or genetic mutations. While our understanding of the role of cilia in cancer remains limited, early data suggests that loss of cilia function may not be a risk factor for lung cancer, as ciliopathy patients rarely present with lung cancer (Inoue et al., 2011). However, genes important for regulating the differentiation of ciliated cells, including FOXJ1 and dynein axonemal intermediate chain 1 (DNAI1) were recently shown to be downregulated in patients with a subset of lung cancer; basal cell-high adenocarcinoma (Fukui et al., 2013). Down regulation of FOXJ1 and DNAI1 was associated with activation of the epithelial-to-mesenchymal transcriptional program (EMT) in these patients. EMT can promote invasion and metastasis, a marker of aggressive disease that usually results in a poor outcome.

Diagnosis in the Clinic

In the clinical setting, impaired lung function associated with disease or environmental insult is typically investigated and diagnosed using a range of tests of breath and lung capacity (spirometry), and measurements of lung volume and gas transfer. Exhaled nitric oxide testing is increasingly used, particularly as a marker of inflammation, and cultures taken from sputum or bronchoalveolar lavage samples can evaluate the nature and level of bacterial infection in the chest, the 'microbiome'. Continued lung function testing aids in monitoring the progression, severity and management of symptoms in patients with chronic respiratory diseases, and these tests also form a key component in any clinical trial.

Lung function and capacity can be measured using different approaches but the widest used is spirometry which measures volume against time, generating the forced expired vital capacity (FVC) (Ranu et al., 2011). Repeated forced breaths are made into a spirometer to measure how much air is being inhaled and exhaled, generating a measure of FVC in litres. From the forced expirogram, forced expired volume in one second (FEV_1 in litres per second) can be calculated as a percentage of FVC, as well as the peak (PEF) and maximal (MEF) expiratory flows. Age, height and sex are taken into account as these influence vital capacity measures. Bronchiectasis which is a feature of cystic fibrosis and motile ciliopathy diseases, as described above, is generally diagnosed by chest X-radiography but this is not highly sensitive, so monitoring is preferably done by high-resolution computed tomography (CT, HRCT). However, since this involves exposure to ionising radiation, chest MRI may be considered as an alternative. The lung clearance index (LCI) is also used as a measure of lung function, recorded using the multiple-breath washout technique (Lucas et al., 2014). Lung volume can be measured to assess obstructive disease, and total lung capacity is often evaluated by whole body plethysmography using a sealed chamber and measures of airway resistance and conductance. Lung gas transfer capacity can also be measured, using carbon monoxide lung diffusion testing.

Worldwide, there is no universally agreed diagnostic pathway for PCD and RGMC motile ciliopathy conditions. European guidelines have been developed that describe a range of clinical tests used to support diagnosis (Barbato et al., 2009; Strippoli et al., 2012; Lucas et al., 2014; Lucas et al., 2017). Individuals referred for diagnostic testing include neonates showing respiratory distress of unknown cause, and people with abnormal situs and respiratory symptoms, daily lifelong wet cough, unexplained bronchiectasis, or airway symptoms accompanied by otitis media or with heterotaxy and dextrocardia. Along with other recessive conditions, the motile ciliopathies are considerably enriched in consanguineous population such as British Asians, and therefore family origin is also taken into account in diagnostic decisions. The clinical recognition of these rare conditions is not optimal and the motile ciliopathies are considered widely underdiagnosed worldwide, complicated by the variability of disease symptoms with congenital recessive and X-linked syndromic forms that are variable in their presentation. The specialized diagnostic testing required and lack of specific clinical diagnostic tests also means patients can remain undiagnosed until adulthood or can be completely missed despite numerous visits to physicians (Lucas et al., 2014). In the case of PCD, diagnostic pick-up is thought to be better in the approximately 50% of cases where left-right body axis defects are apparent in X-rays.

If there is an initial index of suspicion, i.e., characteristic clinical phenotype, this would prompt further clinical testing. The clinical tests in use typically are machine-based measurements of expelled nasal nitric oxide levels, since levels are low in PCD patients. Since this involves the holding of breath, then it is not suitable for younger children (Strippoli et al., 2012). A lower tech saccharine movement test is also possible, but is considered unreliable in children. Diagnosis would then mainly rest upon testing of a biopsy of ciliated cells obtained either by nasal brushing or via bronchoscopy. These samples are used for high resolution imaging to identify abnormal ciliary beating (beat frequency, beat pattern) by light microscopy and high speed video imaging, and also to characterise PCD-related specific ultrastructural ciliary defects and ciliary disorientation that are identified through transmission electron microscopy. PCD is characterized by a range of typical structural defects visible in cilia cross sections, and the ciliary beat frequency and pattern is assessed to determine reduced motility and mucus flow. Culture of ciliated epithelium is possible in order to provide follow up testing and rule out secondary cilary dyskinesia. Increasingly, the use of genetic diagnosis and light microscopic **immunofluoresence** studies of ciliary proteins in ciliated cells is being incorporated into the clinical diagnosis, which can help with difficult cases to further define the defects (Lucas et al., 2017). For example, the low number of cilia that characterize RGMC make these clinically difficult since a lack of cilia can be missed or regarded as poor sample quality, so these patients may have to undergo numerous brush biopsies before the lack of cilia can be verified robustly, whilst a genetic test can be rapidly performed using a blood or saliva sample. As outlined by Lucas and colleagues, these diagnostic investigations are complex, requiring expensive infrastructure. Ideally, an experienced team of clinicians and scientists at tertiary respiratory centres play a key role (Lucas et al., 2014).

Conclusions

Progress is rapidly being made into understanding the biology of lung cilia in development and disease. Recent advances in genetics such as gene editing technology and the Genomics England 100,000 Genomes Project are likely to greatly facilitate future studies of ciliopathies affecting the lungs (Marks 2015). In addition we are likely to see a rapid expansion in our understanding of how the underlying genotype of an affected individual will influence their likely disease course. To date, most research has focused on the multicilia, which are considerably more tractable. However, studies of non-motile, primary lung cilia are beginning to emerge and it will be interesting to compare the biology of primary cilia in the lungs with that from other organs such as the kidney or heart, where only primary cilia are present.

References

Al Alam, D., Danopoulos, S., Schall, K., Sala, F. G., Almohazey, D., Fernandez, G. E., Georgia, S., Frey, M. R., Ford, H. R., Grikscheit, T. et al. 2015. Fibroblast growth factor 10 alters the balance between goblet and Paneth cells in the adult mouse small intestine. AM. J. Physiol. Gastrointest. Liver Physiol. 308(8): G678–90.

Alanis, D. M., Chang, D. R., Akiyama, H., Krasnow, M. A. and Chen, J. 2014. Two nested developmental waves demarcate a compartment boundary in the mouse lung. Nat. Commun. 5: 3923.

Amirav, I., Wallmeier, J., Loges, N. T., Menchen, T., Pennekamp, P., Mussaffi, H., Abitbul, R., Avital, A., Bentur, L., Dougherty, G. W. et al. 2016. Systematic analysis of CCNO variants in a defined population: implications for clinical phenotype and differential diagnosis. Hum. Mutat.

Amitani, R., Wilson, R., Rutman, A., Read, R., Ward, C., Burnett, D., Stockley, R. A. and Cole, P. J. 1991. Effects of human neutrophil elastase and Pseudomonas aeruginosa proteinases on human respiratory epithelium. Am. J. Respir Cell Mol. Biol. 4(1): 26–32.

Austin-Tse, C., Halbritter, J., Zariwala, M. A., Gilberti, R. M., Gee, H. Y., Hellman, N., Pathak, N., Liu, Y., Panizzi, J. R., Patel-King, R. S. et al. 2013. Zebrafish ciliopathy screen plus human mutational analysis identifies C21orf59 and CCDC65 defects as causing primary ciliary dyskinesia. Am. J. Hum. Genet. 93(4): 672–86.

Balder, R., Krunkosky, T. M., Nguyen, C. Q., Feezel, L. and Lafontaine, E. R. 2009. Hag mediates adherence of Moraxella catarrhalis to ciliated human airway cells. Infect. Immun. 77(10): 4597–608.

Barbato, A., Frischer, T., Kuehni, C. E., Snijders, D., Azevedo, I., Baktai, G., Bartoloni, L., Eber, E., Escribano, A., Haarman, E. et al. 2009. Primary ciliary dyskinesia: a consensus statement on diagnostic and treatment approaches in children. Eur. Respir J. 34(6): 1264–76.

Barkauskas, C. E., Cronce, M. J., Rackley, C. R., Bowie, E. J., Keene, D. R., Stripp, B. R., Randell, S. H., Noble, P. W. and Hogan, B. L. 2013. Type 2 alveolar cells are stem cells in adult lung. J. Clin. Invest. 123(7): 3025–36.

Barnes, P. J., Shapiro, S. D. and Pauwels, R. A. 2003. Chronic obstructive pulmonary disease: molecular and cellular mechanisms. Eur. Respir. J. 22(4): 672–88.

Bartalesi, B., Cavarra, E., Fineschi, S., Lucattelli, M., Lunghi, B., Martorana, P. A. and Lungarella, G. 2005. Different lung responses to cigarette smoke in two strains of mice sensitive to oxidants. Eur. Respir. J. 25(1): 15–22.

Bartoloni, L., Blouin, J. L., Pan, Y., Gehrig, C., Maiti, A. K., Scamuffa, N., Rossier, C., Jorissen, M., Armengot, M., Meeks, M. et al. 2002. Mutations in the DNAH11 (axonemal heavy chain dynein type 11) gene cause one form of situs inversus totalis and most likely primary ciliary dyskinesia. Proc. Natl. Acad. Sci. USA 99(16): 10282–6.

Becker-Heck, A., Zohn, I. E., Okabe, N., Pollock, A., Lenhart, K. B., Sullivan-Brown, J., McSheene, J., Loges, N. T., Olbrich, H., Haeffner, K. et al. 2011. The coiled-coil domain containing protein CCDC40 is essential for motile cilia function and left-right axis formation. Nat. Genet. 43(1): 79–84.

Berbari, N. F., O'Connor, A. K., Haycraft, C. J. and Yoder, B. K. 2009. The primary cilium as a complex signaling center. Current Biology CB 19(13): R526–35.

Boon, M., Wallmeier, J., Ma, L., Loges, N. T., Jaspers, M., Olbrich, H., Dougherty, G. W., Raidt, J., Werner, C., Amirav, I. et al. 2014. MCIDAS mutations result in a mucociliary clearance disorder with reduced generation of multiple motile cilia. Nat. Commun 5: 4418.

Budny, B., Chen, W., Omran, H., Fliegauf, M., Tzschach, A., Wisniewska, M., Jensen, L. R., Raynaud, M., Shoichet, S. A., Badura, M. et al. 2006. A novel X-linked recessive mental retardation syndrome comprising macrocephaly and ciliary dysfunction is allelic to oral-facial-digital type I syndrome. Human Genetics 120(2): 171–8.

Bukowy-Bieryllo, Z., Zietkiewicz, E., Loges, N. T., Wittmer, M., Geremek, M., Olbrich, H., Fliegauf, M., Voelkel, K., Rutkiewicz, E., Rutland, J. et al. 2013. RPGR mutations might cause reduced orientation of respiratory cilia. Pediatr Pulmonol. 48(4): 352–63.

Burney, P. G., Patel, J., Newson, R., Minelli, C. and Naghavi, M. 2015. Global and regional trends in COPD mortality, 1990–2010. Eur. Respir. J. 45(5): 1239–47.

Carraro, G., El-Hashash, A., Guidolin, D., Tiozzo, C., Turcatel, G., Young, B. M., De Langhe, S. P., Bellusci, S., Shi, W., Parnigotto, P. P. et al. 2009. miR-17 family of microRNAs controls FGF10-mediated embryonic lung epithelial branching morphogenesis through MAPK14 and STAT3 regulation of E-Cadherin distribution. Dev. Biol. 333(2): 238–50.

Casalino-Matsuda, S. M., Monzon, M. E. and Forteza, R. M. 2006. Epidermal growth factor receptor activation by epidermal growth factor mediates oxidant-induced goblet cell metaplasia in human airway epithelium. Am. J. Respir Cell Mol. Biol. 34(5): 581–91.

Castleman, V. H., Romio, L., Chodhari, R., Hirst, R. A., de Castro, S. C., Parker, K. A., Ybot-Gonzalez, P., Emes, R. D., Wilson, S. W., Wallis, C. et al. 2009. Mutations in radial spoke head protein genes RSPH9 and RSPH4A cause primary ciliary dyskinesia with central-microtubular-pair abnormalities. Am. J. Hum. Genet. 84(2): 197–209.

Chambers, R. C. and Mercer, P. F. 2015. Mechanisms of alveolar epithelial injury, repair, and fibrosis. Annals of the American Thoracic Society 12 Suppl 1: S16–20.

Chang, D. R., Martinez Alanis, D., Miller, R. K., Ji, H., Akiyama, H., McCrea, P. D. and Chen, J. 2013. Lung epithelial branching program antagonizes alveolar differentiation. Proc. Natl. Acad Sci. USA 110(45): 18042–51.

Chao, C. M., El Agha, E., Tiozzo, C., Minoo, P. and Bellusci, S. 2015. A breath of fresh air on the mesenchyme: impact of impaired mesenchymal development on the pathogenesis of bronchopulmonary dysplasia. Front Med. (Lausanne) 2: 27.

Chen, J. and Krasnow, M. A. 2012. Integrin Beta 1 suppresses multilayering of a simple epithelium. PLoS One 7(12): e52886.

Chen, Z. H., Kim, H. P., Sciurba, F. C., Lee, S. J., Feghali-Bostwick, C., Stolz, D. B., Dhir, R., Landreneau, R. J., Schuchert, M. J., Yousem, S. A. et al. 2008. Egr-1 regulates autophagy in cigarette smoke-induced chronic obstructive pulmonary disease. PLoS One 3(10): e3316.

Chilvers, M. A. and O'Callaghan, C. 2000. Local mucociliary defence mechanisms. Paediatric Respiratory Reviews 1(1): 27–34.

Choi, A. M., Ryter, S. W. and Levine, B. 2013. Autophagy in human health and disease. N. Engl. J. Med. 368(19): 1845–6.

Chung, S. and Andrew, D. J. 2008. The formation of epithelial tubes. J. Cell Sci. 121(Pt 21): 3501–4.

Cloonan, S. M., Lam, H. C., Ryter, S. W. and Choi, A. M. 2014. Ciliophagy: The consumption of cilia components by autophagy. Autophagy 10(3): 532–4.

Crystal, R. G., Randell, S. H., Engelhardt, J. F., Voynow, J. and Sunday, M. E. 2008. Airway epithelial cells: current concepts and challenges. Proc. Am. Thorac. Soc. 5(7): 772–7.

Davey, M. G., McTeir, L., Barrie, A. M., Freem, L. J. and Stephen, L. A. 2014. Loss of cilia causes embryonic lung hypoplasia, liver fibrosis, and cholestasis in the talpid3 ciliopathy mutant. Organogenesis 10(2): 177–85.

Davies, A. and Moores, C. 2003. The Respiratory System: Churchill Livingstone.

Davies, J. C. 2002. Pseudomonas aeruginosa in cystic fibrosis: pathogenesis and persistence. Paediatr. Respir. Rev. 3(2): 128–34.

Davis, S. D., Ferkol, T. W., Rosenfeld, M., Lee, H. S., Dell, S. D., Sagel, S. D., Milla, C., Zariwala, M. A., Pittman, J. E., Shapiro, A. J. et al. 2015. Clinical features of childhood primary ciliary dyskinesia by genotype and ultrastructural phenotype. Am. J. Respir Crit. Care. Med. 191(3): 316–24.

De Moerlooze, L., Spencer-Dene, B., Revest, J. M., Hajihosseini, M., Rosewell, I. and Dickson, C. 2000. An important role for the IIIb isoform of fibroblast growth factor receptor 2 (FGFR2) in mesenchymal-epithelial signalling during mouse organogenesis. Development 127(3): 483–92.

Domyan, E. T., Ferretti, E., Throckmorton, K., Mishina, Y., Nicolis, S. K. and Sun, X. 2011. Signaling through BMP receptors promotes respiratory identity in the foregut via repression of Sox2. Development 138(5): 971–81.

Duquesnoy, P., Escudier, E., Vincensini, L., Freshour, J., Bridoux, A. M., Coste, A., Deschildre, A., de Blic, J., Legendre, M., Montantin, G. et al. 2009. Loss-of-function mutations in the human ortholog of Chlamydomonas reinhardtii ODA7 disrupt dynein arm assembly and cause primary ciliary dyskinesia. Am. J. Hum. Genet. 85(6): 890–6.

Ehre, C., Ridley, C. and Thornton, D. J. 2014. Cystic fibrosis: an inherited disease affecting mucin-producing organs. Int. J. Biochem. Cell Biol. 52: 136–45.

Erle, D. J. and Sheppard, D. 2014. The cell biology of asthma. J. Cell Biol. 205(5): 621–31.

Filosto, S., Khan, E. M., Tognon, E., Becker, C., Ashfaq, M., Ravid, T. and Goldkorn, T. 2011. EGF receptor exposed to oxidative stress acquires abnormal phosphorylation and aberrant activated conformation that impairs canonical dimerization. PLoS One 6(8): e23240.

Francis, R. J., Chatterjee, B., Loges, N. T., Zentgraf, H., Omran, H. and Lo, C. W. 2009. Initiation and maturation of cilia-generated flow in newborn and postnatal mouse airway. Am. J. Physiol. Lung Cell Mol. Physiol. 296(6): L1067–75.

Fukui, T., Shaykhiev, R., Agosto-Perez, F., Mezey, J. G., Downey, R. J., Travis, W. D. and Crystal, R. G. 2013. Lung adenocarcinoma subtypes based on expression of human airway basal cell genes. Eur. Respir J. 42(5): 1332–44.

Funk, M. C., Bera, A. N., Menchen, T., Kuales, G., Thriene, K., Lienkamp, S. S., Dengjel, J., Omran, H., Frank, M. and Arnold, S. J. 2015. Cyclin O (Ccno) functions during deuterosome-mediated centriole amplification of multiciliated cells. EMBO J. 34(8): 1078–89.

Gaultier, C., Bourbon, J. and Post, M. 1999. Lung Development: Springer.

Goetz, S. C., Ocbina, P. J. and Anderson, K. V. 2009. The primary cilium as a Hedgehog signal transduction machine. Methods Cell Biol. 94: 199–222.

Goggolidou, P., Stevens, J. L., Agueci, F., Keynton, J., Wheway, G., Grimes, D. T., Patel, S. H., Hilton, H., Morthorst, S. K., DiPaolo, A. et al. 2014. ATMIN is a transcriptional regulator of both lung morphogenesis and ciliogenesis. Development 141(20): 3966–77.

Gomperts, B. N., Kim, L. J., Flaherty, S. A. and Hackett, B. P. 2007. IL-13 regulates cilia loss and foxj1 expression in human airway epithelium. Am. J. Respir. Cell Mol. Biol. 37(3): 339–46.

Goss, A. M., Tian, Y., Tsukiyama, T., Cohen, E. D., Zhou, D., Lu, M. M., Yamaguchi, T. P. and Morrisey, E. E. 2009. Wnt2/2b and beta-catenin signaling are necessary and sufficient to specify lung progenitors in the foregut. Dev. Cell 17(2): 290–8.

Hagood, J. S., Miller, P. J., Lasky, J. A., Tousson, A., Guo, B., Fuller, G. M. and McIntosh, J. C. 1999. Differential expression of platelet-derived growth factor-alpha receptor by Thy-1(−) and Thy-1(+) lung fibroblasts. Am. J. Physiol. 277(1 Pt 1): L218–24.

Halayko, A. J., Tran, T., Ji, S. Y., Yamasaki, A. and Gosens, R. 2006. Airway smooth muscle phenotype and function: interactions with current asthma therapies. Current Drug Targets 7(5): 525–40.

Han, Y. G., Kim, H. J., Dlugosz, A. A., Ellison, D. W., Gilbertson, R. J. and Alvarez-Buylla, A. 2009. Dual and opposing roles of primary cilia in medulloblastoma development. Nat. Med. 15(9): 1062–5.

Hassounah, N. B., Bunch, T. A. and McDermott, K. M. 2012. Molecular pathways: the role of primary cilia in cancer progression and therapeutics with a focus on Hedgehog signaling. Clin. Cancer Res. 18(9): 2429–35.

Haswell, L. E., Hewitt, K., Thorne, D., Richter, A. and Gaca, M. D. 2010. Cigarette smoke total particulate matter increases mucous secreting cell numbers *in vitro*: a potential model of goblet cell hyperplasia. Toxicol. *In Vitro* 24(3): 981–7.

Hessel, J., Heldrich, J., Fuller, J., Staudt, M. R., Radisch, S., Hollmann, C., Harvey, B. G., Kaner, R. J., Salit, J., Yee-Levin, J. et al. 2014. Intraflagellar transport gene expression associated with short cilia in smoking and COPD. PLoS One 9(1): e85453.

Hjeij, R., Lindstrand, A., Francis, R., Zariwala, M. A., Liu, X., Li, Y., Damerla, R., Dougherty, G. W., Abouhamed, M., Olbrich, H. et al. 2013. ARMC4 mutations cause primary ciliary dyskinesia with randomization of left/right body asymmetry. Am. J. Hum. Genet. 93(2): 357–67.

Hjeij, R., Onoufriadis, A., Watson, C. M., Slagle, C. E., Klena, N. T., Dougherty, G. W., Kurkowiak, M., Loges, N. T., Diggle, C. P., Morante, N. F. et al. 2014. CCDC151 mutations cause primary ciliary dyskinesia by disruption of the outer dynein arm docking complex formation. Am. J. Hum. Genet. 95(3): 257–74.

Hogan, B. L., Barkauskas, C. E., Chapman, H. A., Epstein, J. A., Jain, R., Hsia, C. C., Niklason, L., Calle, E., Le, A., Randell, S. H. et al. 2014. Repair and regeneration of the respiratory system: complexity, plasticity, and mechanisms of lung stem cell function. Cell Stem Cell 15(2): 123–38.

Horani, A., Druley, T. E., Zariwala, M. A., Patel, A. C., Levinson, B. T., Van Arendonk, L. G., Thornton, K. C., Giacalone, J. C., Albee, A. J., Wilson, K. S. et al. 2012. Whole-exome capture and sequencing identifies HEATR2 mutation as a cause of primary ciliary dyskinesia. Am. J. Hum. Genet. 91(4): 685–93.

Horani, A., Brody, S. L., Ferkol, T. W., Shoseyov, D., Wasserman, M. G., Ta-shma, A., Wilson, K. S., Bayly, P. V., Amirav, I., Cohen-Cymberknoh, M. et al. 2013. CCDC65 mutation causes primary ciliary dyskinesia with normal ultrastructure and hyperkinetic cilia. PLoS One 8(8): e72299.

Hu, Q., Wu, Y., Tang, J., Zheng, W., Wang, Q., Nahirney, D., Duszyk, M., Wang, S., Tu, J. C. and Chen, X. Z. 2014. Expression of polycystins and fibrocystin on primary cilia of lung cells. Biochemistry and Cell Biology = Biochimie et Biologie Cellulaire 92(6): 547–54.

Hung, C., Linn, G., Chow, Y. H., Kobayashi, A., Mittelsteadt, K., Altemeier, W. A., Gharib, S. A., Schnapp, L. M. and Duffield, J. S. 2013. Role of lung pericytes and resident fibroblasts in the pathogenesis of pulmonary fibrosis. American Journal of Respiratory and Critical Care medicine 188(7): 820–30.

Ibanez-Tallon, I., Pagenstecher, A., Fliegauf, M., Olbrich, H., Kispert, A., Ketelsen, U. P., North, A., Heintz, N. and Omran, H. 2004. Dysfunction of axonemal dynein heavy chain Mdnah5 inhibits ependymal flow and reveals a novel mechanism for hydrocephalus formation. Hum. Mol. Genet. 13(18): 2133–41.

Inoue, Y., Suga, A., Sekido, Y., Yamada, S. and Iwazaki, M. 2011. A case of surgically resected lung cancer in a patient with Kartagener's syndrome. Tokai. J. Exp. Clin. Med. 36(2): 21–4.

Jain, R., Pan, J., Driscoll, J. A., Wisner, J. W., Huang, T., Gunsten, S. P., You, Y. and Brody, S. L. 2010. Temporal relationship between primary and motile ciliogenesis in airway epithelial cells. American Journal of Respiratory Cell and Molecular Biology 43(6): 731–9.

Jeanson, L., Copin, B., Papon, J. F., Dastot-Le Moal, F., Duquesnoy, P., Montantin, G., Cadranel, J., Corvol, H., Coste, A., Desir, J. et al. 2015. RSPH3 mutations cause primary ciliary Dyskinesia with Central-Complex Defects and a Near Absence of Radial Spokes. Am. J. Hum. Genet. 97(1): 153–62.

Kadzik, R. S., Cohen, E. D., Morley, M. P., Stewart, K. M., Lu, M. M. and Morrisey, E. E. 2014. Wnt ligand/Frizzled 2 receptor signaling regulates tube shape and branch-point formation in the lung through control of epithelial cell shape. Proc. Natl. Acad. Sci. USA 111(34): 12444–9.

Kantar, A., Oggiano, N., Giorgi, P. L., Braga, P. C. and Fiorini, R. 1994. Polymorphonuclear leukocyte-generated oxygen metabolites decrease beat frequency of human respiratory cilia. Lung 172(4): 215–22.

King, S. M. and Patel-King, R. S. 2015. The oligomeric outer dynein arm assembly factor CCDC103 is tightly integrated within the ciliary axoneme and exhibits periodic binding to microtubules. J. Biol. Chem. 290(12): 7388–401.

Knowles, M. R., Leigh, M. W., Carson, J. L., Davis, S. D., Dell, S. D., Ferkol, T. W., Olivier, K. N., Sagel, S. D., Rosenfeld, M., Burns, K. A. et al. 2012. Mutations of DNAH11 in patients with primary ciliary dyskinesia with normal ciliary ultrastructure. Thorax. 67(5): 433–41.

Knowles, M. R., Leigh, M. W., Ostrowski, L. E., Huang, L., Carson, J. L., Hazucha, M. J., Yin, W., Berg, J. S., Davis, S. D., Dell, S. D. et al. 2013a. Exome sequencing identifies mutations in CCDC114 as a cause of primary ciliary dyskinesia. Am. J. Hum. Genet. 92(1): 99–106.

Knowles, M. R., Ostrowski, L. E., Loges, N. T., Hurd, T., Leigh, M. W., Huang, L., Wolf, W. E., Carson, J. L., Hazucha, M. J., Yin, W. et al. 2013b. Mutations in SPAG1 cause primary ciliary dyskinesia associated with defective outer and inner dynein arms. Am. J. Hum. Genet. 93(4): 711–20.

Knowles, M. R., Ostrowski, L. E., Leigh, M. W., Sears, P. R., Davis, S. D., Wolf, W. E., Hazucha, M. J., Carson, J. L., Olivier, K. N., Sagel, S. D. et al. 2014. Mutations in RSPH1 cause primary ciliary dyskinesia with a unique clinical and ciliary phenotype. Am. J. Respir. Crit. Care Med. 189(6): 707–17.

Kott, E., Duquesnoy, P., Copin, B., Legendre, M., Dastot-Le Moal, F., Montantin, G., Jeanson, L., Tamalet, A., Papon, J. F., Siffroi, J. P. et al. 2012. Loss-of-function mutations in LRRC6, a gene essential for proper axonemal assembly of inner and outer dynein arms, cause primary ciliary dyskinesia. Am. J. Hum. Genet. 91(5): 958–64.

Kott, E., Legendre, M., Copin, B., Papon, J. F., Dastot-Le Moal, F., Montantin, G., Duquesnoy, P., Piterboth, W., Amram, D., Bassinet, L. et al. 2013. Loss-of-function mutations in RSPH1 cause primary ciliary dyskinesia with central-complex and radial-spoke defects. Am. J. Hum. Genet. 93(3): 561–70.

Kropski, J. A., Blackwell, T. S. and Loyd, J. E. 2015. The genetic basis of idiopathic pulmonary fibrosis. Eur. Respir. J. 45(6): 1717–27.

Kugler, M. C., Joyner, A. L., Loomis, C. A. and Munger, J. S. 2015. Sonic hedgehog signaling in the lung. From development to disease. American Journal of Respiratory Cell and Molecular Biology 52(1): 1–13.

Kuyper, L. M., Pare, P. D., Hogg, J. C., Lambert, R. K., Ionescu, D., Woods, R. and Bai, T. R. 2003. Characterization of airway plugging in fatal asthma. Am. J. Med. 115(1): 6–11.

Lam, H. C., Cloonan, S. M., Bhashyam, A. R., Haspel, J. A., Singh, A., Sathirapongsasuti, J. F., Cervo, M., Yao, H., Chung, A. L., Mizumura, K. et al. 2013. Histone deacetylase 6-mediated selective autophagy regulates COPD-associated cilia dysfunction. J. Clin. Invest. 123(12): 5212–30.

Leopold, P. L., O'Mahony, M. J., Lian, X. J., Tilley, A. E., Harvey, B. G. and Crystal, R. G. 2009. Smoking is associated with shortened airway cilia. PLoS One 4(12): e8157.

Levine, B. and Yuan, J. 2005. Autophagy in cell death: an innocent convict? J. Clin. Invest. 115(10): 2679–88.

Liu, Y., Stein, E., Oliver, T., Li, Y., Brunken, W. J., Koch, M., Tessier-Lavigne, M. and Hogan, B. L. 2004. Novel role for Netrins in regulating epithelial behavior during lung branching morphogenesis. Curr. Biol. 14(10): 897–905.

Liu, Y., Martinez, L., Ebine, K. and Abe, M. K. 2008. Role for mitogen-activated protein kinase p38 alpha in lung epithelial branching morphogenesis. Dev. Biol. 314(1): 224–35.

Loges, N. T., Olbrich, H., Fenske, L., Mussaffi, H., Horvath, J., Fliegauf, M., Kuhl, H., Baktai, G., Peterffy, E., Chodhari, R. et al. 2008. DNAI2 mutations cause primary ciliary dyskinesia with defects in the outer dynein arm. Am. J. Hum. Genet. 83(5): 547–58.

Loges, N. T., Olbrich, H., Becker-Heck, A., Haffner, K., Heer, A., Reinhard, C., Schmidts, M., Kispert, A., Zariwala, M. A., Leigh, M. W. et al. 2009. Deletions and point mutations of LRRC50 cause primary ciliary dyskinesia due to dynein arm defects. Am. J. Hum. Genet. 85(6): 883–9.

Look, D. C., Walter, M. J., Williamson, M. R., Pang, L., You, Y., Sreshta, J. N., Johnson, J. E., Zander, D. S. and Brody, S. L. 2001. Effects of paramyxoviral infection on airway epithelial cell Foxj1 expression, ciliogenesis, and mucociliary function. Am. J. Pathol. 159(6): 2055–69.

Lucas, J. S., Burgess, A., Mitchison, H. M., Moya, E., Williamson, M., Hogg, C. and National Pcd Service, U. K. 2014. Diagnosis and management of primary ciliary dyskinesia. Arch. Dis. Child. 99(9): 850–6.

Lucas, J. S. et al. 2017. Eur. Respir. J. Jan 4; 49(1) pii 1601090.

Maeda, Y., Dave, V. and Whitsett, J. A. 2007. Transcriptional control of lung morphogenesis. Physiol. Rev. 87(1): 219–44.

Mao, P., Wu, S., Li, J., Fu, W., He, W., Liu, X., Slutsky, A. S., Zhang, H. and Li, Y. 2015. Human alveolar epithelial type II cells in primary culture. Physiological Reports 3(2).

Marcet, B., Chevalier, B., Luxardi, G., Coraux, C., Zaragosi, L. E., Cibois, M., Robbe-Sermesant, K., Jolly, T., Cardinaud, B., Moreilhon, C. et al. 2011. Control of vertebrate multiciliogenesis by miR-449 through direct repression of the Delta/Notch pathway. Nat. Cell Biol. 13(6): 693–9.

Marx, V. 2015. The DNA of a nation. Nature 524(7566): 503–5.

Mazor, M., Alkrinawi, S., Chalifa-Caspi, V., Manor, E., Sheffield, V. C., Aviram, M. and Parvari, R. 2011. Primary ciliary dyskinesia caused by homozygous mutation in DNAL1, encoding dynein light chain 1. Am. J. Hum. Genet. 88(5): 599–607.

McCauley, H. A. and Guasch, G. 2015. Three cheers for the goblet cell: maintaining homeostasis in mucosal epithelia. Trends in Molecular Medicine 21(8): 492–503.

McGowan, S. E. 2014. Paracrine cellular and extracellular matrix interactions with mesenchymal progenitors during pulmonary alveolar septation. Birth Defects Research. Part A, Clinical and Molecular Teratology 100(3): 227–39.

McGowan, S. E. and McCoy, D. M. 2014. Regulation of fibroblast lipid storage and myofibroblast phenotypes during alveolar septation in mice. American Journal of Physiology. Lung Cellular and Molecular Physiology 307(8): L618–31.

McQualter, J. L., McCarty, R. C., Van der Velden, J., O'Donoghue, R. J., Asselin-Labat, M. L., Bozinovski, S. and Bertoncello, I. 2013. TGF-beta signaling in stromal cells acts upstream of FGF-10 to regulate epithelial stem cell growth in the adult lung. Stem Cell Research 11(3): 1222–33.

Mendelsohn, C., Lohnes, D., Decimo, D., Lufkin, T., LeMeur, M., Chambon, P. and Mark, M. 1994. Function of the retinoic acid receptors (RARs) during development (II). Multiple abnormalities at various stages of organogenesis in RAR double mutants. Development 120(10): 2749–71.

Mercer, B. A., Lemaitre, V., Powell, C. A. and D'Armiento, J. 2006. The Epithelial Cell in Lung Health and Emphysema Pathogenesis. Curr. Respir. Med. Rev. 2(2): 101–142.

Merveille, A. C., Davis, E. E., Becker-Heck, A., Legendre, M., Amirav, I., Bataille, G., Belmont, J., Beydon, N., Billen, F., Clement, A. et al. 2011. CCDC39 is required for assembly of inner dynein arms and the dynein regulatory complex and for normal ciliary motility in humans and dogs. Nat. Genet. 43(1): 72–8.

Metzger, R. J., Klein, O. D., Martin, G. R. and Krasnow, M. A. 2008. The branching programme of mouse lung development. Nature 453(7196): 745–50.

Min, H., Danilenko, D. M., Scully, S. A., Bolon, B., Ring, B. D., Tarpley, J. E., DeRose, M. and Simonet, W. S. 1998. Fgf-10 is required for both limb and lung development and exhibits striking functional similarity to Drosophila branchless. Genes Dev. 12(20): 3156–61.

Minoo, P., Hamdan, H., Bu, D., Warburton, D., Stepanik, P. and deLemos, R. 1995. TTF-1 regulates lung epithelial morphogenesis. Dev. Biol. 172(2): 694–8.

Mitchison, H. M., Schmidts, M., Loges, N. T., Freshour, J., Dritsoula, A., Hirst, R. A., O'Callaghan, C., Blau, H., Al Dabbagh, M., Olbrich, H. et al. 2012. Mutations in axonemal dynein assembly factor DNAAF3 cause primary ciliary dyskinesia. Nat. Genet. 44(4): 381–9, S1-2.

Moore, A., Escudier, E., Roger, G., Tamalet, A., Pelosse, B., Marlin, S., Clement, A., Geremek, M., Delaisi, B., Bridoux, A. M. et al. 2006. RPGR is mutated in patients with a complex X linked phenotype combining primary ciliary dyskinesia and retinitis pigmentosa. J. Med. Genet. 43(4): 326–33.

Moore, D. J., Onoufriadis, A., Shoemark, A., Simpson, M. A., zur Lage, P. I., de Castro, S. C., Bartoloni, L., Gallone, G., Petridi, S., Woollard, W. J. et al. 2013. Mutations in ZMYND10, a gene essential for proper axonemal assembly of inner and outer dynein arms in humans and flies, cause primary ciliary dyskinesia. Am. J. Hum. Genet. 93(2): 346–56.

Morimoto, M., Liu, Z., Cheng, H. T., Winters, N., Bader, D. and Kopan, R. 2010. Canonical Notch signaling in the developing lung is required for determination of arterial smooth muscle cells and selection of Clara versus ciliated cell fate. J. Cell Sci. 123(Pt 2): 213–24.

Morimoto, M., Nishinakamura, R., Saga, Y. and Kopan, R. 2012. Different assemblies of Notch receptors coordinate the distribution of the major bronchial Clara, ciliated and neuroendocrine cells. Development 139(23): 4365–73.

Morrisey, E. E. and Hogan, B. L. 2010. Preparing for the first breath: genetic and cellular mechanisms in lung development. Dev. Cell 18(1): 8–23.

Nakajima, M., Kawanami, O., Jin, E., Ghazizadeh, M., Honda, M., Asano, G., Horiba, K. and Ferrans, V. J. 1998. Immunohistochemical and ultrastructural studies of basal cells, Clara cells and bronchiolar cuboidal cells in normal human airways. Pathology International 48(12): 944–53.

Narayanan, M., Owers-Bradley, J., Beardsmore, C. S., Mada, M., Ball, I., Garipov, R., Panesar, K. S., Kuehni, C. E., Spycher, B. D., Williams, S. E. et al. 2012. Alveolarization continues during childhood and adolescence: new evidence from helium-3 magnetic resonance. American Journal of Respiratory and Critical Care Medicine 185(2): 186–91.

Nonaka, S., Tanaka, Y., Okada, Y., Takeda, S., Harada, A., Kanai, Y., Kido, M. and Hirokawa, N. 1998. Randomization of left-right asymmetry due to loss of nodal cilia generating leftward flow of extraembryonic fluid in mice lacking KIF3B motor protein. Cell 95(6): 829–37.

Nonaka, S., Shiratori, H., Saijoh, Y. and Hamada, H. 2002. Determination of left-right patterning of the mouse embryo by artificial nodal flow. Nature 418(6893): 96–9.

Oda, T., Yanagisawa, H., Kamiya, R. and Kikkawa, M. 2014. A molecular ruler determines the repeat length in eukaryotic cilia and flagella. Science 346(6211): 857–60.

Olbrich, H., Haffner, K., Kispert, A., Volkel, A., Volz, A., Sasmaz, G., Reinhardt, R., Hennig, S., Lehrach, H., Konietzko, N. et al. 2002. Mutations in DNAH5 cause primary ciliary dyskinesia and randomization of left-right asymmetry. Nat. Genet. 30(2): 143–4.

Olbrich, H., Schmidts, M., Werner, C., Onoufriadis, A., Loges, N. T., Raidt, J., Banki, N. F., Shoemark, A., Burgoyne, T., Al Turki, S. et al. 2012. Recessive HYDIN mutations cause primary ciliary dyskinesia without randomization of left-right body asymmetry. Am. J. Hum. Genet. 91(4): 672–84.

Olbrich, H., Cremers, C., Loges, N. T., Werner, C., Nielsen, K. G., Marthin, J. K., Philipsen, M., Wallmeier, J., Pennekamp, P., Menchen, T. et al. 2015. Loss-of-function GAS8 mutations cause primary ciliary dyskinesia and disrupt the nexin-dynein regulatory complex. Am. J. Hum. Genet. 97(4): 546–54.

Omran, H., Kobayashi, D., Olbrich, H., Tsukahara, T., Loges, N. T., Hagiwara, H., Zhang, Q., Leblond, G., O'Toole, E., Hara, C. et al. 2008. Ktu/PF13 is required for cytoplasmic pre-assembly of axonemal dyneins. Nature 456(7222): 611–6.

Onoufriadis, A., Paff, T., Antony, D., Shoemark, A., Micha, D., Kuyt, B., Schmidts, M., Petridi, S., Dankert-Roelse, J. E., Haarman, E. G. et al. 2013. Splice-site mutations in the axonemal outer dynein arm docking complex gene CCDC114 cause primary ciliary dyskinesia. Am. J. Hum. Genet. 92(1): 88–98.

Onoufriadis, A., Shoemark, A., Munye, M. M., James, C. T., Schmidts, M., Patel, M., Rosser, E. M., Bacchelli, C., Beales, P. L., Scambler, P. J. et al. 2014a. Combined exome and whole-genome sequencing identifies mutations in ARMC4 as a cause of primary ciliary dyskinesia with defects in the outer dynein arm. J. Med. Genet. 51(1): 61–7.

Onoufriadis, A., Shoemark, A., Schmidts, M., Patel, M., Jimenez, G., Liu, H., Thomas, B., Dixon, M., Hirst, R. A., Rutman, A. et al. 2014b. Targeted NGS gene panel identifies mutations in RSPH1 causing primary ciliary dyskinesia and a common mechanism for ciliary central pair agenesis due to radial spoke defects. Hum. Mol. Genet. 23(13): 3362–74.

Panizzi, J. R., Becker-Heck, A., Castleman, V. H., Al-Mutairi, D. A., Liu, Y., Loges, N. T., Pathak, N., Austin-Tse, C., Sheridan, E., Schmidts, M. et al. 2012. CCDC103 mutations cause primary ciliary dyskinesia by disrupting assembly of ciliary dynein arms. Nat. Genet. 44(6): 714–9.

Patel, B. V., Wilson, M. R., O'Dea, K. P. and Takata, M. 2013. TNF-induced death signaling triggers alveolar epithelial dysfunction in acute lung injury. Journal of Immunology 190(8): 4274–82.

Paudyal, A., Damrau, C., Patterson, V. L., Ermakov, A., Formstone, C., Lalanne, Z., Wells, S., Lu, X., Norris, D. P., Dean, C. H. et al. 2010. The novel mouse mutant, chuzhoi, has disruption of Ptk7 protein and exhibits defects in neural tube, heart and lung development and abnormal planar cell polarity in the ear. BMC Dev. Biol. 10: 87.

Pennarun, G., Escudier, E., Chapelin, C., Bridoux, A. M., Cacheux, V., Roger, G., Clement, A., Goossens, M., Amselem, S. and Duriez, B. 1999. Loss-of-function mutations in a human gene related to Chlamydomonas reinhardtii dynein IC78 result in primary ciliary dyskinesia. Am. J. Hum. Genet. 65(6): 1508–19.

Pepicelli, C. V., Lewis, P. M. and McMahon, A. P. 1998. Sonic hedgehog regulates branching morphogenesis in the mammalian lung. Curr. Biol. 8(19): 1083–6.

Piorunek, T., Marszalek, A., Biczysko, W., Gozdzik, J., Cofta, S. and Seget, M. 2008. Correlation between the stage of cystic fibrosis and the level of morphological changes in adult patients. J. Physiol. Pharmacol. 59 Suppl 6: 565–72.

Ramos, E. M., De Toledo, A. C., Xavier, R. F., Fosco, L. C., Vieira, R. P., Ramos, D. and Jardim, J. R. 2011. Reversibility of impaired nasal mucociliary clearance in smokers following a smoking cessation programme. Respirology 16(5): 849–55.

Ranu, H., Wilde, M. and Madden, B. 2011. Pulmonary function tests. Ulster. Med. J. 80(2): 84–90.

Rawlins, E. L. and Hogan, B. L. 2006. Epithelial stem cells of the lung: privileged few or opportunities for many? Development 133(13): 2455–65.

Rehan, V. K., Sugano, S., Wang, Y., Santos, J., Romero, S., Dasgupta, C., Keane, M. P., Stahlman, M. T. and Torday, J. S. 2006. Evidence for the presence of lipofibroblasts in human lung. Experimental Lung Research 32(8): 379–93.

Rock, J. R., Randell, S. H. and Hogan, B. L. 2010. Airway basal stem cells: a perspective on their roles in epithelial homeostasis and remodeling. Disease Models & Mechanisms 3(9-10): 545–56.

Rock, J. R., Gao, X., Xue, Y., Randell, S. H., Kong, Y. Y. and Hogan, B. L. 2011. Notch-dependent differentiation of adult airway basal stem cells. Cell Stem Cell 8(6): 639–48.

Rosenbluth, D. B., Wilson, K., Ferkol, T. and Schuster, D. P. 2004. Lung function decline in cystic fibrosis patients and timing for lung transplantation referral. Chest. 126(2): 412–9.

Sakai, T., Larsen, M. and Yamada, K. M. 2003. Fibronectin requirement in branching morphogenesis. Nature 423(6942): 876–81.

Schittny, J. C., Mund, S. I. and Stampanoni, M. 2008. Evidence and structural mechanism for late lung alveolarization. Am. J. Physiol. Lung Cell Mol. Physiol. 294(2): L246–54.

Schmidt, M. and Mattoli, S. 2013. A mouse model for evaluating the contribution of fibrocytes and myofibroblasts to airway remodeling in allergic asthma. Methods in Molecular Biology 1032: 235–55.

Schwabe, G. C., Hoffmann, K., Loges, N. T., Birker, D., Rossier, C., de Santi, M. M., Olbrich, H., Fliegauf, M., Failly, M., Liebers, U. et al. 2008. Primary ciliary dyskinesia associated with normal axoneme ultrastructure is caused by DNAH11 mutations. Hum. Mutat. 29(2): 289–98.

Shapiro, A. J., Davis, S. D., Ferkol, T., Dell, S. D., Rosenfeld, M., Olivier, K. N., Sagel, S. D., Milla, C., Zariwala, M. A., Wolf, W. et al. 2014. Laterality defects other than situs inversus totalis in primary ciliary dyskinesia: insights into situs ambiguus and heterotaxy. Chest. 146(5): 1176–86.

Sharpe, J., Ahlgren, U., Perry, P., Hill, B., Ross, A., Hecksher-Sorensen, J., Baldock, R. and Davidson, D. 2002. Optical projection tomography as a tool for 3D microscopy and gene expression studies. Science 296(5567): 541–5.

Shaykhiev, R., Zuo, W. L., Chao, I., Fukui, T., Witover, B., Brekman, A. and Crystal, R. G. 2013. EGF shifts human airway basal cell fate toward a smoking-associated airway epithelial phenotype. Proc. Natl. Acad. Sci. USA 110(29): 12102–7.

Short, K., Hodson, M. and Smyth, I. 2013. Spatial mapping and quantification of developmental branching morphogenesis. Development 140(2): 471–8.

Shpak, M., Goldberg, M. M. and Cowperthwaite, M. C. 2014. Cilia gene expression patterns in cancer. Cancer Genomics Proteomics 11(1): 13–24.

Shu, W., Guttentag, S., Wang, Z., Andl, T., Ballard, P., Lu, M. M., Piccolo, S., Birchmeier, W., Whitsett, J. A., Millar, S. E. et al. 2005. Wnt/beta-catenin signaling acts upstream of N-myc, BMP4, and FGF signaling to regulate proximal-distal patterning in the lung. Dev. Biol. 283(1): 226–39.

Sisson, J. H., Papi, A., Beckmann, J. D., Leise, K. L., Wisecarver, J., Brodersen, B. W., Kelling, C. L., Spurzem, J. R. and Rennard, S. I. 1994. Smoke and viral infection cause cilia loss detectable by bronchoalveolar lavage cytology and dynein ELISA. Am. J. Respir. Crit. Care Med. 149(1): 205–13.

Stephen, L. A., Davis, G. M., McTeir, K. E., James, J., McTeir, L., Kierans, M., Bain, A. and Davey, M. G. 2013. Failure of centrosome migration causes a loss of motile cilia in talpid(3) mutants. Developmental Dynamics: An Official Publication of the American Association of Anatomists 242(8): 923–31.

Strippoli, M. P., Frischer, T., Barbato, A., Snijders, D., Maurer, E., Lucas, J. S., Eber, E., Karadag, B., Pohunek, P., Zivkovic, Z. et al. 2012. Management of primary ciliary dyskinesia in European children: recommendations and clinical practice. Eur. Respir. J. 39(6): 1482–91.

Stubbs, J. L., Vladar, E. K., Axelrod, J. D. and Kintner, C. 2012. Multicilin promotes centriole assembly and ciliogenesis during multiciliate cell differentiation. Nat. Cell Biol. 14(2): 140–7.

Swarr, D. T. and Morrisey, E. E. 2015. Lung endoderm morphogenesis: Gasping for form and function. Annu. Rev. Cell Dev. Biol.

Tadokoro, T., Wang, Y., Barak, L. S., Bai, Y., Randell, S. H. and Hogan, B. L. 2014. IL-6/STAT3 promotes regeneration of airway ciliated cells from basal stem cells. Proc. Natl. Acad. Sci. USA 111(35): E3641–9.

Takeyama, K., Dabbagh, K., Jeong Shim, J., Dao-Pick, T., Ueki, I. F. and Nadel, J. A. 2000. Oxidative stress causes mucin synthesis via transactivation of epidermal growth factor receptor: role of neutrophils. J. Immunol. 164(3): 1546–52.

Tarkar, A., Loges, N. T., Slagle, C. E., Francis, R., Dougherty, G. W., Tamayo, J. V., Shook, B., Cantino, M., Schwartz, D., Jahnke, C. et al. 2013. DYX1C1 is required for axonemal dynein assembly and ciliary motility. Nat. Genet. 45(9): 995–1003.

Thomas, B., Rutman, A., Hirst, R. A., Haldar, P., Wardlaw, A. J., Bankart, J., Brightling, C. E. and O'Callaghan, C. 2010. Ciliary dysfunction and ultrastructural abnormalities are features of severe asthma. J. Allergy. Clin. Immunol. 126(4): 722–729 e2.

Tilley, A. E., Walters, M. S., Shaykhiev, R. and Crystal, R. G. 2015. Cilia dysfunction in lung disease. Annu. Rev. Physiol. 77: 379–406.

Treutlein, B., Brownfield, D. G., Wu, A. R., Neff, N. F., Mantalas, G. L., Espinoza, F. H., Desai, T. J., Krasnow, M. A. and Quake, S. R. 2014. Reconstructing lineage hierarchies of the distal lung epithelium using single-cell RNA-seq. Nature 509(7500): 371–5.

Tsao, P. N., Chen, F., Izvolsky, K. I., Walker, J., Kukuruzinska, M. A., Lu, J. and Cardoso, W. V. 2008. Gamma-secretase activation of notch signaling regulates the balance of proximal and distal fates in progenitor cells of the developing lung. J. Biol. Chem. 283(43): 29532–44.

Tsao, P. N., Vasconcelos, M., Izvolsky, K. I., Qian, J., Lu, J. and Cardoso, W. V. 2009. Notch signaling controls the balance of ciliated and secretory cell fates in developing airways. Development 136(13): 2297–307.

Ventura, A., Young, A. G., Winslow, M. M., Lintault, L., Meissner, A., Erkeland, S. J., Newman, J., Bronson, R. T., Crowley, D., Stone, J. R. et al. 2008. Targeted deletion reveals essential and overlapping functions of the miR-17 through 92 family of miRNA clusters. Cell 132(5): 875–86.

Wallmeier, J., Al-Mutairi, D. A., Chen, C. T., Loges, N. T., Pennekamp, P., Menchen, T., Ma, L., Shamseldin, H. E., Olbrich, H., Dougherty, G. W. et al. 2014. Mutations in CCNO result in congenital mucociliary clearance disorder with reduced generation of multiple motile cilia. Nat. Genet. 46(6): 646–51.

Weatherbee, S. D., Niswander, L. A. and Anderson, K. V. 2009. A mouse model for Meckel syndrome reveals Mks1 is required for ciliogenesis and Hedgehog signaling. Human Molecular Genetics 18(23): 4565–75.

Whitsett, J. A. and Weaver, T. E. 2015. Alveolar development and disease. Am. J. Respir. Cell Mol. Biol. 53(1): 1–7.

Wilson, M. S. and Wynn, T. A. 2009. Pulmonary fibrosis: pathogenesis, etiology and regulation. Mucosal. Immunol. 2(2): 103–21.

Wirschell, M., Olbrich, H., Werner, C., Tritschler, D., Bower, R., Sale, W. S., Loges, N. T., Pennekamp, P., Lindberg, S., Stenram, U. et al. 2013. The nexin-dynein regulatory complex subunit DRC1 is essential for motile cilia function in algae and humans. Nat. Genet. 45(3): 262–8.

Wistuba, II and Gazdar, A. F. 2006. Lung cancer preneoplasia. Annu. Rev. Pathol. 1: 331–48.

Wong, S. Y., Seol, A. D., So, P. L., Ermilov, A. N., Bichakjian, C. K., Epstein, E. H., Jr., Dlugosz, A. A. and Reiter, J. F. 2009. Primary cilia can both mediate and suppress Hedgehog pathway-dependent tumorigenesis. Nat. Med. 15(9): 1055–61.

Yaghi, A., Zaman, A., Cox, G. and Dolovich, M. B. 2012. Ciliary beating is depressed in nasal cilia from chronic obstructive pulmonary disease subjects. Respir. Med. 106(8): 1139–47.

Yang, I. V., Coldren, C. D., Leach, S. M., Seibold, M. A., Murphy, E., Lin, J., Rosen, R., Neidermyer, A. J., McKean, D. F., Groshong, S. D. et al. 2013. Expression of cilium-associated genes defines novel molecular subtypes of idiopathic pulmonary fibrosis. Thorax. 68(12): 1114–21.

Yang, J., Hernandez, B. J., Martinez Alanis, D., Narvaez Del Pilar, O., Vila-Ellis, L., Akiyama, H., Evans, S. E., Ostrin, E. J. and Chen, J. 2016. The development and plasticity of alveolar type 1 cells. Development 143(1): 54–65.

Yates, L. L., Schnatwinkel, C., Murdoch, J. N., Bogani, D., Formstone, C. J., Townsend, S., Greenfield, A., Niswander, L. A. and Dean, C. H. 2010. The PCP genes Celsr1 and Vangl2 are required for normal lung branching morphogenesis. Hum. Mol. Genet. 19(11): 2251–67.

Yates, L. L., Schnatwinkel, C., Hazelwood, L., Chessum, L., Paudyal, A., Hilton, H., Romero, M. R., Wilde, J., Bogani, D., Sanderson, J. et al. 2013. Scribble is required for normal epithelial cell-cell contacts and lumen morphogenesis in the mammalian lung. Dev. Biol. 373(2): 267–80.

Zariwala, M. A., Gee, H. Y., Kurkowiak, M., Al-Mutairi, D. A., Leigh, M. W., Hurd, T. W., Hjeij, R., Dell, S. D., Chaki, M., Dougherty, G. W. et al. 2013. ZMYND10 is mutated in primary ciliary dyskinesia and interacts with LRRC6. Am. J. Hum. Genet. 93(2): 336–45.

CHAPTER 3

The Role of Cilia in Heart Development and Disease

Sigolène M. Meilhac

INTRODUCTION

Cilia are small organelles functioning as a type of antenna at the cell surface. Since the demonstration of their fundamental role in embryonic development (Pazour et al., 2000), cilia have been extensively studied for their structure and function. Motile cilia can generate fluid flow, whereas primary (non-motile) cilia are involved in sensing and transducing cell and mechanical signalling (see Goetz and Anderson, 2010). Defects in the structure or function of cilia give rise to rare diseases, referred to as ciliopathies, which are pleiotropic genetic diseases, affecting the formation of different organs including the heart (see Baker and Beales, 2009). Yet, the role of cilia in heart development remains poorly understood.

Although cilia are present at the surface of most vertebrate cells, recent advances in correlative microscopy (Goetz et al., 2014), super-resolution microscopy (Lambacher et al., 2016) and in the reconstruction of transcriptional networks regulating ciliogenesis (Choksi et al., 2014) have highlighted extensive variations in shapes and sizes, ultrastructural details, numbers per cell, motility patterns and sensory capabilities of cilia between cell types. This emphasises the need to dissect the function of cilia in a specific cellular context, beyond the textbook picture of the cilium as a signalling hub. Formation of the heart involves the coordination of several cell types contributing to the organ and it also depends on signalling from neighbouring tissues, such as the endoderm or the left-right organiser.

Imagine – Institut Pasteur, Laboratory of Heart Morphogenesis, 75015 Paris, France; and INSERM UMR1163, Université Paris Descartes, 75015 Paris, France.
E-mail: sigolene.meilhac@institutimagine.org

In this chapter, we will discuss current knowledge on cilia and signalling potentially transduced through cilia, in the context of the formation of the heart. After a brief summary of heart development and physiology, we will review the cardiopathies associated with ciliopathies in patients and take the mouse as the main experimental model to assess the presence and function of cilia during heart development.

Development and Physiology of the Heart

The heart is the first organ to form in development, from pools of progenitors, heart fields, sequentially added to the heart (Meilhac et al., 2014) (Figure 1). The primordium of the heart, at embryonic day (E) 8.5 in the mouse, is a tube, with a double layer of endocardium, inside, and myocardium, outside. Cardiac muscle cells (cardiomyocytes), which underlie the contractile function of the heart, are also essential for the morphogenesis of the heart. Cardiomyocytes, which are differentiated cells and beat, still have a proliferative potential *in utero* underlying the increase in size of the heart that accompanies the hemodynamic demand of the growing organism. The myocardium progressively matures in architecture, including oriented myofibres and trabeculations, which are important components of the efficient contraction. After birth, mature cardiomyocytes exit the cell cycle (Li et al., 1996). The structure and development of the myocardium is regulated in space, with variations in the four cardiac chambers, as well as in their inlets and outlets. In addition to the endocardium and myocardium, a third tissue layer, the epicardium, covers the embryonic heart, from stage E9.5 in the mouse. From E12.5, epicardial derived cells are produced: they invade the ventricular walls and become either interstitial fibroblasts, which provide growth factors and secrete the extracellular matrix, or smooth muscle cells of the coronary blood vessels. In addition to providing support and trophic factors, the endocardium (Grego-Bessa et al., 2007) and epicardium (Shen et al., 2015) are involved in reciprocal signalling with the myocardium for the regulation of myocardial growth.

Another function of the heart is the establishment of double blood circulation, which completely separates the blood that is oxygenated in the lung from the blood that is carbonated in the organs. The correct plumbing of the heart depends on the alignment of the cardiac chambers as well as their septation. In the early cardiac tube, the prospective cardiac chambers are positioned as an array along the axis of the tube. It is after heart looping, that the cardiac chambers are correctly aligned. Rightward looping of the cardiac tube is the first morphological sign of left-right asymmetry in the embryo, lying downstream of the left-right organiser (Brennan et al., 2002). However, the embryonic heart remains a tube, with ventricles connected in series and a single outflow tract. Septation at fetal stages corresponds to the partitioning of cardiac regions, including the aorta and pulmonary artery

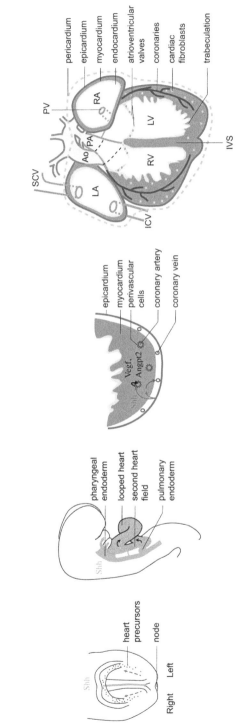

Figure 1. Potential role of cilia at different steps of heart formation. At embryonic day (E)7.5, cilia are important in the node, the left-right organiser, for the determination of left/right patterning of the embryo. Heart precursors (blue) migrate (dashed arrows) within the lateral plate mesoderm towards the anterior. Shh is expressed in the midline (green) and received by heart precursors. At E8.5, looping of the heart tube is a readout of left/right patterning. At this stage, secretion of Shh by endoderm cells activates signalling (green arrows) in the second heart field precursors, controlling the septation of the atria and the great arteries. Black arrows indicate the progressive differentiation and recruitment of heart precursors at the poles of the heart tube. At E13.5, a section of the ventricular wall depicts the different tissues of the heart. Shh, secreted by the epicardium (blue arrows), is received by perivascular (black) or myocardial (orange) cells, which in response produce angiogenic factors (Vascular Endothelial Growth Factor and Angiopoietin) stimulating the growth of the coronary vessels (purple). At E16.5, the structure of the mature heart is shown, including the organisation of the different cardiac tissues (colour coded) as well as the alignment of cardiac chambers and the great vessels. Ao: aorta; IVS: interventricular septum; ICV: inferior caval vein; LA: left atrium; LV: left ventricle; PA: pulmonary artery; PV: pulmonary vein; RA: right atrium; RV: right ventricle; SCV: superior caval vein.

from the outflow tract, the two atria and the two ventricles. The outflow tract and atria, which are connected to the body, are colonised by progenitor cells for their septation, from the neural crest lineage (Waldo et al., 1998) and the second heart field (see Rochais et al., 2009). Finally, the unidirectional blood flow depends on the formation of valves, between the atria and the ventricles or between the ventricles and the great arteries, deriving from the endocardium (Ramsdell and Markwald, 1997) and epicardium (Wessels et al., 2012). The formation of the valves and the completion of heart septation in the atrio-ventricular canal are intertwined, as they derive from the same transient mesenchymal structures, the endocardial cushions (Lin et al., 2012). We will see that Hedgehog signalling, from extra-cardiac tissues such as the endoderm, is important for the septation of the heart.

An indication that cilia play a role in the formation of the heart comes from clinical observations.

Ciliopathies and Cardiopathies

Ciliopathies are rare diseases, for which several genetic variations have already been identified (Baker and Beales 2009). They are classified as different syndromes, according to their clinical features. These severe diseases have been associated with cardiac malformations (Table 1), with important consequences for the quality of life and the mortality of patients. About 10% of patients with Bardet-Biedl syndrome display heart malformations, including valvular stenosis, bicuspid aortic valve or defective septation of the atria. The Meckel or McKusick—Kaufman syndrome, caused by mutations in proteins required for ciliogenesis, exhibits cardiac malformations in about 15% of cases. Alström syndrome, characterised by more than 80 different mutations in the unique gene *ALMS1* of unknown function, is associated with transient dilated cardiomyopathy in two thirds of the patients (Louw et al., 2014). Recently, it has been suggested that mutations of *ALMS1* are responsible for mitogenic cardiomyopathies, in which cardiomyocyte proliferation fails to arrest after birth (Shenje et al., 2014). Ellis–van Creveld (EVC) syndrome, caused by mutations in the *EVC* or *EVC2* genes, is associated with defective septation of the heart in 60% of cases, leading to a common atrium or atrioventricular canal. Autosomal dominant Polycystic Kidney Diseases (PKD) results from mutations in *PKD1* and *PKD2*, which encode proteins called polycystins that frequently localise to the primary cilium. PKD is thus considered as a ciliopathy in which primary cilia are present but dysfunctional (Tobin and Beales, 2009). Patients affected by PKD have a higher probability to develop anomalies of cardiac valves and hypertrophy of the left ventricle (Perrone, 1997). In addition to these ciliopathies in which primary cilia are malformed or dysfunctional, there are ciliopathies arising from defects in motile cilia. This is the case for Primary Ciliary Dyskinesia, which is associated with cardiac defects (see Harrison et al., 2016). In *situs inversus totalis* such as

Table 1. Cardiopathies associated with ciliopathies.

	Syndrome	Mutated gene	Cardiopathy	OMIM reference	Prevalence (orphanet)
Motile cilium	Primary Ciliary Dyskinesia	ARMC4, C21ORF59, CCDC39/40/114/151/164, DNAAF2/3/4, DNAH5/11, DNAI1, HEATR2♦	Interatrial communication, transposition of the great arteries, heterotaxy, ventricular septal defects, anomalous pulmonary venous return	244400	1/16 000*
	Situs inversus totalis	DNAI1	Situs inversus		
	Heterotaxy	NPHP4, CFAP52/53✧	Complex cardiac malformations		1/10 000**
Primary cilium	Alström	ALMS1	Transient dilated cardiomyopathy	203800	450 cases worldwide
	Bardet-Biedl	BBS1-19	Cardiac malformations	209900	1/125,000 à 1/175,000
	Ellis-Van Creveld	EVC, EVC2	Interatrial communication	225500	150 cases worldwide
	McKusick-Kaufman	MKKS	Cardiac malformations	236700	1% in Amish population
	Meckel-Gruber	MKS1-11, IFT88, RPGRIP1, BBS15	Cardiac malformations	249000	1/9,000 in Finland
	Autosomal dominant polycystic kidney disease	PKD1-2	Hypertrophy of the left ventricle, mitral or aortic valve prolapse	173900	1/1,000

Gene function: ARMC4, C21ORF59, CCDC39/40/114/151/164, DNAAF2/3/4, HEATR2 are required for the assembly of dynein arms in the axoneme, DNAH5/11, DNAI1 are components of dynein arms, MKS1/3/5 are involved in the formation of the basal centriole, RPGRIP1, NPHP4 are component of the transition zone at the base of cilia, IFT88, BBS19 are components of intraflagellar transport, CFAP53 is a centriolar satellite protein, ALMS1 is a centrosomal protein, EVC/EVC2 are dominant negative partners of the co-receptor Smoothened, PKD2 is a calcium channel, PKD1 is a mecanoreceptor, MKKS, BBS12 are chaperone proteins, BBS1/2/4/5/7/8/9/17/18 form the BBSome that recruits the GTPase RAB8A for vesicular transport towards the cilium, BBS3/14 are required for vesicular transport, BBS11 is a ubiquitin ligase E3, BBS15 is involved in planar polarity, CFAP52 functions in protein-protein interactions.

♦see Harrison et al., 2015 and Li et al., 2015 ✧ see Guimier et al., 2015 * see Harrison et al., 2015 ** see Lin et al., 2014 in Kartagener syndrome, the structure of the heart is mirror-imaged, and the heart is located on the right side of the thoracic cavity. In other cases, the determination of the right and left of the body is incomplete, and not coherent between visceral organs, leading to a spectrum of defects referred to as heterotaxy, including complex cardiac malformations such as transposition of the great arteries, double outlet right ventricle (a heart malformation in which the aorta and pulmonary artery both connect to the right ventricle), isomerism of the atrial appendages, atrioventricular canal and anomalous venous return (Brueckner, 2007). The cases of heterotaxy associated with Primary Ciliary Dyskinesia, are due to defective cilium motility in the left-right organiser (Nonaka et al., 1998), as well as in other tissues such as the lung and sperm. In addition, there are cases of heterotaxy with mutations in other ciliary genes, not involved in cilium motility in the lung or sperm (see Sutherland and Ware, 2009). The broad spectrum of cardiac malformations associated with ciliopathies suggests that cilia are required at different steps of heart morphogenesis.

There are numerous mouse models of ciliopathies, some of which will be presented in the following sections. In a large-scale genetic screen, aimed at recovering chemically-induced recessive mutations of congenital heart defects in fetuses, mutations in 61 genes were identified. Strikingly, 56% of the mutations (34) were in cilia-related genes and other genes (25%) were potentially involved in signalling transduced by the primary cilium (Li et al., 2015). In contrast to the rarity of cardiopathies associated with ciliopathies in patients, the frequency, in the mouse, of cardiac malformations associated with a dysfunction of cilia suggests an important role of cilia in heart development, which might be masked by lethality *in utero* in humans. This raises the questions of which cell types require cilia for the formation of the heart and what the mechanisms involved are.

Cilia of the Left-Right Organiser are Important for the Alignment of Cardiac Chambers

Motile and primary cilia are essential for the function of the left-right organiser, the node (Figure 2A–B), in the embryo at E7.5, by generating and sensing a leftward fluid flow (Nonaka et al., 1998; see Babu and Roy, 2013). Cells of the node do not contribute directly to the heart. However, the establishment of left-right asymmetry in the node has important consequences for embryonic development and for the morphogenesis of visceral organs. In particular, left-right signalling affects several steps in heart morphogenesis. Two fields of precursors are localised on either side of the midline (Figure 1 E7.5), thus receiving either left or right signals. They are added progressively to the growing heart tube, which undergoes looping. Such looping explains the twisted map of the contribution of left/

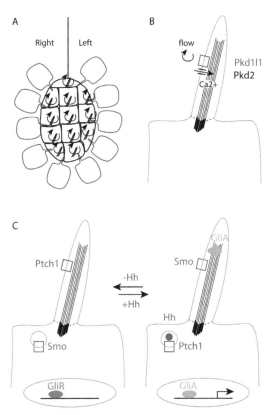

Figure 2. Ciliary structure and signal transduction in cilia. (A) In the centre of the node (in black), cells have motile cilia that rotate, thus generating a fluid flow towards the left of the embryo. The signal is received at the periphery (in red), by cells with primary cilia. **(B)** These primary cilia are enriched in polycystins, Pkd1l1 and Pkd2. Primary cilia are membrane protrusions (grey), built on a central scaffold of microtubules, the axoneme (stripes), anchored on a modified centriole, the basal body (thick stripes). Primary cilia are isolated from the cytosol by a filter (dashed line). **(C)** In many cell types, such as heart precursors, primary cilia are required for the transduction of Hedgehog (Hh) signalling. In the absence of the ligand (left panel), the inhibitory receptor Patched-1 (Ptch1) is localised to the primary cilium, whereas the co-receptor Smoothend (Smo) is internalised in the cell. The transcription factors Gli are in their repressive forms (GliR). In the presence of the ligand (right panel), Ptch1 is internalised and Smo can accumulate in the primary cilium, leading to the activation of Gli (GliA) and hence of transcriptional targets.

right precursors to heart regions. Left precursors contribute to the dorsal left atrium, superior atrioventricular canal (Dominguez et al., 2012), left superior caval vein, pulmonary vein and pulmonary trunk (Lescroart et al., 2012), whereas right precursors contribute to the right atrium, ventral left atrium, inferior atrioventricular canal (Dominguez et al., 2012), right superior caval vein and aorta (Lescroart et al., 2012). Thus, left-right signalling impacts the identity of regions of the heart. For example, the left and right atria differ in their morphology, venous drainage and in the nature of the valves which connect to the ventricles. Left-right signalling is also required for the process of rightward looping of the heart tube (Figure 1 E8.5), resulting in the relative positioning of the left and right ventricles, with respect to the great arteries and atria. Thus, heart looping is important for the alignment of cardiac chambers, as well as for the localisation of the heart on the left of the thoracic cavity (levocardia). Another process controlled by left-right signalling is the rotation of the outflow tract, such that the base of the aorta, emerging from the left ventricle, spirals around the base of the pulmonary artery, emerging from the right ventricle. The structure of the heart is therefore sensitive to perturbations in the determination of the left and right of the body. Defects in left-right signalling, with a failure of cardiac chambers to align properly and thus to be compartmentalised properly, have dramatic consequences for the plumbing of the blood circulation, resulting in incomplete separation of carbonated and oxygenated bloods.

In several mouse models, the absence or dysfunction of cilia affect heart morphogenesis, and specifically the processes which depend on left-right signalling. Absence of cilia, following a defect in the basal centriole in the $Ofd1^{-/-}$ mutants (Ferrante et al., 2006) or a defect in ciliary transport in the $Ift46^{-/-}$, $Ift88^{-/-}$, $Ift172^{-/-}$, $Kif3a^{-/-}$ and $Kif3b^{-/-}$ mutants (Nonaka et al., 1998; Takeda et al., 1999; Murcia et al., 2000; Huangfu et al., 2003; Lee et al., 2015), induces abnormal determination of the left side of the body, with randomised heart looping towards the right or left. Sometimes, ciliogenesis is partial, with shorter or less numerous cilia. In $Fuz^{-/-}$ mutants (Gray et al., 2009), in which an effector of Planar Cell Polarity, involved in membrane trafficking and ciliogenesis is mutated, septation defects of the great arteries and ventricles are observed, whereas in $Mks1^{-/-}$ mutants (Cui et al., 2011), characterised by the alteration of a regulator of the basal centriole, the direction of heart looping is randomised, with transposition of the great arteries observed at a later stage. $Nphp4$ (French et al., 2012) and $Cfap53$ (Silva et al., 2016), which are associated with human heterotaxy, have been shown to be required for cilium assembly in the left-right organiser of zebrafish, with consequences for heart looping. After depletion of the axonemal dynein Dnaic1 (Francis et al., 2012) or Dnah11 (Supp et al., 1999), which are models of human Primary Ciliary Dyskinesia, ciliogenesis is not affected but cilia motility is impaired. Abnormal heart looping results in misposition of the heart (dextrocardia) if the loop is reversed, or in heterotaxy if looping is incomplete. When the ciliary kinase $Nek8$ is mutated (Manning et al., 2013), cilia form normally. However, the direction of heart looping is randomised and later defects of cardiac chamber septation and double outlet right ventricle become apparent. Invalidation of $Pifo$ (Kinzel et al., 2010), which is involved in cilium resorption, generates abnormally duplicated cilia accompanied by double outlet right ventricle. However, the determination of left identity and the direction of heart looping are normal in $Pifo^{-/-}$ mutants, indicating a role downstream of the left-right organiser. *Nek2* is another example of a gene involved in cilium resorption, by promoting centrosome splitting. *Nek2* is required for left-right determination and heart looping, as shown in *Xenopus* (Endicott et al., 2015). Thus, mutations in genes involved in cilium assembly or function impair the determination of left-right asymmetry, often by affecting the left-right organiser, in which cilia play a central role. To better understand the spectrum of cardiac malformations resulting from abnormal left-right signalling, it will be important in the future to investigate the mechanism of heart looping and characterise how the shape of the heart loop determines the structure of the mature heart.

In these animal models of ciliopathies, the genetic inactivation is zygotic, i.e., ubiquitous, which does not permit to understand in which cell types cilia are required for heart morphogenesis. It is likely that the phenotypes of heart looping and heterotaxy result from an indirect ciliary defect, in cells of the left-right organiser which do not contribute to the heart. In addition to this indirect role in the determination of the left-right of the body, cilia may be required directly for signal transduction in cells which participate in the heart.

Localisation of Primary Cilia during Heart Development

The primary cilium is an organelle present in most cell types of vertebrates, except during mitosis, since it sequesters a centriole for anchoring. In cardiac cells, observations by electron microscopy have established the presence of primary cilia, i.e., non-motile cilia devoid of a central doublet of microtubules. Apical monocilia have been detected in cardiac precursor cells at E9.5, in the dorsal pericardial wall referred to as the second heart field, at cell surfaces facing the pericardial cavity (Francou et al., 2014). A high proportion of ciliated cells has been reported in the cardiac muscle at embryonic stages, with cilia frequently invaginated in a ciliary pocket (Rash et al., 1969; Myklebust et al., 1977). Cilia have been detected in the embryonic heart between E9.5 and E12.5 (Slough et al., 2008; Gerhardt et al., 2013). However, at later and adult stages, primary cilia are mainly detected in interstitial cells, rather than in cardiomyocytes (Myklebust et al., 1977; Diguet et al., 2015). As cardiomyocytes mature, with the increased organisation of the contractile apparatus and specification of its membranes, a loss of primary cilia has been observed. In neonatal cardiomyocytes, centrioles adopt a particular configuration: they are split and pericentriolar proteins such as pericentrin and Pcm1 are relocalised to the nuclear membrane. Experiments of RNA interference in cultured cardiomyocytes suggested that such disassembly of the centrosome is associated with loss of function at the microtubule-organising centre and hence promoted cell cycle arrest. Split centrioles are also less potent for ciliogenesis in culture conditions (Zebrowski et al., 2015). In addition to these temporal changes, primary cilia tend to be regionalised in the different compartments of the heart (Figure 1 E16.5). In the embryonic heart, cardiomyocytes of the atria and at the base of the ventricles are more frequently ciliated (Gerhardt et al., 2013). It is unclear whether this is associated with the state of maturation of cardiomyocytes. In the epicardium of the embryonic heart, acetylated-tubulin, present in stabilised microtubules such as these of the ciliary axoneme, is enriched at the cell surface facing the pericardial cavity (Slough et al., 2008). In the endocardium of the embryonic heart, cilia point towards the blood cavity and are directly exposed to hemodynamic flow. Cilia are preferentially localised to endocardial cells exposed to lower mechanical stress, for example in the atria, the base of ventricular trabeculations or the base of the cushion primordia of atrioventricular valves. This regionalisation would suggest that primary cilia may play a role as mechanosensors of hemodynamic forces (Van der Heiden et al., 2006). Cilia have also been observed in the mesenchyme of valve cushions with no apparent orientation (Slough et al., 2008), or in the smooth muscle cells of coronaries at fetal stages (Diguet et al., 2015), in keeping with a role of cilia in the smooth muscle of other vessels (Lu et al., 2008).

Cells with primary cilia are distinct from cells with motile cilia, which express a specific set of genes providing motility, such as genes involved in the production and regulation of the nexin/dynein complex. Recent advances in the biology of cilia have shown that primary cilia may also have specific roles in different cell types or at different stages (see Choksi et al., 2014), depending for example on the ciliary localization of a receptor. It is possible that different populations of cardiac precursors or cells, involved in different aspects of heart morphogenesis, express a specific set of ciliary genes. Cilia in the mesenchyme of valve cushions (Slough et al., 2008) and in the coronary smooth muscle (Diguet et al., 2015) have specific antigens, such as the small GTPase Arl13b. Mature cardiomyocytes, after birth, but not other cardiac cells, express specifically the marker of centriolar satellite Pcm1 (Bergmann et al., 2011). Transcriptomic analysis of second heart field cells has documented the set of ciliary genes in heart precursors (Burnicka-Turek et al., 2016). Differential expression of ciliary genes in different cell populations may impact the classification of human diseases as proposed for the concordance between Atrio-Ventricular Septal Defects (AVSD) and heterotaxy. Analyses of the ciliary genes mutated in ciliopathy cases or models, led to the proposal of 3 classes of AVSD: AVSD in combination with heterotaxy are associated with mutations in genes involved in cilium motility, AVSD without heterotaxy are associated with mutations in ciliary genes expressed in the second heart field and not in the left-right organiser, AVSD irrespective of heterotaxy are associated with mutations in ciliary genes expressed in the second heart field as well as in the left-right organiser (Burnicka-Turek et al., 2016).

Although many cell types have a cilium, the presence of a primary cilium is regulated in space and time. In addition, the differential composition of cilia in distinct cell types raises the question of which signalling pathways are required in primary cilia for impacting heart morphogenesis. We will take the example of two signalling pathways, which have been clearly associated with primary cilia, hedgehog signalling and polycystins.

Hedgehog Signalling in Heart Development

The Hedgehog (Hh) signalling pathway, that regulates growth and patterning of tissues, critically depends on the primary cilium (Huangfu et al., 2003; see Nozawa et al., 2013). In the presence of a ligand (Sonic hedgehog (Shh), Indian hedgehog (Ihh) or Desert hedgehog (Dhh)), the co-receptor Smoothened (Smo) is localised to the cilium, where it can trigger the activation of the Gli transcription factors (Figure 2C). In the early embryo, Shh and Ihh play a role in the determination of laterality of the body (Zhang et al., 2001), whereas Smo and Ihh are required for the establishment of blood circulation by vascularisation of the yolk sac (Byrd et al., 2002). These two processes have indirect effects on the formation of the heart. In a direct way, Shh is required for the growth and septation of the heart as well as for the formation of coronaries, the vessels which irrigate the cardiac muscle.

Hh signalling in left-right patterning

At the time of the establishment of left-right patterning (2 somite stage, E8.0), two Hh ligands have been shown to play redundant functions: Shh is expressed in the midline and Ihh in the definitive endoderm. The double mutants $Ihh^{-/-}$; $Shh^{-/-}$ or mutants for the co-receptor Smo fail to specify a left identity in the lateral plate mesoderm, which contains heart precursors. The left markers *Nodal*, *Lefty2* or *Pitx2* are absent. The heart does not loop and remains linear (Zhang et al., 2001; Tsiairis and McMahon, 2009). $Shh^{-/-}$ single mutants display left isomerism of the atria and outflow tract, as depicted by the bilateral expression of *Pitx2* (Hildreth et al., 2009). Inactivation of *Sufu*, which encodes a negative regulator of Hh signalling, leads to randomisation of the direction of heart looping (Cooper, 2005). Given that the components of Hh signalling are not asymmetrically expressed in the embryo, it is intriguing how they can influence left-right patterning. An elegant series of experiments has shown that Hh signalling is required for the competence of the lateral plate mesoderm to respond to asymmetric Nodal signalling (Tsiairis and McMahon, 2009). A transgene expressing *Smo* in the lateral plate mesoderm rescued left-right patterning in $Smo^{-/-}$ mutants, indicating that reception of Hh signalling occurs in the lateral plate mesoderm. Therefore, in ciliary mutants, defective left-right patterning may not be solely due to the function of cilia in the node, but potentially also to defective Hh signalling in the lateral plate mesoderm.

Hh signalling in myocardial growth

$Shh^{-/-}$ mutants display additional defects, with a reduced size of the right ventricle (Tsukui et al., 1999; Washington Smoak et al., 2005). In contrast, depletion of the negative regulator of Hh signalling, Sufu, leads to hyperplasia of the ventricular cardiac muscle (myocardium) (Cooper 2005). Nkx2-5, a transcription factor required for the differentiation of the myocardium, is downregulated in mutants for the co-receptor Smo at the early stages of cardiac differentiation (cardiac crescent stage) and upregulated in mutants for the inhibitory receptor Patched1 (Zhang et al., 2001). In addition to cell differentiation, Shh has been proposed to regulate proliferation of second heart field cells (Dyer and Kirby, 2009). The fact that the right, rather than the left, ventricle is preferentially affected upon reduction of Hh signalling might suggest that it is received by cells of the second heart field. However, defects of the right ventricle have not been reported in mutants with a second heart field-specific depletion of Smo ($Mef2c$-AHF-Cre; $Smo^{flox/-}$), which nonetheless display precocious cardiomyocyte differentiation (Goddeeris et al., 2008). $Rpgrip1l^{-/-}$ mutants (Gerhardt et al., 2013), which have shorter cilia, show defects in the maturation of the transcription factor Gli3, an effector of the Hh pathway. A phenotype of these mutants is the thinning of the ventricular wall associated with decreased cell proliferation. As Gli3 has been observed in primary cilia of the embryonic ventricles, it is possible that Hh signalling may play a role in the cells of cardiac chambers, and not just in the precursor cells of the second heart field. In other mouse models of ciliopathy, such as $Kif3a^{-/-}$, $Ofd1^{-/-}$, $Pifo^{-/-}$ (Takeda et al., 1999; Ferrante et al., 2006; Kinzel et al., 2010), a thinner myocardium has also been observed. However, it has not been demonstrated that this is due to defective Hh signalling in the cardiac lineage.

Hh signalling in heart septation

For the survival of the organism, it is essential that the blood that is oxygenated in the lungs, does not mix with the carbonated blood, which returns from the other organs. Thus, in the embryonic heart, which is a tube, the cardiac chambers (atria, ventricles) and the great arteries (aorta, pulmonary artery) have to be separated. Among the cells which contribute to this process of septation, there are mesoderm cells of the second heart field, which express the receptor Patched1 or the transcription factor Gli1, indicating that the cells are sensitive to the secretion of Shh by the neighbouring endoderm cells (Figure 1 E8.5) (Goddeeris et al., 2008; Hoffmann et al., 2009). By genetic tracing with an inducible Cre recombinase, expressed under the control of the regulatory sequences of $Gli1$, it was shown that these second heart field cells participate in the septum (dividing wall) of the atria and in the pulmonary artery (Hoffmann et al., 2009). Thus, when Shh is ubiquitously inactivated (Washington Smoak et al., 2005) or when the co-receptor Smo is specifically inactivated in the second heart field precursors (Lin et al., 2006; Goddeeris et al., 2008), the interatrial septum does not form, the pulmonary artery is atrophied and the interventricular septum is incomplete. These animals do not survive after birth, because of the inability to feed organs on sufficiently oxygenated blood.

Because Hh signalling requires a primary cilium for activation of the downstream pathway, mutations affecting ciliogenesis impair heart septation. This is the case of the *Ift88$^{cbs/cbs}$* hypomorphic mutants, which reproduce defects of *Shh$^{-/-}$* mutants (Willaredt et al., 2012). Shh has been proposed to regulate the survival of cells in the pharyngeal arches, which are important for the septation of the great arteries (Washington Smoak et al., 2005). *Rpgrip1l$^{-/-}$* mutants, which present defects in the maturation of the effector Gli3, display impaired ventricular septation (Gerhardt et al., 2013). Shh signalling is modulated by other components of the ciliary transport machinery, such as Ift25 (Keady et al., 2012), which are thus required for heart septation. In humans, septation defects are observed in the ciliopathy syndrome of Ellis-van Creveld (EVC). The genes involved in the disease, *EVC* and *EVC2*, encode dominant negative partners of the co-receptor Smo, the localisation of which is regulated by a ciliary complex that was recently identified (Pusapati et al., 2014).

Hh signalling for the formation of coronaries

When the cardiac muscle thickens, it has to be irrigated by a network of blood vessels, referred to as coronaries. The Shh pathway is necessary for the formation of this network, implemented as paracrine signalling between different cardiac cell types. Shh is expressed in the epicardium, the tissue covering the myocardium, in a wave from the base of the ventricles at E12.5 to the apex at E13.5 (Lavine et al., 2006). Depletion of the co-receptor Smo, specifically in the myocardium, arrests the growth of the coronary veins, whereas the same depletion targeted to perivascular cells reduces the growth of the coronary arteries (Lavine et al., 2008b). Thus, secretion of Shh by epicardial cells is sensed by neighbouring, myocardial or perivascular cells (Figure 1 E13.5), which respond by expressing vascular growth factors including Vascular Endothelial Growth Factor (Vegf) and Angiopoietin (Angpt)2. In the adult heart, Shh signalling is required for the maintenance of the coronary microvasculature: it is required in the myocardium, and not in the smooth muscle, specifically for the maintenance of small arteries (Lavine et al., 2008a). It is intriguing that adult cardiomyocytes in this study, or neonatal cardiomyocytes in culture (Carbe et al., 2014) respond to Hh signalling, since mature cardiomyocytes tend to be deprived of primary cilia

(Diguet et al., 2015; Zebrowski et al., 2015). Given the role of Hh signalling in the formation of coronary vessels, it will be interesting to investigate coronary development in ciliary mutants. No such defect has yet been described in ciliopathy patients or animal models.

We have seen that transduction of Hh signalling is important in cells that will contribute to the heart. Such transduction very probably occurs in primary cilia, given also that mutations in ciliary components lead to similar phenotypes of defective septation and ventricular growth. Another example of a signalling pathway observed in primary cilia that is important for the formation of the heart is the polycystin pathway.

The Role of Polycystins in Heart Development

Polycystins are membrane proteins often localised to primary cilia. In humans, mutations in polycystins cause Autosomal Dominant Polycystic Kidney disease (ADPKD). Pkd2 functions as a calcium channel (Figure 2B), whereas Pkd1 is considered as a (G-protein-coupled) mechanoreceptor of cell-cell or cell-matrix interactions (see Zhou, 2009). In the mouse, inactivation of *Pkd1* or *Pkd2* leads to malformations of the cardiac valves (Boulter et al., 2001), as well as defects of cardiac chamber septation (Wu et al., 2000; Boulter et al., 2001). Additional growth defects have been observed, with a thinner myocardium detected before birth upon loss of Pkd function (Boulter et al., 2001; Slough et al., 2008). In contrast, myocardium hypertrophy manifests upon 2–15-fold overexpression of Pkd1 (Kurbegovic et al., 2010), which is reminiscent of the human disease. When inactivation of *Pkd1* is restricted to endothelial cells (Garcia-Gonzalez et al., 2010) or to both the smooth muscle and myocardium (Hassane et al., 2011), there is no heart defect, which is different from the heart defects seen in the ubiquitous loss-of-function models described above. This difference in phenotype between constitutive (ubiquitous) and conditional (restricted) mutants is unclear: it could be due to variations in the DNA sequence associated with Pkd1 inactivation, or due to the role of Pkd1 in another cell type, for example in interstitial fibroblasts. Some growth defects may also be due to compensatory myocardial remodelling, secondary to the hypertension induced by kidney or vascular dysfunction (Alam and Perrone, 2013). Indeed, in the adult, *Pkd1* is required for stress-induced cardiac hypertrophy (Pedrozo et al., 2015). Thus the mechanism of action of Pkd1 and Pkd2 remains to be determined. Are they directly involved in cardiac cells? Which cell behaviour is affected and via which signalling pathway? Similarly, the link between Pkd and cardiac primary cilia has not yet been demonstrated.

Other signalling pathways have been linked with primary cilia, such as the canonical Wnt pathway (Lancaster et al., 2011), the Wnt/planar cell polarity pathway, the PDGFRα (see Goetz and Anderson, 2010) or TGFβ (Clement et al., 2013) pathways, which are important for aspects of heart morphogenesis. However, mutations in ciliary components result in less severe phenotypes than the inactivation of the Wnt or TGFβ pathways (e.g., Haegel et al., 1995; Dunn, 2004), supporting the view that primary cilia are not necessarily required for the transduction of these pathways, but involved in modulating them. For these signalling pathways, the requirement of primary cilia in cardiac cells *in vivo* thus remains to be elucidated.

Conclusion

In conclusion, the rare incidence of cardiopathies associated with ciliopathies in patients is balanced by a higher frequency in the mouse model, which is amenable to the study of defects resulting in lethality *in utero*. The broad spectrum of cardiac malformations associated with ciliopathies suggests that cilia may play specific roles at different stages of heart development. Experiments in the mouse have shown that motile cilia in the node, which are important for left-right signalling, have indirect but dramatic consequences for the formation of the heart, by controlling the alignment and identity of cardiac compartments. Within the cardiac lineage, primary cilia are found in most cell types at embryonic stages, but not maintained in mature cells such as cardiomyocytes, which have exited the cell cycle. The transcriptional regulation of ciliogenesis in the cardiac lineage remains unknown. It will be interesting to investigate how it relates to the regulation of cell proliferation and differentiation. Detailed analysis of the spatio-temporal composition of cardiac primary cilia, as well as finer genetic analyses will be required to uncouple the many potential functions of cilia and understand which cell type, at which developmental stage requires a primary cilium to regulate the septation and growth of the heart or the formation of valves and coronary vessels. Mutants with impaired Hh signalling appear to phenocopy many of the defects observed in ciliopathies. However, it will have to be formally demonstrated whether Hh signalling is the only pathway transduced in primary cilia of the cardiac lineage or whether other pathways, including polycystins, PDGFRα, TGFβ or Wnt, require primary cilia for their implementation. In addition, ciliary proteins such as Ift88 and Ift20 have been shown to localise also to the centrosome and thus have cellular functions not related to cilia (Delaval et al., 2011). The study of cardiac primary cilia will not only provide insight into the fundamental mechanisms of heart morphogenesis, but will also increase our knowledge of the origin of the congenital heart defects associated with ciliopathies.

Acknowledgements

We are grateful to L. Houyel and P. Bouvagnet for comments on the manuscript. Work in the group of S. Meilhac is supported by the Institut *Imagine*, the Institut Pasteur, the Institut National de la Santé Et de la

Recherche Médicale, the Université Paris Descartes and the Association Française contre les Myopathies.

References

Alam, A. and Perrone, R. D. 2013. Left ventricular hypertrophy in ADPKD: changing demographics. Current Hypertension Reviews 9: 27–31.

Babu, D. and Roy, S. 2013. Left-right asymmetry: cilia stir up new surprises in the node. Open Biol. 3: 130052.

Baker, K. and Beales, P. L. 2009. Making sense of cilia in disease: the human ciliopathies. Am. J. Med. Genet. C. Semin. Med. Genet. 151C: 281–95.

Bergmann, O., Zdunek, S., Alkass, K., Druid, H., Bernard, S. and Frisén, J. 2011. Identification of cardiomyocyte nuclei and assessment of ploidy for the analysis of cell turnover. Experimental Cell Research 317: 188–194.

Boulter, C., Mulroy, S., Webb, S., Fleming, S., Brindle, K. and Sandford, R. 2001. Cardiovascular, skeletal, and renal defects in mice with a targeted disruption of the Pkd1 gene. Proc. Natl. Acad. Sci. USA 98: 12174–9.

Brennan, J., Norris, D. P. and Robertson, E. J. 2002. Nodal activity in the node governs left-right asymmetry. Genes Dev. 16: 2339–44.

Brueckner, M. 2007. Heterotaxia, congenital heart disease, and primary ciliary dyskinesia. Circulation 115: 2793–5.

Burnicka-Turek, O., Steimle, J. D., Huang, W., Felker, L., Kamp, A., Kweon, J., Peterson, M., Reeves, R. H., Maslen, C. L., Gruber, P. J. et al. 2016. Cilia gene mutations cause atrioventricular septal defects by multiple mechanisms. Human Molecular Genetics 25: 3011–3028.

Byrd, N., Becker, S., Maye, P., Narasimhaiah, R., St-Jacques, B., Zhang, X., McMahon, J., McMahon, A. and Grabel, L. 2002. Hedgehog is required for murine yolk sac angiogenesis. Development 129: 361–72.

Carbe, C. J., Cheng, L., Addya, S., Gold, J. I., Gao, E., Koch, W. J. and Riobo, N. A. 2014. Gi proteins mediate activation of the canonical hedgehog pathway in the myocardium. AJP: Heart and Circulatory Physiology 307: H66–H72.

Choksi, S. P., Lauter, G., Swoboda, P. and Roy, S. 2014. Switching on cilia: transcriptional networks regulating ciliogenesis. Development 141: 1427–41.

Clement, C. A., Ajbro, K. D., Koefoed, K., Vestergaard, M. L., Veland, I. R., Henriques de Jesus, M. P., Pedersen, L. B., Benmerah, A., Andersen, C. Y., Larsen, L. A. et al. 2013. TGF-beta signaling is associated with endocytosis at the pocket region of the primary cilium. Cell Rep. 3: 1806–14.

Cooper, A. F. 2005. Cardiac and CNS defects in a mouse with targeted disruption of suppressor of fused. Development 132: 4407–4417.

Cui, C., Chatterjee, B., Francis, D., Yu, Q., SanAgustin, J. T., Francis, R., Tansey, T., Henry, C., Wang, B., Lemley, B. et al. 2011. Disruption of Mks1 localization to the mother centriole causes cilia defects and developmental malformations in Meckel-Gruber syndrome. Dis. Model Mech. 4: 43–56.

Delaval, B., Bright, A., Lawson, N. D. and Doxsey, S. 2011. The cilia protein IFT88 is required for spindle orientation in mitosis. Nat. Cell Biol. 13: 461–8.

Diguet, N., Le Garrec, J.-F., Lucchesi, T. and Meilhac, S. M. 2015. Imaging and analyzing primary cilia in cardiac cells. Methods in Cilia & amp; Flagella, vol. 127: Elsevier.

Dominguez, J. N., Meilhac, S. M., Bland, Y. S., Buckingham, M. E. and Brown, N. A. 2012. Asymmetric fate of the posterior part of the second heart field results in unexpected left/right contributions to both poles of the heart. Circ. Res. 111: 1323–35.

Dunn, N. R. 2004. Combinatorial activities of Smad2 and Smad3 regulate mesoderm formation and patterning in the mouse embryo. Development 131: 1717–1728.

Dyer, L. A. and Kirby, M. L. 2009. Sonic hedgehog maintains proliferation in secondary heart field progenitors and is required for normal arterial pole formation. Dev. Biol. 330: 305–17.

Endicott, S. J., Basu, B., Khokha, M. and Brueckner, M. 2015. The NIMA-like kinase Nek2 is a key switch balancing cilia biogenesis and resorption in the development of left-right asymmetry. Development 142: 4068–4079.

Ferrante, M. I., Zullo, A., Barra, A., Bimonte, S., Messaddeq, N., Studer, M., Dolle, P. and Franco, B. 2006. Oral-facial-digital type I protein is required for primary cilia formation and left-right axis specification. Nat. Genet. 38: 112–7.

Francis, R. J., Christopher, A., Devine, W. A., Ostrowski, L. and Lo, C. 2012. Congenital heart disease and the specification of left-right asymmetry. Am. J. Physiol. Heart. Circ. Physiol. 302: H2102–11.

Francou, A., Saint-Michel, E., Mesbah, K. and Kelly, R. G. 2014. TBX1 regulates epithelial polarity and dynamic basal filopodia in the second heart field. Development 141: 4320–31.

French, V. M., van de Laar, I. M. B. H., Wessels, M. W., Rohe, C., Roos-Hesselink, J. W., Wang, G., Frohn-Mulder, I. M. E., Severijnen, L. A., de Graaf, B. M., Schot, R. et al. 2012. NPHP4 variants are associated with pleiotropic heart malformations. Circulation Research 110: 1564–1574.

Garcia-Gonzalez, M. A., Outeda, P., Zhou, Q., Zhou, F., Menezes, L. F., Qian, F., Huso, D. L., Germino, G. G., Piontek, K. B. and Watnick, T. 2010. Pkd1 and Pkd2 are required for normal placental development. PLoS One 5.

Gerhardt, C., Lier, J. M., Kuschel, S. and Ruther, U. 2013. The ciliary protein Ftm is required for ventricular wall and septal development. PLoS One 8: e57545.

Goddeeris, M. M., Rho, S., Petiet, A., Davenport, C. L., Johnson, G. A., Meyers, E. N. and Klingensmith, J. 2008. Intracardiac septation requires hedgehog-dependent cellular contributions from outside the heart. Development 135: 1887–95.

Goetz, J. G., Steed, E., Ferreira, R. R., Roth, S., Ramspacher, C., Boselli, F., Charvin, G., Liebling, M., Wyart, C., Schwab, Y. et al. 2014. Endothelial cilia mediate low flow sensing during zebrafish vascular development. Cell Rep. 6: 799–808.

Goetz, S. C. and Anderson, K. V. 2010. The primary cilium: a signalling centre during vertebrate development. Nat. Rev. Genet. 11: 331–44.

Gray, R. S., Abitua, P. B., Wlodarczyk, B. J., Szabo-Rogers, H. L., Blanchard, O., Lee, I., Weiss, G. S., Liu, K. J., Marcotte, E. M., Wallingford, J. B. et al. 2009. The planar cell polarity effector Fuz is essential for targeted membrane trafficking, ciliogenesis and mouse embryonic development. Nat. Cell Biol. 11: 1225–32.

Grego-Bessa, J., Luna-Zurita, L., del Monte, G., Bolos, V., Melgar, P., Arandilla, A., Garratt, A. N., Zang, H., Mukouyama, Y. S., Chen, H. et al. 2007. Notch signaling is essential for ventricular chamber development. Dev. Cell 12: 415–29.

Haegel, H., Larue, L., Ohsugi, M., Fedorov, L., Herrenknecht, K. and Kemler, R. 1995. Lack of beta-catenin affects mouse development at gastrulation. Development 121: 3529–3537.

Harrison, M. J., Shapiro, A. J. and Kennedy, M. P. 2016. Congenital heart disease and primary ciliary dyskinesia. Paediatr. Respir. Rev. 18: 25–32.

Hassane, S., Claij, N., Jodar, M., Dedman, A., Lauritzen, I., Duprat, F., Koenderman, J. S., van der Wal, A., Breuning, M. H., de Heer, E. et al. 2011. Pkd1-inactivation in vascular smooth muscle cells and adaptation to hypertension. Lab. Invest. 91: 24–32.

Hildreth, V., Webb, S., Chaudhry, B., Peat, J. D., Phillips, H. M., Brown, N., Anderson, R. H. and Henderson, D. J. 2009. Left cardiac isomerism in the Sonic hedgehog null mouse. Journal of Anatomy 214: 894–904.

Hoffmann, A. D., Peterson, M. A., Friedland-Little, J. M., Anderson, S. A. and Moskowitz, I. P. 2009. Sonic hedgehog is required in pulmonary endoderm for atrial septation. Development 136: 1761–70.

Huangfu, D., Liu, A., Rakeman, A. S., Murcia, N. S., Niswander, L. and Anderson, K. V. 2003. Hedgehog signalling in the mouse requires intraflagellar transport proteins. Nature 426: 83–7.

Keady, B. T., Samtani, R., Tobita, K., Tsuchya, M., San Agustin, J. T., Follit, J. A., Jonassen, J. A., Subramanian, R., Lo, C. W. and Pazour, G. J. 2012. IFT25 links the signal-dependent movement of Hedgehog components to intraflagellar transport. Dev. Cell 22: 940–51.

Kinzel, D., Boldt, K., Davis, E. E., Burtscher, I., Trumbach, D., Diplas, B., Attie-Bitach, T., Wurst, W., Katsanis, N., Ueffing, M. et al. 2010. Pitchfork regulates primary cilia disassembly and left-right asymmetry. Dev. Cell 19: 66–77.

Kurbegovic, A., Cote, O., Couillard, M., Ward, C. J., Harris, P. C. and Trudel, M. 2010. Pkd1 transgenic mice: adult model of polycystic kidney disease with extrarenal and renal phenotypes. Hum. Mol. Genet. 19: 1174–89.

Lambacher, N. J., Bruel, A. L., van Dam, T. J., Szymanska, K., Slaats, G. G., Kuhns, S., McManus, G. J., Kennedy, J. E., Gaff, K., Wu, K. M. et al. 2016. TMEM107 recruits ciliopathy proteins to subdomains of the ciliary transition zone and causes Joubert syndrome. Nat. Cell Biol. 18: 122–31.

Lancaster, M. A., Schroth, J. and Gleeson, J. G. 2011. Subcellular spatial regulation of canonical Wnt signalling at the primary cilium. Nat. Cell Biol. 13: 700–7.

Lavine, K. J., White, A. C., Park, C., Smith, C. S., Choi, K., Long, F., Hui, C. C. and Ornitz, D. M. 2006. Fibroblast growth factor signals regulate a wave of Hedgehog activation that is essential for coronary vascular development. Genes Dev. 20: 1651–66.

Lavine, K. J., Kovacs, A. and Ornitz, D. M. 2008a. Hedgehog signaling is critical for maintenance of the adult coronary vasculature in mice. Journal of Clinical Investigation 1–11.

Lavine, K. J., Long, F., Choi, K., Smith, C. and Ornitz, D. M. 2008b. Hedgehog signaling to distinct cell types differentially regulates coronary artery and vein development. Development 135: 3161–3171.

Lee, M.-S., Hwang, K.-S., Oh, H.-W., Ji-Ae, K., Kim, H.-T., Cho, H.-S., Lee, J.-J., Ko, J. Y., Choi, J.-H., Jeong, Y.-M. et al. 2015. IFT46 plays an essential role in cilia development. Developmental Biology 400: 248–257.

Lescroart, F., Mohun, T., Meilhac, S. M., Bennett, M. and Buckingham, M. 2012. Lineage tree for the venous pole of the heart: clonal analysis clarifies controversial genealogy based on genetic tracing. Circulation Research 111: 1313–1322.

Li, F., Wang, X., Capasso, J. M. and Gerdes, A. M. 1996. Rapid transition of cardiac myocytes from hyperplasia to hypertrophy during postnatal development. J. Mol. Cell Cardiol. 28: 1737–46.

Li, Y., Klena, N. T., Gabriel, G. C., Liu, X., Kim, A. J., Lemke, K., Chen, Y., Chatterjee, B., Devine, W., Damerla, R. R. et al. 2015. Global genetic analysis in mice unveils central role for cilia in congenital heart disease. Nature 521: 520–4.

Lin, C. J., Lin, C. Y., Chen, C. H., Zhou, B. and Chang, C. P. 2012. Partitioning the heart: mechanisms of cardiac septation and valve development. Development 139: 3277–99.

Lin, L., Bu, L., Cai, C. L., Zhang, X. and Evans, S. 2006. Isl1 is upstream of sonic hedgehog in a pathway required for cardiac morphogenesis. Dev. Biol. 295: 756–63.

Louw, J. J., Corveleyn, A., Jia, Y., Iqbal, S., Boshoff, D., Gewillig, M., Peeters, H., Moerman, P. and Devriendt, K. 2014. Homozygous loss-of-function mutation in ALMS1 causes the lethal disorder mitogenic cardiomyopathy in two siblings. Eur. J. Med. Genet. 57: 532–5.

Lu, C. J., Du, H., Wu, J., Jansen, D. A., Jordan, K. L., Xu, N., Sieck, G. C. and Qian, Q. 2008. Non-random distribution and sensory functions of primary cilia in vascular smooth muscle cells. Kidney Blood Press Res. 31: 171–84.

Manning, D. K., Sergeev, M., van Heesbeen, R. G., Wong, M. D., Oh, J. H., Liu, Y., Henkelman, R. M., Drummond, I., Shah, J. V. and Beier, D. R. 2013. Loss of the ciliary kinase Nek8 causes left-right asymmetry defects. J. Am. Soc. Nephrol. 24: 100–12.

Meilhac, S. M., Lescroart, F. , Blanpain, C. and Buckingham, M. E. 2014. Cardiac cell lineages that form the heart. Cold Spring Harb Perspect Med. 4: a013888.

Murcia, N. S., Richards, W. G., Yoder, B. K., Mucenski, M. L., Dunlap, J. R. and Woychik, R. P. 2000. The Oak Ridge Polycystic Kidney (orpk) disease gene is required for left-right axis determination. Development 127: 2347–2355.

Myklebust, R., Engedal, H., Saetersdal, T. S. and Ulstein, M. 1977. Primary 9 + 0 cilia in the embryonic and the adult human heart. Anat. Embryol. 151: 127–39.

Nonaka, S., Tanaka, Y., Okada, Y., Takeda, S., Harada, A., Kanai, Y., Kido, M. and Hirokawa, N. 1998. Randomization of left-right asymmetry due to loss of nodal cilia generating leftward flow of extraembryonic fluid in mice lacking KIF3B motor protein. Cell 95: 829–37.

Nozawa, Y. I., Lin, C. and Chuang, P. T. 2013. Hedgehog signaling from the primary cilium to the nucleus: an emerging picture of ciliary localization, trafficking and transduction. Curr. Opin. Genet. Dev. 23: 429–37.

Pazour, G. J., Dickert, B. L., Vucica, Y., Seeley, E. S., Rosenbaum, J. L., Witman, G. B. and Cole, D. G. 2000. Chlamydomonas IFT88 and its mouse homologue, polycystic kidney disease gene tg737, are required for assembly of cilia and flagella. J. Cell Biol. 151: 709–18.

Pedrozo, Z., Criollo, A., Battiprolu, P. K., Morales, C. R., Contreras-Ferrat, A., Fernandez, C., Jiang, N., Luo, X., Caplan, M. J., Somlo, S. et al. 2015. Polycystin-1 is a cardiomyocyte mechanosensor that governs L-type Ca2+ channel protein stability. Circulation 131: 2131–2142.

Perrone, R. D. 1997. Extrarenal manifestations of ADPKD. Kidney Int. 51: 2022–36.

Ramsdell, A. F. and Markwald, R. R. 1997. Induction of endocardial cushion tissue in the avian heart is regulated, in part, by TGFbeta-3-mediated autocrine signaling. Dev. Biol. 188: 64–74.

Rash, J. E., Shay, J. W. and Biesele, J. J. 1969. Cilia in cardiac differentiation. J. Ultrastruct. Res. 29: 470–84.

Rochais, F., Mesbah, K. and Kelly, R. G. 2009. Signaling pathways controlling second heart field development. Circ. Res. 104: 933–42.

Shen, H., Cavallero, S., Estrada, K. D., Sandovici, I., Kumar, S. R., Makita, T., Lien, C. L., Constancia, M. and Sucov, H. M. 2015. Extracardiac control of embryonic cardiomyocyte proliferation and ventricular wall expansion. Cardiovasc. Res. 105: 271–8.

Shenje, L. T., Andersen, P., Halushka, M. K., Lui, C., Fernandez, L., Collin, G. B., Amat-Alarcon, N., Meschino, W., Cutz, E., Chang, K. et al. 2014. Mutations in Alstrom protein impair terminal differentiation of cardiomyocytes. Nat. Commun. 5: 3416.

Silva, E., Betleja, E., John, E., Spear, P., Moresco, J. J., Zhang, S., Yates, J. R. r., Mitchell, B. J. and Mahjoub, M. R. 2016. Ccdc11 is a novel centriolar satellite protein essential for ciliogenesis and establishment of left-right asymmetry. Molecular Biology of the Cell 27: 48–63.

Slough, J., Cooney, L. and Brueckner, M. 2008. Monocilia in the embryonic mouse heart suggest a direct role for cilia in cardiac morphogenesis. Dev. Dyn. 237: 2304–14.

Supp, D. M., Brueckner, M., Kuehn, M. R., Witte, D. P., Lowe, L. A., McGrath, J., Corrales, J. and Potter, S. S. 1999. Targeted deletion of the ATP binding domain of left-right dynein confirms its role in specifying development of left-right asymmetries. Development 126: 5495–5504.

Sutherland, M. J. and Ware, S. M. 2009. Disorders of left-right asymmetry: Heterotaxy and situs inversus. American Journal of Medical Genetics Part C: Seminars in Medical Genetics 151C: 307–317.

Takeda, S., Yonekawa, Y., Tanaka, Y., Okada, Y., Nonaka, S. and Hirokawa, N. 1999. Left-right asymmetry and kinesin superfamily protein KIF3A: new insights in determination of laterality and mesoderm induction by kif3A–/– mice analysis. J. Cell Biol. 145: 825–36.

Tobin, J. L. and Beales, P. L. 2009. The nonmotile ciliopathies. Genet. Med. 11: 386–402.

Tsiairis, C. D. and McMahon, A. P. 2009. An Hh-dependent pathway in lateral plate mesoderm enables the generation of left/right asymmetry. Current Biology 19: 1912–1917.

Tsukui, T., Capdevila, J., Tamura, K., Ruiz-Lozano, P., Rodriguez-Esteban, C., Yonei-Tamura, S., Magallón, J., Chandraratna, R. A. S., Chien, K., Blumberg, B. et al. 1999. Multiple left-right asymmetry defects in Shh–/– mutant mice unveil a convergence of the Shh and retinoic acid pathways in the control of Lefty-1. Proceedings of the National Academy of Sciences of the United States of America 96: 11376–6.

Van der Heiden, K., Groenendijk, B. C., Hierck, B. P., Hogers, B., Koerten, H. K., Mommaas, A. M., Gittenberger-de Groot, A. C. and Poelmann, R. E. 2006. Monocilia on chicken embryonic endocardium in low shear stress areas. Dev. Dyn. 235: 19–28.

Waldo, K., Miyagawa-Tomita, S., Kumiski, D. and Kirby, M. L. 1998. Cardiac neural crest cells provide new insight into septation of the cardiac outflow tract: aortic sac to ventricular septal closure. Dev. Biol. 196: 129–44.

Washington Smoak, I., Byrd, N. A., Abu-Issa, R., Goddeeris, M. M., Anderson, R., Morris, J., Yamamura, K., Klingensmith, J. and Meyers, E. N. 2005. Sonic hedgehog is required for cardiac outflow tract and neural crest cell development. Dev. Biol. 283: 357–72.

Wessels, A., van den Hoff, M. J., Adamo, R. F., Phelps, A. L., Lockhart, M. M., Sauls, K., Briggs, L. E., Norris, R. A., van Wijk, B., Perez-Pomares, J. M. et al. 2012. Epicardially derived fibroblasts preferentially contribute to the parietal leaflets of the atrioventricular valves in the murine heart. Dev. Biol. 366: 111–24.

Willaredt, M. A., Gorgas, K., Gardner, H. A. and Tucker, K. L. 2012. Multiple essential roles for primary cilia in heart development. Cilia 1: 23.

Wu, G., Markowitz, G. S., Li, L., D'Agati, V. D., Factor, S. M., Geng, L., Tibara, S., Tuchman, J., Cai, Y., Park, J. H. et al. 2000. Cardiac defects and renal failure in mice with targeted mutations in Pkd2. Nat. Genet. 24: 75–8.

Zebrowski, D. C., Vergarajauregui, S., Wu, C. C., Piatkowski, T., Becker, R., Leone, M., Hirth, S., Ricciardi, F., Falk, N., Giessl, A. et al. 2015. Developmental alterations in centrosome integrity contribute to the post-mitotic state of mammalian cardiomyocytes. Elife 4.

Zhang, X. M., Ramalho-Santos, M. and McMahon, A. P. 2001. Smoothened mutants reveal redundant roles for Shh and Ihh signaling including regulation of L/R symmetry by the mouse node. Cell 106: 781–92.

Zhou, J. 2009. Polycystins and primary cilia: primers for cell cycle progression. Annu. Rev. Physiol. 71: 83–113.

CHAPTER 4

Cilia in Kidney Development and Disease

Paraskevi Goggolidou

INTRODUCTION

The kidney is a fascinating organ, consisting of a combination of variable cell types that work together to achieve fluid homeostasis. An important structure in the kidney is the nephron, comprising of the glomerulus, the tubule and the collecting duct. The collecting duct is the most distal part of the nephron, the glomerulus is the most proximal part, while the tubule is the fragment in between. The number of nephrons found in each kidney varies from a single nephron in zebrafish pronephros to more than a million nephrons in human metanephros (Schneider et al., 2015). The significance of this organ is manifested by the diversity of the functions that it performs; it secretes hormones, maintains blood pressure, controls pH and regulates red blood cells. The mammalian kidney in particular, is derived from cells of the intermediate mesoderm that gradually give rise to the metanephric mesenchyme. Although nephrogenesis in mammals and avians stops at the end of gestation or after birth, it occurs through the animal's lifetime in fish, amphibians and reptiles (Romagnani et al., 2013). As a result, the limited regenerative properties of the kidney are being exploited as a means of treating kidney disease. In addition, research over the last twenty years has highlighted the significance of a unique cellular structure, the primary cilium in correct kidney development and function. This chapter will focus on the

School of Biomedical Science and Physiology, Faculty of Science and Engineering, University of Wolverhampton, UK; and Centre for Nephrology, UCL Medical Campus, Royal Free Hospital, UK.
E-mail: p.goggolidou@wlv.ac.uk.

role of the cilium in the kidney, dissect the signalling pathways associated with normal kidney structure and function and discuss the inherited renal diseases that have been reclassified as renal ciliopathies.

Kidney Development and the Adult Human Kidney

The kidney is a bean-shaped organ located on either side of the spine, just below the diaphragm. Its characteristic shape consists of endothelial and epithelial cells. These cells sit within a stroma composed of a population of matrix-producing fibroblasts that surround adjacent nephrogenic structures and the collecting duct (Goggolidou, 2014). The interactions between the various cell types found in the kidney allow it to perform its crucial physiological function. The primary function of the kidney is to remove waste from the body through the production of urine. Kidneys also assist in the regulation of blood pressure, blood volume and electrolyte composition in the blood. In doing so, they regulate the body's sodium, potassium, phosphorus and calcium levels, they remove drugs and toxins from the body and they produce and release erythropoietin, renin and calcitriol into the bloodstream.

The formation of a fully functional kidney initiates with kidney morphogenesis, a three-stage process involving the pronephric, mesonephric and metanephric kidney structures. During development, three types of organs are derived from the intermediate mesoderm: pronephros and mesonephros disappear during embryogenesis and metanephros remains throughout adulthood. Pronephros consists of few tubule-like structures, but it is important for correct pronephric and mesonephric duct formation. The transcription factor Pax2 regulates mesonephric and metanephric duct formation and its expression is controlled by Bone Morphogenetic Protein-4 (BMP-4) (Sainio and Raatikainen-Ahokas, 1999). The mesonephric kidney serves as an embryonic secretory organ that is thereafter removed by programmed cell death. It is the source of multiple stem cells, including somatic cells in the male gonad, vascular endothelial cells and haematopoietic stem cells. Inactivation of another transcription factor, Wilms Tumour-1 (WT-1) causes Wilm's tumour and WT-1 mutant mice lack gonads, mesonephric tubules and metanephros (Sainio and Raatikainen-Ahokas, 1999).

Kidney development is well conserved between mouse and human, with metanephric kidney development initiating at embryonic day (E) 10 in mouse and resulting in the distinctly branched kidney with the many nephron segments. The process of metanephric kidney morphogenesis commences when the ureteric bud branches within the metanephric mesenchyme. Branching of the ureteric bud provokes the proliferation of the nephron progenitors within the metanephric mesenchyme. At the same time, the nephron progenitors undergo a mesenchymal to epithelial transition (MET) and they form the renal vesicles, precursors of the adult nephron

(Goggolidou, 2014). During MET and the generation of renal vesicles, sustained signalling from the mesenchyme allows the ureteric bud to branch further and this will eventually lead to the creation of the adult collecting duct and ureter.

While the kidney is developing to form the functional, adult kidney, renal vesicles arrange into subsequent distinct shapes, initially referred to as comma-shaped and then S-shaped bodies (Figure 1A). Specific cell populations of these structures will upon further development and branching of the kidney produce the proximal, distal and collecting tubules, the renal corpuscle and the loop of Henle. Once renal vesicles have formed, they arrange into recognisable shapes, first the comma- and later the S-shaped bodies. Distinct cell populations of these structures will then give rise to the renal corpuscle, the proximal, distal and collecting tubules and the loop of Henle. It is important to note that although kidney tubules increase in length during the process of kidney development, their diameter always remains constant. Defects in kidney diameter associated with convergent extension have been shown to result in dysfunctional kidneys in both animal models and human kidney tissues (Goggolidou and Wilson, 2016).

In addition, it has been demonstrated that absent or malformed kidney cilia also impact on kidney morphogenesis and affect renal function, both in animal models and human patients. It is thus crucial to investigate the role of the primary cilium in the kidney and dissect how its function influences renal disease.

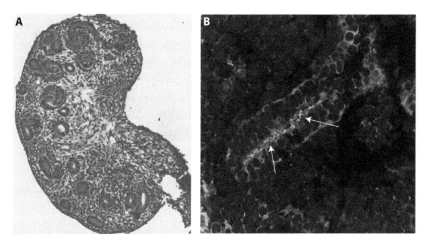

Figure 1. The mouse embryonic kidney with its specialised structures and primary cilia. (**A**) Periodic-acid Schiff (PAS) staining of embryonic (E) 13.5 mouse kidney highlights the differentiating renal vesicles and comma shaped bodies (dashed yellow lines). (**B**) The branching ureteric bud (E13.5) contains primary cilia across its whole length (see white arrows). Cilia were detected by immunofluorescence staining with acetylated tubulin (green), while nuclei were visualised by DAPI (blue).

Cilia in the Kidney

The cilium, the cell's antenna, is an evolutionarily-conserved, microtubule-based organelle whose function remained elusive until recently. Cilia are found in almost every cell type and depending on the cell type, cells can be mono- or multi-ciliated. Discovered more than 200 years ago, cilia contain similarities with the prokaryotic flagella and they are apical, cellular protrusions that can be motile or immotile, consistent with the role they perform (Hildebrandt et al., 2011). Protein synthesis does not take place in cilia, which typically form in the G0 or G1 stages of the cell cycle, hence a network of proteins called intraflagellar transport (IFT) proteins, as well as the kinesin and dynein motor proteins, are required for ciliary formation and resorption. When IFT trafficking is disrupted, defective ciliogenesis or deficient cilia function ensues (Oh and Katsanis, 2012).

The epithelial, non-proliferating cells of the kidney are mono-ciliated and are thought to use the cilium as a sensory organelle, either for signal transduction or to sense calcium flow (Figure 1B). The length of cilia in the kidney varies according to the developmental stage of the kidney, with shorter cilia found in renal vesicles and the longest cilia found in post-natal nephrons (Goggolidou and Wilson, 2016). It is important to emphasise that many proteins that are crucial for correct kidney development and function have been shown to localise to kidney cilia or to require renal cilia in order to transduce downstream signals. In addition, gene inactivation of the anterograde motor protein *Kif3a* or the IFT proteins *Ift88* and *Ift20* that disrupt ciliogenesis, all result in the formation of kidney cysts (Davenport et al., 2007; Jonassen et al., 2008; Lin et al., 2003). It is thus important to better understand the links between cilia, signalling and cystic renal disease and in the next section, we will concentrate on Wnt signalling, a pathway whose function has been associated with the primary cilium in the kidney.

The Wnt Signalling Pathway

The Wnt signalling pathway is an ancient, evolutionarily conserved pathway, with diverse roles in mammalian tissues and human diseases. Wnts are small glycoproteins that are the ligands of Wnt signalling. More than 19 Wnts exist in mammals and a few of them have been shown to play a role in the structure and function of the kidney. For example, overexpression of Wnt1 was recently shown to accelerate the progression from acute kidney injury to chronic kidney disease (Xiao et al., 2016), while the Wnt ligand was able to drive interstitial fibrosis in a Wnt1 inducible proximal tubule mouse model (Maarouf et al., 2016). Mutations in *WNT4* were identified in families suffering from familial, non-syndromic renal hypodysplasias (Vivante et al., 2013) and Wnt4 was lately proposed as an early biomarker of kidney tubular injury (Zhao et al., 2016). Loss of Wnt5a results in dilated renal tubules and cyst formation in zebrafish and kidney agenesis or duplex kidneys in mice (Huang et al., 2014). Further, in a mouse model, Wnt5a controls the expression of aquaporin-2 (Ando et al., 2016), suggesting that it might play a role in nephrogenic diabetes insipidus. Interestingly, Wnt6 was found to be expressed in developing mouse kidney mesonephros and differential WNT6 expression manifested in human renal biopsies (Beaton et al., 2016), potentially providing an indicator of disease severity. Similarly, Wnt9b was demonstrated to be essential for mouse metanephric kidney development, with mesenchyme deficient in Wnt9b, failing to undergo tubulogenesis (Carroll et al., 2005). Further work showed that loss of Wnt9b resulted in increased kidney tubule diameter and randomly oriented cell divisions (Karner et al., 2009), which are both indicators of abnormal planar cell polarity in the kidney epithelium. *Wnt11* mutant mouse kidneys were hypoplastic and had reduced ureteric bud tips (Majumdar et al., 2003), with genetic intercrosses manifesting a reciprocal interaction between Wnt11 and Ret/Gdnf in the regulation of kidney branching morphogenesis.

It is hypothesised that the type of Wnt ligand can influence the downstream response of the Wnt signalling pathway. Hence, Wnt signalling is an umbrella that includes three separate branches of Wnt signalling that all initiate when a Wnt ligand binds to a receptor. The three different types of Wnt signalling can be distinguished into the canonical Wnt, non-canonical Wnt/Planar Cell Polarity (PCP) and non-canonical Wnt/calcium (Ca^{+2}) pathways (Figure 2). Canonical Wnt signalling (that involves β catenin) initiates when a canonical Wnt ligand binds to a Frizzled (Fz) receptor in the presence of lipoprotein receptor-related protein 5 or 6 (LRP5 or 6). The outcome of binding is the activation of Dishevelled (Dvl) and the inhibition of the β catenin destruction complex, consisting of Glycogen synthase kinase 3β (GSK3β), adenomatous polyposis coli (APC), Axin and casein kinase 1 (CK1). Upon canonical Wnt signalling stimulation, β catenin accumulates in the cytosol driving the activation of *Wnt* target genes. When the canonical Wnt pathway is inactive, β catenin is targeted through phosphorylation for proteosomal degradation, hence switching the canonical Wnt signalling pathway off.

The non-canonical branches of Wnt signalling involve more protein players and show greater complexity in their responses, interactions and activation than the canonical Wnt signalling pathway. The non-canonical Wnt/ PCP pathway also initiates when a Wnt ligand behinds to a Fz receptor, but in this case, activation of Dvl leads to either downstream cytoskeletal rearrangements or transcriptional activation through the stimulation of *RhoA* or *Rac1*. It is believed that some Wnt ligands activate only canonical Wnt signalling, whereas others are more specific to non-canonical Wnt signalling activation, nevertheless, since most Wnt ligands have been shown to activate both pathways, what leads to the stimulation of one pathway over the other remains elusive.

Nonetheless, the significance of components of the canonical and non-canonical Wnt signalling cascades has been demonstrated both in animal

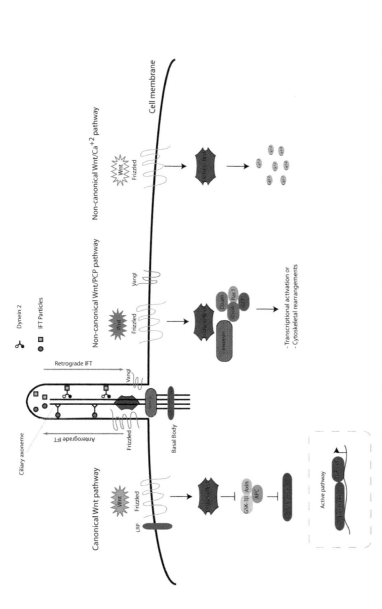

Figure 2. The various branches of Wnt signalling. Wnt signalling initiates when a Wnt ligand binds to a Frizzled receptor. This results in the activation of Dishevelled (purple box), triggering either the canonical Wnt pathway (involving β catenin, left hand side of figure), the Planar Cell Polarity (PCP) pathway (causing the transcriptional activation of downstream genes or cytoskeletal rearrangements, middle of figure) or the calcium pathway, resulting in increased intracellular calcium levels (right hand side of figure). Many Wnt-related proteins localise to the primary cilium (Frizzled, Vangl, Dishevelled, Inversin, β-catenin), which uses anterograde and retrograde intraflaggelar transport (IFT) for protein trafficking.

models and in human patients. Loss of Fz4 and Fz8 that mediate non-canonical Wnt signalling via Wnt11 resulted in reduced ureteric bud growth in embryonic mouse kidneys (Ye et al., 2011). Elevated secreted frizzled-related protein 4 (sFRP4) expression was observed in Autosomal Dominant Polycystic Kidney Disease (ADPKD) kidneys compared to normal (Romaker et al., 2009) and recently elevated sFRP4 levels at baseline were observed in ADPKD patients that also showed rapid kidney decline (Zschiedrich et al., 2016), presenting the possibility that sFRP4 can be used as a predictor of failure of renal function in ADPKD. Dishevelled was initially shown to associate with actin fibers and focal adhesion plates in mouse metanephric mesenchymal cells (Torres and Nelson, 2000). Interestingly, abnormal DVL-1 expression was detected in an NPHP2 family that carried novel inversin mutations, further supporting the role of DVL in the kidney (Bellavia et al., 2010).

So far, the most compelling evidence on the role of canonical Wnt signalling in renal disease stems from experiments relating kidney abnormalities to modulations of β catenin expression. Recent studies have demonstrated increased expression of β catenin in the renal stroma, both in human and mouse kidneys, which lead to renal dysplasia (Boivin et al., 2016). Ectopic overexpression of a canonical Wnt, Wnt4, was also observed in stroma-like cells and it caused a disruption in endothelial cell migration and vessel formation in the kidneys. Another study has proposed aquaporin-1 as a modulator of canonical Wnt signalling, since its overexpression decreased β catenin expression and aquaporin-1 co-immunoprecipitated with β catenin, GSK3β, Axin1 and LRP6, all components of canonical Wnt signalling (Wang et al., 2015). Significantly, alterations in normal β catenin expression levels and correct localisation patterns have been observed in all three kidney cell populations (epithelial, mesenchymal and stromal) both in animal mutants and human dysplastic kidney tissue (Boivin et al., 2015). Besides canonical Wnt signalling, however, β catenin is also involved in adherens cell junctions, TGFβ/Bmp signalling and the Gdnf/Ret pathway, making it difficult to accurately determine which and how many of these functions of β catenin are involved in the normal and pathogenic states at each stage of kidney development.

In order to firmly establish how Wnt signalling alone impacts kidney development and function, it might be worth to thoroughly investigate the upstream components of the pathway and define how receptor proteins affect the Wnt pathway and hence might be critical for renal disease. Besides the Fz receptor, there are additional receptors referred to as core PCP proteins that are crucial for non-canonical Wnt/PCP pathway signal transduction. Examples of these include the core PCP proteins Van Gogh-like (Vangl) and Celserus (Celsr), both of which have been shown to impact kidney development and disease. In particular, Vangl2 was shown to be required for mouse kidney branching morphogenesis and glomerular maturation and *Vangl2^{Lp}* null mice displayed poor cortico-medullary definition, mildly dysmorphic tubules and reduced glomeruli (Yates et al., 2010). In addition, activation of the non-canonical Wnt/PCP pathway in cultured human podocytes resulted in the endocytosis of nephrin, a core slit diaphragm protein, while loss of Vangl2 caused an increase of nephrin at the cell surface (Babayeva et al., 2013). Podocyte-specific ablation of the Vangl2 gene resulted in glomerular maturation defects in fetal kidneys and in significantly smaller glomeruli in aging mice (Rocque et al., 2015). Interestingly, Vangl2 co-immunoprecipitates with E-cadherin in embryonic kidneys and its overexpression results in E-cadherin internalisation (Nagaoka et al., 2014). Recently, Vangl2 was also demonstrated to interact with Celsr1 in mouse embryonic kidneys and to contribute to ureteric tree growth and glomeruli maturation (Brzoska et al., 2016), advancing the evidence of an increasing role of core PCP proteins in kidney development and function. It should be noted though that although mutations in PCP genes lead to loss of planar organisation, they leave cell identity and apical-basal polarity intact.

In addition to core PCP proteins, effector PCP molecules such as Fat4 and Atmin have been shown to impact non-canonical Wnt/PCP and thus affect the kidney. Loss of an atypical cadherin, Fat4, disrupted oriented cell division and tubule elongation in the developing kidney; Fat4 interacts with Vangl2 and through PCP disruption leads to cystic kidney disease (Saburi et al., 2008). Atmin, a transcription factor with diverse roles in DNA damage repair and ciliogenesis is also an effector PCP molecule, whose loss of expression leads to defective kidney morphogenesis (Goggolidou et al., 2014). In mice deficient of Atmin, which resulted in embryonic lethality, kidney development was severely impaired with reduced numbers of differentiated, epithelial structures and ureteric buds. As this defect was not caused by reduced proliferation or apoptosis, it was demonstrated that loss of Atmin incited changes in expression of non-canonical Wnt components (*Wnt9b, Wnt11*, Vangl2, *Daam2*) that influenced oriented cell division and the cytoskeletal organisation of the kidney. Further, disturbances in non-canonical Wnt signalling resulted in abnormal canonical Wnt signalling, with greatly reduced levels of total and active β catenin, supporting the theory that a balance between these two pathways is required for normal kidney morphogenesis and function.

Nevertheless, how Wnt ligands trigger the canonical and non-canonical Wnt signalling pathways, what leads to their activation or how timely regulation of canonical and non-canonical Wnt signalling is achieved still remains elusive. Inversin, the protein encoded by *Nephronophthisis-2* (*NPHP2*), mutations in which cause a cystic renal disease called nephronophthisis, has been proposed to be key for the shift between canonical Wnt and non-canonical Wnt/PCP signalling. The interaction between inversin and Dvl is proposed to act as the molecular switch between canonical Wnt and non-canonical Wnt/PCP signalling, by inhibiting canonical Wnt and promoting the non-canonical Wnt/PCP pathway (Simons et al., 2005). The primary cilium is key in this hypothesis, with an active cilium transducing non-canonical Wnt signalling and acting as a suppressor of canonical Wnt signalling, enabling the switch between the two pathways. Inversin was demonstrated to target Dvl1 for degradation, inhibiting the canonical Wnt pathway; further, fluid flow caused an increase in inversin levels in ciliated, tubular epithelial cells, linking the cilium to the non-canonical Wnt pathway (Simons et al., 2005). Interestingly, another study on a family with homozygous *NPHP2* mutations and cystic, end-stage kidneys found increased DVL1 and β catenin expression, supporting increased canonical Wnt signalling (Bellavia et al., 2010). Unfortunately, this study did not explore non-canonical Wnt signalling or cilia in these patients and hence a conclusion on the role of Inversin in maintaining the balance between these two pathways cannot safely be reached. In addition, mutant inversin mice demonstrated random oriented cell division but normal canonical Wnt signalling, contradicting the previous findings (Sugiyama et al., 2011). Nevertheless, Kif3a, a crucial protein in the primary cilium's anterograde IFT has been shown to block canonical Wnt signalling by restricting Dvl phosphorylation (Corbit et al., 2008), providing concrete evidence for the cilium's preference for non-canonical Wnt signalling. Since inversin also modulates the actin cytoskeleton (Werner et al., 2013), further work on its role in non-canonical Wnt signalling will help elucidate the mechanism by which it may act as a molecular switch.

It should also be noted that there is a third component to Wnt signalling, the non-canonical Wnt/Ca^{+2} signalling that upon Dvl stimulation results in increased intracellular calcium concentration and activated calcineurin, protein kinase C and Ca^{+2}/calmodulin-dependent protein kinase II (CAMKII). The specific receptors and channels involved in the activation of this pathway are poorly defined, nevertheless, recent work has identified a fascinating role for the Polycystic Kidney Disease (PKD) complex in this pathway. WNT9B was demonstrated to bind to the PKD1 extracellular domain and activate PKD1/PKD2 Ca^{+2} currents (Kim et al., 2016). In addition, PKD1 forms a complex with DVL1 and DVL2, linking the PKD proteins with the upstream components of Wnt signalling. Further work will shed light on whether these proteins might also affect additional components of Wnt signalling, potentially shifting the balance amongst the various Wnt branches.

It becomes obvious from the above that there are links between the primary cilium in the kidney and the regulation of Wnt signalling. Although most Wnt ligands are able to activate all three branches of Wnt signalling, the primary cilium is considered to be important for the timely regulation of the Wnt signalling pathways. With regards to the non-canonical Wnt signalling pathway, the interaction between inversin and Dvl is believed to be crucial for activation of the pathway. In addition, an active cilium is thought to suppress canonical Wnt signalling and permit the activation of the non-canonical Wnt branches. The discovery that many proteins produced by cystic kidney disease genes localise to the primary cilium further strengthens the evidence for the role of the cilium in kidney development and disease. This has led to the ciliary hypothesis of cystic kidney disease and has prompted the re-classification of some cystic renal diseases into ciliopathies. The following section will concentrate on discussing the symptoms of these diseases, dissecting the links between their manifestations and the role of the cilium and shedding light on how enhancing our knowledge of ciliary biology might help with discovering novel treatments for these disorders.

Kidney Ciliopathies

Polycystic kidney disease encompasses a spectrum of disorders that are associated with the presentation of cysts in the kidney, cause loss of kidney function and will eventually result in end stage kidney disease. These disorders might lead to the formation of larger kidneys, such as in Autosomal Dominant Polycystic Kidney Disease (ADPKD) and Autosomal Recessive Polycystic Kidney Disease (ARPKD) or to kidney hypoplasia, such as in Nephronophthisis (NPHP).

Autosomal dominant polycystic kidney disease (ADPKD)

ADPKD is the most common inherited cystic kidney disease, with a prevalence of 1:500–1:1,000 live births (Wilson, 2008). It is an autosomal dominant condition that affects 12.5 million people worldwide, with symptoms most commonly arising in adulthood. It is usually diagnosed by ultrasound of the kidneys that display bilateral cysts; other symptoms include cysts in the liver and pancreas, intracranial aneurisms and mitral valve prolapse (Wilson and Goilav, 2007). Kidney cysts give rise to renal pain, hypertension and renal insufficiency, with half of the ADPKD individuals manifesting end stage renal disease (ESRD) by 60 years of age and having a need for a kidney transplant. It should be noted that polycystic liver disease is the most common extrarenal manifestation in ADPKD, with liver cysts affecting 75% of ADPKD patients in their 60s. Although most of ADPKD patients do not show symptoms until adulthood, about 2% of ADPKD patients present with clinical manifestations before 15 years of age, with some displaying significant perinatal morbidity and mortality, rendering them indistinguishable from Autosomal Recessive Polycystic Kidney Disease (ARPKD) patients. More than 90% of ADPKD patients have an affected parent, with only 10% of cases arising due to *de novo* mutations (Cornec-Le Gall et al., 2014). These mutations that usually give rise to truncated proteins are mostly detected in one of two loci, *Polycystic Kidney Disease (PKD)1* and *PKD2*, with 85% of mutations identified in *PKD1*, while 15% of ADPKD patients have mutations in *PKD2*. Patients with *PKD1* mutations develop ESRD 20 years earlier than individuals with mutations in *PKD2*, with the median ages of ESRD onset at 58.1 and 79.9 years respectively (Bergmann, 2015). Mutations in the *PKD2* locus are associated with smaller cystic kidneys and milder renal impairment (Bacallao and McNeill, 2009). Strikingly, ADPKD patients manifest great phenotypic variability, even within related individuals, demonstrating that besides mutations in causative genes, modifying genes, epigenetic mechanisms and stochastic and environmental factors also impact symptom severity and manifestation.

Nevertheless, mouse studies have demonstrated that fully inactivating *Pkd1* or *Pkd2* is not compatible with life (Geng et al., 1997; Wu and Somlo, 2000), while a missense *Pkd1* mutation in mice showed its involvement in early onset ADPKD (Hopp et al., 2012). Additive effects of mutations in more than one cystogenic genes (*HNF1B*, *PKHD1*) have also been suggested (Bergmann et al., 2011) and it is possible that ADPKD severity is associated with the levels of the polycystin proteins. Further, *Pkd* mutations before postnatal (P) day 13 in mice have been shown to result in more severe cystic kidney disease compared to loss of polycystin function after this time point (Lantinga-van Leeuwen et al., 2007; Piontek et al., 2007).

It is thus important to have a thorough understanding of the structure and function of the polycystin proteins. Polycystin (PC)1 is a 4,303 amino acid (aa) receptor-like protein with a great number of domains that are involved in protein-protein and protein-carbohydrate interactions, 11 transmembrane domains and a cytoplasmic tail. PC1 is expressed in most nephron segments, although the precise special pattern remains unclear. During kidney development, PC1 localisation is observed in the distal portion of the ureteric bud and the portion of the ureteric bud tip within the nephrogenic zone (Wilson and Goilav, 2007). PC1 expression generally declines after birth and low expression confined to the cortical and medullary collecting duct is seen in adults. PC2 is a 968 aa six-transmembrane protein, with a role in calcium response. PC2 is expressed later in development compared to PC1 and it is also more highly expressed in the adult kidney, although the *PKD2* transcript is detectable before *PKD1* in human embryos (Chauvet et al., 2002).

The spatial and temporal differences between PC1 and PC2 hint at their differing roles, which support the need for them to form a protein complex *in vivo*. PC1 and PC2 interact through their C-terminal tails and PC2 localises to the primary cilium. Experiments involving induced pluripotent stem cells derived from fibroblasts of ADPKD and ARPKD patients showed that cilia formed normally in the ADPKD and ARPKD cells, exhibiting no ciliogenesis defect (Freedman et al., 2013). However, lower levels of ciliary PC2 were detected in ADPKD cells, with ectopic expression and overexpression of PC1 increasing the expression levels of PC2 in the cilium and hinting that PC1 regulates the localisation of PC2 to the primary cilium. It should be emphasised that PC2 has been shown to be a calcium-permeable calcium channel, belonging to the transient receptor potential (TRP) family (Petri et al., 2010; Koulen et al., 2002). It is also an integral membrane protein with six transmembrane domains and trafficking motifs that permit the ciliary localisation of the protein (Geng et al., 2006). PC2 is expressed in all nephron segments in the kidney, but for the glomerulus (Lee and Somlo, 2014). Hence, the PC1-PC2 interaction is crucial for kidney function, as PC1 is believed to act as a mechanosensor at the primary cilium, coordinating the activity of the PC2 calcium channel (Celic et al., 2012; Ferreira et al., 2011).

It is interesting to note that the cilia-mediated role of polycystins also manifests through their interaction with other proteins. The retinitis pigmentosa (RP)2 protein localises to the primary cilium and its loss results in developmental defects caused by ciliary dysfunction such as kidney cysts, situs inversus and hydrocephalus (Hurd et al., 2010). Interestingly, RP2 forms a complex with polycystin 2 and it may regulate PC2 trafficking in the cilium; work in zebrafish demonstrated that the two genes regulate common developmental processes controlled by ciliary function, highlighting the network of interactions exhibited by ciliopathy proteins. In addition, the extra-renal expression of polycystins has emphasized their ciliary role in other organs. Ablation of *Pkd1* in the developing mouse brain causes hydrocephalus (Wodarczyk et al., 2009), a phenotype commonly associated with defective cilia. PC1 and PC2 localise to the primary cilia of radial glia cells and their loss disrupts PCP of these cells (Ohata et al., 2015). Further, loss of polycystin expression disrupts the asymmetric localisation of Vangl2 in brain ependymal cells, with a genetic interaction also demonstrated between *Pkd1* and *Vangl2*, linking the role of polycystins to cilia and PCP.

Hence PC1 and PC2 are not only the protein products of the genes mutated in ADPKD, but also two of the proteins that localise to cilia by IFT, leading to the conclusion that PC1 and PC2 form a receptor-channel complex in the kidney cilium that allow it to serve as a sensory organelle. Hence, a synergistic relationship was hypothesised between the loss of cilia and polycystins, but unexpectedly, recent experiments have demonstrated that structurally intact cilia promote cyst growth, subsequent to PC1 or PC2 inactivation (Ma et al., 2013). Although independent ablation of kidney cilia or loss of PC1/PC2 alone both led to rapid, severe cyst formation, the combined inactivation of *Pkd1* or *Pkd2* together with loss of *Kif3a* and *Ift20* resulted in less severe cystic disease. The implication of this work places the exciting possibility that PC1/PC2 play a role in the regulation of a yet unidentified ciliary signalling cascade, relating polycystins and ciliary signalling in ADPKD that could be highly beneficial for the identification of novel therapeutic approaches for this disease.

Autosomal recessive polycystic kidney disease (ARPKD)

ARPKD is a rare, autosomal recessive disease with an incidence of 1 in 20,000 and a carrier frequency of 1 in 70 in the Caucasian population (Bergmann, 2015). It is a disease of variable severity that manifests equally in men and women and where no ethnicity bias has been detected, with 30–50% of affected individuals dying in the neonatal period. The median age of ARPKD patients is 2–5 years, however, severe cases will display symptoms around birth and recently, a small group of ARPKD individuals in their early 30s with no need for renal dialysis has been discovered. Nevertheless, greater than 50% of ARPKD patients will progress to ESRD within the first decade of life, signifying the need for kidney dialysis and transplantation.

The common symptom in all ARPKD individuals is enlarged, echogenic kidneys with bilateral renal cysts, arising only from the collecting duct and being diagnosed by ultrasound. ARPKD foetuses might also display oligo- or anhydramnios, in which case they are treated by the addition of amniotic fluid. Approximately 30–50% of affected ARPKD neonates are underdeveloped and die shortly after birth from respiratory failure due to pulmonary hypoplasia. In those that survive the neonatal period, besides renal cysts, the associated hypertension and poor kidney function, extra-renal manifestations include poor growth, liver fibrosis, portal hypertension, variceal bleeding, cholangitis and frequent urinary tract infections. It should be noted, however, that great variation is observed in the severity of the symptoms, with some patients displaying single organ-specific phenotypes (exclusively renal or liver symptoms), whereas others manifested both liver and kidney phenotypes.

ARPKD arises due to mutations in *Polycystic Kidney and Hepatic Disease 1 (PKHD1)*, a 472,941 bp gene found on chromosome 6. The protein it encodes, fibrocystin/polyductin, is 4074 aa long and it contains a single transmembrane-spanning domain and multiple immunoglobulin-like plexin-transcription-factor domains (Ward et al., 2003). The type of mutations in *PKHD1* affect disease severity, with two different truncating mutations observed in severe ARPKD cases, while milder cases have either a missense mutation and a truncation or two missense mutations in *PKHD1* (Rossetti and Harris, 2007). Patients with two truncating *PKHD1* mutations display a more severe perinatal or neonatal phenotype, while hypomorphic (missense) mutations confer better prognosis. The *Pkhd1* transcript has got many splice variants and the protein they encode, fibrocystin, is expressed in a tissue-specific manner and manifests a complex pattern of proteolytic processing.

In the developing kidney, fibrocystin expression is observed in the ureteric bud branches and in the adult kidney fibrocystin is detected in the cortical and medullary collecting tubule epithelia and the thick ascending limbs of Henle. The cytoplasmic tail of fibrocystin contains an 18-residue motif which is required for targeting the protein to the primary cilium (Follit et al., 2010). Rab8 binds this ciliary targeting motif, potentially offering an insight into how trafficking through the endomembrane system and into the cilium is regulated. Intriguingly, fibrocystin was also shown to be secreted from exosome-like vesicles that bind to the primary cilium (Bakeberg et al., 2011), however, this was a rare event and hence its significance remains unclear. Experiments in clonal cell lines derived from normal and ARPKD kidneys showed similar resting and peak intracellular calcium concentrations, although the flow induced intracellular calcium peak was greater in ARPKD (Rohatgi et al., 2008). Further, cilia were 20% shorter in ARPKD cells compared to normal foetal collecting tubule cells, however, no difference in ciliary localisation or total expression of polycystin-2 was detected. In the PCK rat, a well-established animal model of ARPKD, cilia were found to be shorter and dysmorphic (Masyuk et al., 2003). Further, siRNA-mediated knockdown of *Pkhd1* in cholangiocytes resulted in shorter cilia. Subsequent characterisation of the liver of the PCK rats supported the finding that cilia were shorter when fibrocystin is not expressed, its loss also resulting in ciliary malformations, with cilia displaying bulbous extensions at the ciliary tip (Masyuk et al., 2004). The Oak Ridge Polycystic Kidney (ORPK) mouse, which is an IFT88 hypomorph that impairs intraflagellar transport, has long been used as another mouse model to study ARPKD. A stunted cilium is observed in these mice (Liu et al., 2005) and abnormal apical flow sensing was detected in cells extracted from the ORPK mouse (Siroky et al., 2006). Fibrocystin was also suggested to participate in intracellular calcium regulation through ciliary mechanosensing (Wang et al., 2007). In addition, fibrocystin was demonstrated to interact with calcium modulating cyclophilin ligand related protein (Nagano et al., 2005), which is involved in calcium signalling, providing further clues into fibrocystin's potential involvement in calcium signalling. It thus becomes apparent from the above that the role of cilia in ARPKD can pose exciting possibilities with regards to discovering novel cilia-related approaches that could impact disease prognosis and treatment.

It should be noted that PC1, PC2 and fibrocystin are also expressed in other tissues besides the kidney, such as the liver, pancreas and lungs. Work on the lung epithelial cell line A549 demonstrated polycystic expression in the mitotic spindle, while fibrocystin was expressed in the centrosome (Hu et al., 2014). Further, all these cystoproteins were also detected in the primary or motile cilia of lung epithelial, sub-mucosal and bronchiol cells, opening exciting possibilities on the function that they might serve in the lung.

Nephronophthisis (NPHP)

NPHP is a tubule-interstitial autosomal recessive cystic kidney disease. It is genetically heterogeneous, with more than twenty mutated genes identified. NPHP is characterised by polyuria, polydipsia, secondary enuresis and anaemia and it can be classified into three types: juventile, infantile and adolescent. Juvenile or type-1 NPHP is the most common form of NPHP, with patients manifesting ESRD at a mean age of 13 years (Wolf, 2015). At the other extreme, infantile or type 2 NPHP is very rare with an early onset of ESRD, prior to 4 years of age. Finally, adolescent or type 3 NPHP has a mean age of ESRD onset of 19 years. Although mutations in more than 20 genes have been shown to be causative of NPHP, these mutations are detected in only one third of the patients, with two thirds of the NPHP cases still carrying mutations in unknown genes.

NPHP presents with small or normal size fibrotic kidneys, displaying cysts at the corticomedullary region. About 70% of NPHP patients show kidney cysts, however, intestinal fibrosis and not cyst development is more closely associated with disease progression. No gender or ethnic bias has been observed in NPHP patients. Besides the kidney manifestation, NPHP has extra-renal manifestations, such as retinal degeneration, hepatic fibrosis, skeletal anomalies, brain malformations, cerebellar vermis hypoplasia and neurological impairment. It displays an autosomal recessive mode of inheritance and more than 90 genes have so far been shown to be mutated in NPHP (Braun and Hildebrandt, 2017). The most frequent mutations associated with 25–30% of NPHP cases are homozygous deletions in the *NPHP1* gene (Halbritter et al., 2013). Around 85% of the *NPHP1* mutations are homozygous deletions of the whole gene (Hildebrandt et al., 2001). Other clinical symptoms include polyuria, polydipsia, anemia and growth retardation and situs inversus and congenital heart defects occur in infantile NPHP patients. Significantly, hypertension is not observed until there is severe reduction in kidney function and under ultrasound, NPHP patients' kidneys are of normal or smaller size and display increased echogenicity and loss of cortico-medullary differentiation.

Further, the renal histology of NPHP patients shows a thickened and disintegrated tubular basement membrane, renal tubular atrophy and tubulointestitial fibrosis. Transmission electron microscopy (TEM) is sometimes useful in accurately diagnosing NPHP, as it permits the study of alterations in the tubular membrane. Serum chemistry also provides information on the stage of renal impairment, however, genetic testing is the most definite way of diagnosing NPHP with confidence. Disappointingly and despite the great developments in NPHP diagnosis, no pharmacological treatment exists for this disease, whereupon when patients reach ESRD, their only option is renal replacement therapy by kidney dialysis or transplantation.

Nevertheless, NPHP is a widely studied ciliopathy, since many of the NPHP protein products localise to the primary cilium. NPHP1 and NPHP4 are expressed in ciliated sensory neurons (Wolf et al., 2005) and NPHP2 localises to the kidney cilium and is thought to facilitate the switch between canonical and non-canonical Wnt signalling (Simons et al., 2005). Interestingly, the *inv* mouse, in which *Nphp2* is mutated, strongly resembles the human phenotype, with mice displaying cystic kidneys, liver malformations, an abnormal left-right axis and congenital heart defects (Phillips et al., 2004). Significantly, recessive truncating mutations in *Doublecortin domain containing 2 (DCDC2)* were detected in a cohort of NPHP patients (Schueler et al., 2015). DCDC2 was demonstrated to co-localise with acetylated α-tubulin to kidney and liver primary cilia and to multiciliated ependymal cells. DCDC2 interacts with DVL1, DVL2 and DVL3 and its loss disrupted ciliation and constitutively activated β catenin in 3D spheroid mouse kidney cells. Further, the ciliation defects observed in the mouse kidney cell line and in zebrafish were rescued by treatment with a Wnt inhibitor, demonstrating the failure of DCDC2 mutants to control canonical Wnt signalling and maintain normal ciliogenesis.

Although NPHP mostly occurs as an isolated kidney disease, about 15% of the patients have extrarenal symptoms (affected organs include the eyes, liver, bones and central nervous system), possibly due to the NPHP protein localisation to primary cilia that are found in almost every cell type. As a result, Joubert, Meckel-Gruber and Jeune syndromes are considered to be sub-categories of NPHP and are complex ciliopathies with renal and extra-renal manifestations. Joubert syndrome is a complex, developmental disorder involving various organs, such as the nervous system, the kidney, the retina and the liver. Its most prominent symptom is the agenesis of the cerebellar vermis that appears as a characteristic molar tooth sign on radiography. This genetically heterogeneous, monogenic condition has been associated with mutations in around 27 genes. The greatest studied of these proteins, Jouberin, interacts with NPHP1, NPHP3 and NPHP4 and localises to the transition zone of the cilium (Omran, 2010). Meckel-Gruber is a severe, perinatally lethal ciliopathy that manifests with cystic kidneys, central nervous system malformations and polydactyly (Barker et al., 2013). It shows great genetic heterogeneity and many of the genes that are causative of Meckel-Gruber syndrome localise to cilia. The most widely studied of these genes, TMEM67, localises both to the basal body of the cilium and the actin cystoskeleton (Dawe et al., 2007; Dawe et al., 2009) and as a result, Meckel-Gruber patients show both cytoskeletal and ciliary defects (Valente et al., 2010; Dawe et al., 2009). Similarly, Jeune syndrome patients' display skeletal defects along with kidney, liver, brain and heart defects (Huber and Cormier-Daire, 2012). The genes that when mutated cause Jeune syndrome either are part of IFT or are associated with the centriole, both of which are key for ciliary formation and function (Halbritter et al., 2013; Shaheen et al., 2014; Schmidts et al., 2015) and explain the complex phenotype observed in Jeune syndrome.

The symptoms and manifestation of NPHP support a unifying theory for cystogenesis, where most proteins altered in cystic kidney disease are expressed in primary cilia, the basal body or centrosomes of renal epithelial cells. Interestingly, a paediatric patient with a novel, complex phenotype consisting of end stage renal failure, skeletal abnormalities, deafness and hepatosplenomegaly was identified to be carrying mutations in *WDR19* that encodes the intraflagellar trafficking protein IFT144 (Fehrenbach et al., 2014). Adding further insight into the significance of cilia in cystic kidney disease was the discovery of a nonsense mutation resulting in truncated inversin protein in a family manifesting a severe case of NPHP (Oud et al., 2014). The result of this mutation was mislocalised inversin protein that instead of localising in the ciliary axoneme and the basal body, was only present at the basal body.

Cystic Kidney Disease Management and Treatment

Although a great understanding of the genes mutated in kidney ciliopathies and the molecular pathways that they affect has been achieved, there is a lag between discoveries in the research lab and translating these discoveries into effective treatments. For example, treating the symptoms of ADPKD is a complex situation, where all the extrarenal symptoms have to be addressed.

Treatment for hypertension includes ACE inhibitors and angiotensin II receptor blockers and dietary modification is advisable (Harris and Torres, 2002). Symptomatic intracranial aneurysms are treated by surgery and polycystic liver disease is usually asymptomatic. ADPKD women who are pregnant are monitored for hypertension, urinary tract infections, oligohydramnios and pre-eclampsia and the correct development of the foetus is also closely inspected.

The only pharmacological treatment that has been approved in the field of cystic renal diseases involves the use of Tolvaptan. Tolvaptan reduces kidney cyst growth as it inhibits intracellular cyclic AMP and it is currently being prescribed to rapidly progressing adult ADPKD patients. Published reports show that subsequent to tolvaptan administration, a reduction is observed both in the increase of total kidney volume and in glomerular filtration rate decrease, suggesting that tolvaptan can successfully slow down renal cystic progression decline in ADPKD (Kai et al., CEN case report 2016; Uchiyama et al., 2016). Nevertheless, an increase in liver enzyme levels has been observed in some ADPKD patients taking tolvaptan (Uchiyama et al., 2016) and a case was recently reported of a patient manifesting pulmonary thrombosis subsequent to tolvaptan administration (Morimoto et al., 2017), requiring the constant monitoring of patients and limiting the numbers of patients to which this drug can be prescribed. Hence, kidney dialysis and transplantation remain the most common approaches of addressing ADPKD.

In the case of ARPKD patients, it should be noted that some may also require a liver transplant, which is transplanted before or together with the kidney transplant, but nevertheless conferring a good clinical outcome and long-term survival. With regards to NPHP, drugs such as paclitaxel have been suggested as a way to address the interstitial fibrosis element of NPHP, given that its effect in rodent models of renal fibrosis proved promising (Slaats et al., 2016). In any event, administering it to paediatric patients and calculating the optimal dosage and efficacy might prove challenging. It is thus important to focus on managing cystic renal diseases in a way that improves the quality of life of the patients and increases their lifespan. Current and future research developments will assist in this direction and will significantly aid in discovering novel treatments.

Conclusion

It becomes apparent from the above that inherited cystic renal diseases have the primary cilium as the common denominator, an observation that could be useful in finding common ways to treat these diseases. It should also be noted that other diseases that phenocopy PKD also exist and one of them is Medullary cystic kidney disease (MCKD). MCKD has a late onset adult renal manifestation, with renal salt wasting. It is the autosomal dominant counterpart of NPHP, caused most commonly by mutations in *UMOD* that encodes uromodulin, the most abundant protein in the urine of healthy individuals. UMOD localises to the primary cilium and co-localises with nephrocystin (Zaucke et al., 2010). In addition, mutations in *Hepatocyte Nuclear Factor 1β (HNF1β)* cause cystic renal dysplasia due to early embryonic mal-development. The disease is inherited in an autosomal dominant manner and variations in symptom severity have been recorded. The transcriptional network between *HNF1β*, *PKHD1*, *PKD2* and *UMOD* (Gresh et al., 2004) clearly demonstrates the inter-relationship between the various cystic kidney diseases. What remains now is the connection of all the dots that will show the order of events that lead to cystic renal disease. Future work will demonstrate how cellular signalling and the primary cilium can help in diminishing the impact these diseases have on patients and whether interventions in these pathways can be applied to treat cystic renal diseases.

References

Ando, F., Sohara, E., Morimoto, T., Yui, N., Nomura, N., Kikuchi, E., Takahashi, D., Mori, T., Vandewalle, A., Rai, T. et al. 2016. Wnt5a induces renal AQP2 expression by activating calcineurin signalling pathway. Nat. Commun. 7: 13636.

Babayeva, S., Rocque, B., Aoudjit, L., Zilber, Y., Li, J., Baldwin, C., Kawachi, H., Takano, T. and Torban, E. 2013. Planar cell polarity pathway regulates nephrin endocytosis in developing podocytes. J. Biol. Chem. 288: 24035–24048.

Bacallao, R. L. and McNeill, H. 2009. Cystic kidney diseases and planar cell polarity signaling. Clin. Genet. 75: 107–117.

Bakeberg, J. L., Tammachote, R., Woollard, J. R., Hogan, M. C., Tuan, H. F., Li, M., van Deursen, J. M., Wu, Y., Huang, B. Q., Torres, V. E. et al. 2011. Epitope-tagged Pkhd1 tracks the processing, secretion, and localization of fibrocystin. J. Am. Soc. Nephrol. 22: 2266–2277.

Beaton, H., Andrews, D., Parsons, M., Murphy, M., Gaffney, A., Kavanagh, D., McKay, G. J., Maxwell, A. P., Taylor, C. T., Cummins, E. P. et al. 2016. Wnt6 regulates epithelial cell differentiation and is dysregulated in renal fibrosis. Am. J. Physiol. Renal Physiol. 311: F35–45.

Bellavia, S., Dahan, K., Terryn, S., Cosyns, J. P., Devuyst, O. and Pirson, Y. 2010. A homozygous mutation in INVS causing juvenile nephronophthisis with abnormal reactivity of the Wnt/beta-catenin pathway. Nephrol. Dial. Transplant. 25: 4097–4102.

Bergmann, C., von Bothmer, J., Ortiz Bruchle, N., Venghaus, A., Frank, V., Fehrenbach, H., Hampel, T., Pape, L., Buske, A., Jonsson, J. et al. 2011. Mutations in multiple PKD genes may explain early and severe polycystic kidney disease. J. Am. Soc. Nephrol. 22: 2047–2056.

Bergmann, C. 2015. ARPKD and early manifestations of ADPKD: the original polycystic kidney disease and phenocopies. Pediatr. Nephrol. 30: 15–30.

Boivin, F. J., Sarin, S., Lim, J., Javidan, A., Svajger, B., Khalili, H. and Bridgewater, D. 2015. Stromally expressed beta-catenin modulates Wnt9b signaling in the ureteric epithelium. PLoS One 10: e0120347.

Boivin, F. J., Sarin, S., Dabas, P., Karolak, M., Oxburgh, L. and Bridgewater, D. 2016. Stromal beta-catenin overexpression contributes to the pathogenesis of renal dysplasia. J. Pathol. 239: 174–185.

Braun, D. A. and Hildebrandt, F. 2017. Ciliopathies. Cold Spring Harb Perspect. Biol. 9: 10.1101/cshperspect.a028191.

Brzoska, H. L., d'Esposito, A. M., Kolatsi-Joannou, M., Patel, V., Igarashi, P., Lei, Y., Finnell, R. H., Lythgoe, M. F., Woolf, A. S., Papakrivopoulou, E. et al. 2016. Planar cell polarity genes Celsr1 and Vangl2 are necessary for kidney growth, differentiation, and rostrocaudal patterning. Kidney Int. 90: 1274–1284.

Carroll, T. J., Park, J. S., Hayashi, S., Majumdar, A. and McMahon, A. P. 2005. Wnt9b Plays a central role in the regulation of mesenchymal to epithelial transitions underlying organogenesis of the mammalian urogenital system. Dev. Cell. 9: 283–292.

Chauvet, V., Qian, F., Boute, N., Cai, Y., Phakdeekitacharoen, B., Onuchic, L. F., Attie-Bitach, T., Guicharnaud, L., Devuyst, O., Germino, G. G. et al. 2002. Expression of PKD1 and PKD2 transcripts and proteins in human embryo and during normal kidney development. Am. J. Pathol. 160: 973–983.

Corbit, K. C., Shyer, A. E., Dowdle, W. E., Gaulden, J., Singla, V., Chen, M. H., Chuang, P. T. and Reiter, J. F. 2008. Kif3a constrains beta-catenin-dependent Wnt signalling through dual ciliary and non-ciliary mechanisms. Nat. Cell Biol. 10: 70–76.

Cornec-Le Gall, E., Audrezet, M. P., Le Meur, Y., Chen, J. M. and Ferec, C. 2014. Genetics and pathogenesis of autosomal dominant polycystic kidney disease: 20 years on. Hum. Mutat. 35: 1393–1406.

Davenport, J. R., Watts, A. J., Roper, V. C., Croyle, M. J., van Groen, T., Wyss, J. M., Nagy, T. R., Kesterson, R. A. and Yoder, B. K. 2007. Disruption of intraflagellar transport in adult mice leads to obesity and slow-onset cystic kidney disease. Curr. Biol. 17: 1586–1594.

Fehrenbach, H., Decker, C., Eisenberger, T., Frank, V., Hampel, T., Walden, U., Amann, K. U., Kruger-Stollfuss, I., Bolz, H. J., Haffner, K. et al. 2014. Mutations in WDR19 encoding the intraflagellar transport component IFT144 cause a broad spectrum of ciliopathies. Pediatr. Nephrol. 29: 1451–1456.

Follit, J. A., Li, L., Vucica, Y. and Pazour, G. J. 2010. The cytoplasmic tail of fibrocystin contains a ciliary targeting sequence. J. Cell Biol. 188: 21–28.

Freedman, B. S., Lam, A. Q., Sundsbak, J. L., Iatrino, R., Su, X., Koon, S. J., Wu, M., Daheron, L., Harris, P. C., Zhou, J. et al. 2013. Reduced ciliary polycystin-2 in induced pluripotent stem cells from polycystic kidney disease patients with PKD1 mutations. J. Am. Soc. Nephrol. 24: 1571–1586.

Geng, L., Segal, Y., Pavlova, A., Barros, E. J., Lohning, C., Lu, W., Nigam, S. K., Frischauf, A. M., Reeders, S. T. and Zhou, J. 1997. Distribution and developmentally regulated expression of murine polycystin. Am. J. Physiol. 272: F451–9.

Goggolidou, P. 2014. Wnt and planar cell polarity signaling in cystic renal disease. Organogenesis 10: 86–95.

Goggolidou, P., Hadjirin, N. F., Bak, A., Papakrivopoulou, E., Hilton, H., Norris, D. P. and Dean, C. H. 2014. Atmin mediates kidney morphogenesis by modulating Wnt signaling. Hum. Mol. Genet. 23: 5303–5316.

Goggolidou, P. and Wilson, P. D. 2016. Novel biomarkers in kidney disease: roles for cilia, Wnt signalling and ATMIN in polycystic kidney disease. Biochem. Soc. Trans. 44: 1745–1751.

Gresh, L., Fischer, E., Reimann, A., Tanguy, M., Garbay, S., Shao, X., Hiesberger, T., Fiette, L., Igarashi, P., Yaniv, M. et al. 2004. A transcriptional network in polycystic kidney disease. EMBO J. 23: 1657–1668.

Halbritter, J., Porath, J. D., Diaz, K. A., Braun, D. A., Kohl, S., Chaki, M., Allen, S. J., Soliman, N. A., Hildebrandt, F., Otto, E. A. et al. 2013. Identification of 99 novel mutations in a worldwide cohort of 1,056 patients with a nephronophthisis-related ciliopathy. Hum. Genet. 132: 865–884.

Hildebrandt, F., Rensing, C., Betz, R., Sommer, U., Birnbaum, S., Imm, A., Omran, H., Leipoldt, M., Otto, E. and Arbeitsgemeinschaft fur Paedatrische Nephrologie (APN) Study Group. 2001. Establishing an algorithm for molecular genetic diagnostics in 127 families with juvenile nephronophthisis. Kidney Int. 59: 434–445.

Hildebrandt, F., Benzing, T. and Katsanis, N. 2011. Ciliopathies. N. Engl. J. Med. 364: 1533–1543.

Hopp, K., Ward, C. J., Hommerding, C. J., Nasr, S. H., Tuan, H. F., Gainullin, V. G., Rossetti, S., Torres, V. E. and Harris, P. C. 2012. Functional polycystin-1 dosage governs autosomal dominant polycystic kidney disease severity. J. Clin. Invest. 122: 4257–4273.

Hu, Q., Wu, Y., Tang, J., Zheng, W., Wang, Q., Nahirney, D., Duszyk, M., Wang, S., Tu, J. C. and Chen, X. Z. 2014. Expression of polycystins and fibrocystin on primary cilia of lung cells. Biochem. Cell Biol. 92: 547–554.

Huang, L., Xiao, A., Choi, S. Y., Kan, Q., Zhou, W., Chacon-Heszele, M. F., Ryu, Y. K., McKenna, S., Zuo, X., Kuruvilla, R. et al. 2014. Wnt5a is necessary for normal kidney development in zebrafish and mice. Nephron. Exp. Nephrol. 128: 80–88.

Hurd, T., Zhou, W., Jenkins, P., Liu, C. J., Swaroop, A., Khanna, H., Martens, J., Hildebrandt, F. and Margolis, B. 2010. The retinitis pigmentosa protein RP2 interacts with polycystin 2 and regulates cilia-mediated vertebrate development. Hum. Mol. Genet. 19: 4330–4344.

Jonassen, J. A., San Agustin, J., Follit, J. A. and Pazour, G. J. 2008. Deletion of IFT20 in the mouse kidney causes misorientation of the mitotic spindle and cystic kidney disease. J. Cell Biol. 183: 377–384.

Kai et al. https://www.ncbi.nlm.nih.gov/pubmed/?term=kai+et+al+2016+tolvaptan.

Karner, C. M., Chirumamilla, R., Aoki, S., Igarashi, P., Wallingford, J. B. and Carroll, T. J. 2009. Wnt9b signaling regulates planar cell polarity and kidney tubule morphogenesis. Nat. Genet. 41: 793–799.

Kim, S., Nie, H., Nesin, V., Tran, U., Outeda, P., Bai, C. X., Keeling, J., Maskey, D., Watnick, T., Wessely, O. et al. 2016. The polycystin complex mediates Wnt/Ca(2+) signalling. Nat. Cell Biol. 18: 752–764.

Lantinga-van Leeuwen, I. S., Leonhard, W. N., van der Wal, A., Breuning, M. H., de Heer, E. and Peters, D. J. 2007. Kidney-specific inactivation of the Pkd1 gene induces rapid cyst formation in developing kidneys and a slow onset of disease in adult mice. Hum. Mol. Genet. 16: 3188–3196.

Lin, F., Hiesberger, T., Cordes, K., Sinclair, A. M., Goldstein, L. S., Somlo, S. and Igarashi, P. 2003. Kidney-specific inactivation of the KIF3A subunit of kinesin-II inhibits renal ciliogenesis and produces polycystic kidney disease. Proc. Natl. Acad. Sci. USA 100: 5286–5291.

Liu, W., Murcia, N. S., Duan, Y., Weinbaum, S., Yoder, B. K., Schwiebert, E. and Satlin, L. M. 2005. Mechanoregulation of intracellular Ca2+ concentration is attenuated in collecting duct of monocilium-impaired orpk mice. Am. J. Physiol. Renal Physiol. 289: F978–88.

Ma, M., Tian, X., Igarashi, P., Pazour, G. J. and Somlo, S. 2013. Loss of cilia suppresses cyst growth in genetic models of autosomal dominant polycystic kidney disease. Nat. Genet. 45: 1004–1012.

Maarouf, O. H., Aravamudhan, A., Rangarajan, D., Kusaba, T., Zhang, V., Welborn, J., Gauvin, D., Hou, X., Kramann, R. and Humphreys, B. D. 2016. Paracrine Wnt1 drives interstitial fibrosis without inflammation by tubulointerstitial cross-talk. J. Am. Soc. Nephrol. 27: 781–790.

Majumdar, A., Vainio, S., Kispert, A., McMahon, J. and McMahon, A. P. 2003. Wnt11 and Ret/Gdnf pathways cooperate in regulating ureteric branching during metanephric kidney development. Development 130: 3175–3185.

Masyuk, T. V., Huang, B. Q., Ward, C. J., Masyuk, A. I., Yuan, D., Splinter, P. L., Punyashthiti, R., Ritman, E. L., Torres, V. E., Harris, P. C. et al. 2003. Defects in cholangiocyte fibrocystin expression and ciliary structure in the PCK rat. Gastroenterology 125: 1303–1310.

Masyuk, T. V., Huang, B. Q., Masyuk, A. I., Ritman, E. L., Torres, V. E., Wang, X., Harris, P. C. and Larusso, N. F. 2004. Biliary dysgenesis in the PCK rat, an orthologous model of autosomal recessive polycystic kidney disease. Am. J. Pathol. 165: 1719–1730.

Nagano, J., Kitamura, K., Hujer, K. M., Ward, C. J., Bram, R. J., Hopfer, U., Tomita, K., Huang, C. and Miller, R. T. 2005. Fibrocystin interacts with CAML, a protein involved in Ca2+ signaling. Biochem. Biophys. Res. Commun. 338: 880–889.

Nagaoka, T., Inutsuka, A., Begum, K., Bin hafiz, K. and Kishi, M. 2014. Vangl2 regulates E-cadherin in epithelial cells. Sci. Rep. 4: 6940.

Oh, E. C. and Katsanis, N. 2012. Cilia in vertebrate development and disease. Development 139: 443–448.

Ohata, S., Herranz-Perez, V., Nakatani, J., Boletta, A., Garcia-Verdugo, J. M. and Alvarez-Buylla, A. 2015. Mechanosensory genes Pkd1 and Pkd2 contribute to the planar polarization of brain ventricular epithelium. J. Neurosci. 35: 11153–11168.

Omran, H. 2010. NPHP proteins: gatekeepers of the ciliary compartment. J. Cell Biol. 190: 715–717.

Oud, M. M., van Bon, B. W., Bongers, E. M., Hoischen, A., Marcelis, C. L., de Leeuw, N., Mol, S. J., Mortier, G., Knoers, N. V., Brunner, H. G. et al. 2014. Early presentation of cystic kidneys in a family with a homozygous INVS mutation. Am. J. Med. Genet. A. 164A: 1627–1634.

Phillips, C. L., Miller, K. J., Filson, A. J., Nurnberger, J., Clendenon, J. L., Cook, G. W., Dunn, K. W., Overbeek, P. A., Gattone, V. H., 2nd and Bacallao, R. L. 2004. Renal cysts of Inv/Inv mice resemble early infantile nephronophthisis. J. Am. Soc. Nephrol. 15: 1744–1755.

Piontek, K., Menezes, L. F., Garcia-Gonzalez, M. A., Huso, D. L. and Germino, G. G. 2007. A critical developmental switch defines the kinetics of kidney cyst formation after loss of Pkd1. Nat. Med. 13: 1490–1495.

Rocque, B. L., Babayeva, S., Li, J., Leung, V., Nezvitsky, L., Cybulsky, A. V., Gros, P. and Torban, E. 2015. Deficiency of the planar cell polarity protein vangl2 in podocytes affects glomerular morphogenesis and increases susceptibility to injury. J. Am. Soc. Nephrol. 26: 576–586.

Rohatgi, R., Battini, L., Kim, P., Israeli, S., Wilson, P. D., Gusella, G. L. and Satlin, L. M. 2008. Mechanoregulation of intracellular Ca2+ in human autosomal recessive polycystic kidney disease cyst-lining renal epithelial cells. Am. J. Physiol. Renal Physiol. 294: F890–9.

Romagnani, P., Lasagni, L. and Remuzzi, G. 2013. Renal progenitors: an evolutionary conserved strategy for kidney regeneration. Nat. Rev. Nephrol. 9: 137–146.

Romaker, D., Puetz, M., Teschner, S., Donauer, J., Geyer, M., Gerke, P., Rumberger, B., Dworniczak, B., Pennekamp, P., Buchholz, B. et al. 2009. Increased expression of secreted frizzled-related protein 4 in polycystic kidneys. J. Am. Soc. Nephrol. 20: 48–56.

Rossetti, S. and Harris, P. C. 2007. Genotype-phenotype correlations in autosomal dominant and autosomal recessive polycystic kidney disease. J. Am. Soc. Nephrol. 18: 1374–1380.

Saburi, S., Hester, I., Fischer, E., Pontoglio, M., Eremina, V., Gessler, M., Quaggin, S. E., Harrison, R., Mount, R. and McNeill, H. 2008. Loss of Fat4 disrupts PCP signaling and oriented cell division and leads to cystic kidney disease. Nat. Genet. 40: 1010–1015.

Sainio, K. and Raatikainen-Ahokas, A. 1999. Mesonephric kidney—a stem cell factory? Int. J. Dev. Biol. 43: 435–439.

Schneider, J., Arraf, A. A., Grinstein, M., Yelin, R. and Schultheiss, T. M. 2015. Wnt signaling orients the proximal-distal axis of chick kidney nephrons. Development 142: 2686–2695.

Schueler, M., Braun, D. A., Chandrasekar, G., Gee, H. Y., Klasson, T. D., Halbritter, J., Bieder, A., Porath, J. D., Airik, R., Zhou, W. et al. 2015. DCDC2 mutations cause a renal-hepatic ciliopathy by disrupting Wnt signaling. Am. J. Hum. Genet. 96: 81–92.

Simons, M., Gloy, J., Ganner, A., Bullerkotte, A., Bashkurov, M., Kronig, C., Schermer, B., Benzing, T., Cabello, O. A., Jenny, A. et al. 2005. Inversin, the gene product mutated in nephronophthisis type II, functions as a molecular switch between Wnt signaling pathways. Nat. Genet. 37: 537–543.

Siroky, B. J., Ferguson, W. B., Fuson, A. L., Xie, Y., Fintha, A., Komlosi, P., Yoder, B. K., Schwiebert, E. M., Guay-Woodford, L. M. and Bell, P. D. 2006. Loss of primary cilia results in deregulated and unabated apical calcium entry in ARPKD collecting duct cells. Am. J. Physiol. Renal Physiol. 290: F1320–8.

Sugiyama, N., Tsukiyama, T., Yamaguchi, T. P. and Yokoyama, T. 2011. The canonical Wnt signaling pathway is not involved in renal cyst development in the kidneys of Inv mutant mice. Kidney Int. 79: 957–965.

Torres, M. A. and Nelson, W. J. 2000. Colocalization and redistribution of dishevelled and actin during Wnt-induced mesenchymal morphogenesis. J. Cell Biol. 149: 1433–1442.

Vivante, A., Mark-Danieli, M., Davidovits, M., Harari-Steinberg, O., Omer, D., Gnatek, Y., Cleper, R., Landau, D., Kovalski, Y., Weissman, I. et al. 2013. Renal hypodysplasia associates with a WNT4 variant that causes aberrant canonical WNT signaling. J. Am. Soc. Nephrol. 24: 550–558.

Wang, S., Zhang, J., Nauli, S. M., Li, X., Starremans, P. G., Luo, Y., Roberts, K. A. and Zhou, J. 2007. Fibrocystin/polyductin, found in the same protein complex with polycystin-2, regulates calcium responses in kidney epithelia. Mol. Cell. Biol. 27: 3241–3252.

Wang, W., Li, F., Sun, Y., Lei, L., Zhou, H., Lei, T., Xia, Y., Verkman, A. S. and Yang, B. 2015. Aquaporin-1 retards renal cyst development in polycystic kidney disease by inhibition of Wnt signaling. FASEB J. 29: 1551–1563.

Ward, C. J., Yuan, D., Masyuk, T. V., Wang, X., Punyashthiti, R., Whelan, S., Bacallao, R., Torra, R., LaRusso, N. F., Torres, V. E. et al. 2003. Cellular and subcellular localization of the ARPKD protein; fibrocystin is expressed on primary cilia. Hum. Mol. Genet. 12: 2703–2710.

Werner, M. E., Ward, H. H., Phillips, C. L., Miller, C., Gattone, V. H. and Bacallao, R. L. 2013. Inversin modulates the cortical actin network during mitosis. Am. J. Physiol. Cell. Physiol. 305: C36–47.

Wilson, P. D. and Goilav, B. 2007. Cystic disease of the kidney. Annu. Rev. Pathol. 2: 341–368.

Wilson, P. D. 2008. Mouse models of polycystic kidney disease. Curr. Top. Dev. Biol. 84: 311–350.

Wodarczyk, C., Rowe, I., Chiaravalli, M., Pema, M., Qian, F. and Boletta, A. 2009. A novel mouse model reveals that polycystin-1 deficiency in ependyma and choroid plexus results in dysfunctional cilia and hydrocephalus. PLoS One 4: e7137.

Wolf, M. T., Lee, J., Panther, F., Otto, E. A., Guan, K. L. and Hildebrandt, F. 2005. Expression and phenotype analysis of the nephrocystin-1 and nephrocystin-4 homologs in caenorhabditis elegans. J. Am. Soc. Nephrol. 16: 676–687.

Wolf, M. T. 2015. Nephronophthisis and related syndromes. Curr. Opin. Pediatr. 27: 201–211.

Wu, G. and Somlo, S. 2000. Molecular genetics and mechanism of autosomal dominant polycystic kidney disease. Mol. Genet. Metab. 69: 1–15.

Xiao, L., Zhou, D., Tan, R. J., Fu, H., Zhou, L., Hou, F. F. and Liu, Y. 2016. Sustained activation of Wnt/beta-catenin signaling drives AKI to CKD progression. J. Am. Soc. Nephrol. 27: 1727–1740.

Yates, L. L., Papakrivopoulou, J., Long, D. A., Goggolidou, P., Connolly, J. O., Woolf, A. S. and Dean, C. H. 2010. The planar cell polarity gene vangl2 is required for mammalian kidney-branching morphogenesis and glomerular maturation. Hum. Mol. Genet. 19: 4663–4676.

Ye, X., Wang, Y., Rattner, A. and Nathans, J. 2011. Genetic mosaic analysis reveals a major role for frizzled 4 and frizzled 8 in controlling ureteric growth in the developing kidney. Development 138: 1161–1172.

Zhao, S. L., Wei, S. Y., Wang, Y. X., Diao, T. T., Li, J. S., He, Y. X., Yu, J., Jiang, X. Y., Cao, Y., Mao, X. Y. et al. 2016. Wnt4 is a novel biomarker for the early detection of kidney tubular injury after ischemia/reperfusion injury. Sci. Rep. 6: 32610.

Zschiedrich, S., Budde, K., Nurnberger, J., Wanner, C., Sommerer, C., Kunzendorf, U., Banas, B., Hoerl, W. H., Obermuller, N., Arns, W. et al. 2016. Secreted frizzled-related protein 4 predicts progression of autosomal dominant polycystic kidney disease. Nephrol. Dial. Transplant. 31: 284–289.

CHAPTER 5

The Role of Cilia in Pancreatic Development and Disease

Jantje M. Gerdes

INTRODUCTION

Basal body and ciliary dysfunction can give rise to a multitude of different developmental and degenerative disorders that are referred to as ciliopathies (Fliegauf et al., 2007; Hildebrandt et al., 2011; Gerdes et al., 2009). Although cilia are virtually ubiquitously present in vertebrates and mammals, the first human diseases linked to ciliary dysfunction mainly affect the renal epithelium progressing to end stage renal disease. This has led to early research efforts towards understanding the role of ciliary function in the kidney. As many of the syndromic ciliopathies such as Bardet Biedl Syndrome (BBS), Joubert Syndrome or Senior-Loken Syndrome present with symptoms in several organs, more recent studies have addressed the role of ciliary function in other organs. This chapter summarizes the information gained from spontaneous and targeted ciliary mutants such as the Oregon Ridge Polycystic Kidney (orpk) mouse or pancreas specific deletion of core ciliary genes. In addition, it discusses the evidence of signalling pathways compromised by impaired ciliary function in the pancreas. Although ciliary function has been implicated in many pathways, the majority of these links have been observed in tissues other than the pancreas or in cultured cells. Furthermore, recent studies demonstrate that tissue specificity is an important factor in the analysis of ciliary function. Therefore, transferring insights gained from other tissues onto pancreatic cilia could be an oversimplification of the complex regulatory network of ciliary signalling pathways. This is

Institute for Diabetes and Regeneration Research, HelmholtzZentrum München, Parkring 11, 85748 Garching Hochbrück, Germany; and German Center for Diabetes Research (DZD).
  E-mail: jantje.gerdes@helmholtz-muenchen.de

especially true for cell culture conditions that might differ significantly from the physiological situation.

This overview focuses on the phenotypes caused by ciliary impairment in the pancreas and attempts to reach conclusions about the role of cilia in this context. The role of ciliary function in pancreatic cancer is difficult to assess because of conflicting evidence that will need to be addressed in additional studies. To avoid speculation, the discussion of cilia and pancreatic cancer will remain perfunctory. Instead, the emphasis of this chapter will be placed on metabolic disease that at least partially arises from ciliary dysfunction in the pancreas.

Pancreas Development and Physiology

The adult pancreas consists of two main compartments—the exocrine portion that includes bulbous formations of acinar cells and the ductal epithelia and the endocrine part consisting of the pancreatic islets of Langerhans that secrete hormones into the blood supply (Figure 1). Both the exocrine and the endocrine portion of the pancreas facilitate nutrient metabolism: acini produce digestive enzymes such as amylase, lipase or trypsinogen, the inactive precursor of trypsin. These digestive enzymes are drained into the duodenum, where the proenzymes are cleaved by enteropeptidase and thus activated. Islet cells produce glucagon (α-cells), insulin (β-cells), somatostatin (δ-cells), ghrelin (ε-cells) and pancreatic polypeptide (PP cells). These peptide hormones are secreted into the blood stream and act on other tissues such as liver, adipose tissue or the brain.

Organogenesis is preceded by regionalization into distinct organ fields and the definitive endoderm germ layer gives rise to the endocrine, exocrine and ductal components of the pancreas (for a more detailed overview over rodent pancreas development, please refer to Pan and Wright 2011); *Pancreatic homeobox 1* (*Pdx1*) is expressed early in the posterior fore- and midgut and later in the pancreatic islets of Langerhans (Ohlsson et al., 1993) and marks the region where the pancreatic bud begins to form at embryonic day 9.5 (E9.5) (Wells and Melton, 1999). In mice, two buds are formed from ventral and dorsal endoderm respectively. Until E9.5, dorsal endoderm is in contact with the notochord and this contact is required for pancreatic gene expression. The ventral component of the pancreas derives from endoderm that is directly adjacent to presumptive liver endoderm. In rodents, pancreas formation is classically divided into two waves of development: primary transition between E9.5 and E12.5 and secondary transition starting at E13.5.

During the primary transition, there is massive proliferation of pancreatic progenitors, formation of microlumens and branching of the epithelial cell layer (around E11.5). In the dorsal pancreas, the first differentiated endocrine cells, predominantly glucagon-producing α-cells, start to appear. Coiling movements of the gut tube induce a rotation that positions the dorsal and ventral pancreatic buds in proximity of each other, facilitating conjoining of

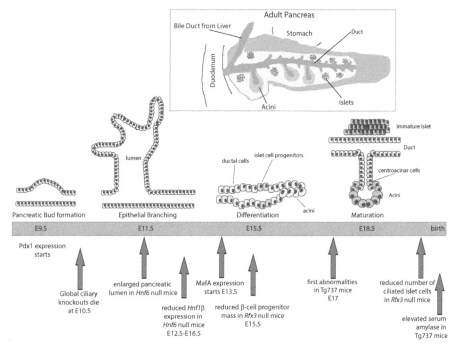

Figure 1. Overview of ciliary and cilia-related phenotypes in murine pancreatic development. The adult pancreas is situated between the stomach, the duodenum and the liver. Precursors of digestive enzymes are produced by the acinar cells and secreted into the ducts that drain into the duodenum. The first pancreatic progenitors start to form at embryonic day (E) 7.75, preceding bud formation by more than 24 hours. Epithelial branching can be observed at later stages, around E11.5. Until E17, the mutant phenotypes of *Hnf6* and *Rfx3* null mice might not be directly cilia-related, because ciliary hypomorphs manifest a phenotype at E17, while Pdx-1-Cre driven conditional ciliary mutants display a phenotype around birth.

the two structures into a single organ. Approximately at E12.5, the densely packed epithelium undergoes extensive remodelling and expansion and, at the same time, compartmentalization of the pancreatic branches starts to segregate distinct "tip" and "trunk" domains. During the secondary transition, the epithelium then differentiates into acini (derived from the "tip"), ductal cells and islet cell progenitors (both derived from the "trunk"epithelial regions; E15.5), which undergo epithelial-mesenchymal transition (EMT) to form immature islets in the developing pancreas. Secondary transition processes are completed around E16.5 and are followed by pancreas maturation at E18.5, during which time the duct, acini and immature islets are already formed but undergo further differentiation into different islet cell types (α-, β- and δ-cells). Importantly, islets are not fully matured at birth but they expand during the first three to four weeks after birth (around weaning) and then form mature, round islets with β-cells at the center surrounded by α-, δ- and PP cells. ε-cells, on the other hand, are dispersed throughout the islet.

Mice lacking functional *Pdx1* do not develop a pancreas and mutations in *PDX1* lead to diabetes in human patients (mature-onset diabetes of the young (MODY)), suggesting that there is significant overlap in the transcriptional machinery of mouse and humans.

The severely limited availability of early (less than 8 weeks gestation) and later gestational-aged (more than 22 weeks) human pancreatic tissues, make it difficult to outline similarities and differences between mouse and human pancreas development. As a consequence, the development of stem-cell based differentiation protocols to produce single hormone-expressing, glucose responsive human β-cells was largely informed by studies of mouse rather than human islet development. For an in-depth review of human pancreas development, please refer to Pan and Brissova, 2014. Pancreas formation is first evident at gestational day 26 (G26d) with the appearance of a dorsal bud followed by two ventral buds at G30d; while the left ventral bud regresses, the right one migrates posteriorly around G35d, conjoining with the dorsal pancreatic bud upon gut rotation at six to seven weeks of gestation (G6-7w). Notably, PDX1 expression does not define the endodermal region that will give rise to the dorsal bud unlike in mouse development, while at the same time, sonic hedgehog (SHH) expression is excluded from this area which is in good agreement with mouse data. Overall, molecular markers of early pancreatic differentiation seem to be highly similar in mouse and human.

In contrast to the presence of glucagon-positive cells in early mouse pancreas formation, hormone-expressing endocrine cells are not detected until G7.5-8w in humans, roughly 3 weeks after the initial outgrowth of the dorsal bud. The first hormone-expressing cells that appear in the human pancreas are insulin-positive cells. These lines of evidence make it unlikely that human pancreas development is characterized by the primary transition processes that are observed in mice. It is also uncertain if human endocrine cells derive exclusively from the trunk region of the pancreatic epithelium, as described in mouse. The spatiotemporal expression pattern of transcription factors, such as Neurogenin 3 (NGN3) and NKX6.2, suggest that similar transcriptional regulatory mechanisms operate in humans and mice. However, the presence of—albeit few—β-cells in patients with homozygous *NGN3* mutations indicate an NGN3-independent pathway for endocrine cell development in the human pancreas, unlike in mice.

Initially, islet clusters contain three- to four-fold more β- than α-cells at G9-13w, before reaching a 1:1 ratio around G14-16w which is maintained through birth. Importantly, adult human islets are characterized by a higher α- to β-cell ratio (approx. 1:1) than mouse islets (approximately 1:4) and a different architecture with a more even distribution of endocrine islet cells throughout the islet, as opposed to a core of β-cells surrounded by a mantle of other cell types as is observed in mouse islets.

Genetic disease such as MODY can inform us about important similarities and differences between mouse and human pancreas development and therefore is an important source of information for the development of stem-cell based islet cell replacement therapies.

Spontaneous and Targeted Mutants of Ciliary Proteins

Ift88 hypomorph

Polycystic kidney disease (PKD) comprises a group of both autosomal dominant (ADPKD) and autosomal recessive (ARPKD) disorders that present with cysts in the kidney and other organs, such as the liver (50% frequency) in ADPKD or liver fibrosis in ARPKD (Hildebrandt et al., 2011). ADPKD is usually caused by mutations in *PKD1* or *PKD2*; with an incidence rate of 1:1,000 it is the most common genetic cause of end stage renal disease. ARPKD is less frequent and caused by mutations in *PKHD1*. Typical pathologies of PKD are observed in mouse models harbouring mutations in PKD disease genes, as well as the *Orpk* mouse model (Moyer et al., 1994). *Orpk* mice arose through a random insertional mutation into the murine *Intraflagellar transport 88* (*Ift88*) gene, a core component of ciliary transport, resulting in an *Ift88* hypomorph. Homozygous animals usually die of an unknown cause within the first two weeks after birth. In addition to hepatic cysts, pancreatic cysts are a complication that occurs in roughly 10% of ADPKD patients. *Orpk* mice have severely damaged pancreata, including extensive loss of acinar cells, formation of abnormal tubular structures and an appearance of endocrine cells in the ducts (Cano et al., 2004; Zhang et al., 2005). *Orpk* embryos develop dilated pancreatic ducts as early as embryonic day E17, while both acini and immature islets appear normal (Zhang et al., 2005) (Figure 1). After postnatal (P) day 2, the acinar morphology becomes disorganized and the ductal lumen continues to expand. At P14 or thereafter, collagen deposition and fibrosis is seen throughout the acinar portion of the organ but not in islets. Interestingly, *Ift88* is only expressed in islet cells and epithelial cells of the duct; yet, pancreata of *Orpk* mice show a 5–10 fold increase in the number of apoptotic cells in the acini, alongside an upregulation of proliferating cells in the duct. Amylase expression levels were comparable between *Orpk* and wildtype embryos at E18.5, but significantly reduced in *Orpk* pups shortly after birth (Cano et al., 2004). At P1, the mutant mice display elevated serum levels of amylase suggesting irregular activation of digestive enzymes in the pancreas during the first days of life (Zhang et al., 2005) (Figure 1). Normally, activation occurs outside the pancreas and only the precursor forms are released. Over time, amylase levels decrease compared to wildtype controls, consistent with the destructive effect of activated digestive enzymes on the acinar portion of the pancreas. Given that centro-acinar cells are also ciliated, epithelial in nature and form the transition between ductal tissue and acini, it seems likely that loss of ciliary function in these cells contributes to dysregulated release of active enzymes. Some proteins, including inversin, localize to cilia and cell-cell junctions, suggesting a link between these two structures

(Nurnberger et al., 2002). Possibly, the structural stability of the junction between centroacinar and acinar cells is disturbed due to ciliary dysfunction itself or impairment of cellular polarization that has been reported for renal cysts of orpk mice (Taulman et al., 2001). Interestingly, endocrine cells that express insulin and other markers of β-cell maturity are found in expanding duct-like structures (Cano et al., 2004). Recent evidence suggests that islet progenitor cells reside in the duct and can undergo epithelial-mesenchymal transition to form β-like cells (Al-Hasani et al., 2013). Potentially, ciliary function is involved in this process. Overall, the orpk phenotype suggests an important role for ciliogenesis, ciliary maintenance and function in proper pancreatic tissue organization.

Pancreas-specific ciliary mutants

To better understand the role of cilia in pancreatic development and tissue homeostasis, Hebrok and co-workers conditionally deleted *Kif3a*, a component of the cilia-specific kinesin-II motor protein, under the control of the *pancreatic homeobox domain 1* (*Pdx1*) promoter (Cano et al., 2006). *Pdx1* or *insulin promoting factor 1* (*Ipf1*) is one of the earliest genes expressed within the pancreatic epithelium (Ohlsson et al., 1993; Gu et al., 2002). Shortly after birth, the pancreatic phenotype resembled that of *Orpk* mice: the characteristic berry-like shape of the exocrine acini was lost and the ductal lumen dilated. At P14, the pancreata were severely affected, resembling those of orpk mice. Strikingly, localized periductal fibrosis with increased deposition of extracellular matrix was observed at this stage, reminiscent of human chronic pancreatitis (Cano et al., 2006). Transforming growth factor-β (TGF-β) and Matrix Metalloprotease (MMP) signalling pathways are dysregulated in ciliary mutants, possibly explaining the fibrotic phenotype in *Kif3a* null pancreata. Similar to orpk mice, acini undergo apoptosis, while ductal cells proliferate in *Pdx1-Cre; Kif3a$^{D/D}$* pancreata (Cano et al., 2004; Cano et al., 2006). Because no cells were observed that expressed both ductal and acinar markers, transdifferentiation of acinar into ductal cells was ruled out; a combination of apoptosis in one compartment and hyperproliferation in the other is the likely cause of ductal expansion and loss of acini in *Pdx1-Cre; Kif3a$^{D/D}$* pancreata. A link between polycystic kidney disease and pancreatic cancer has long been suggested (Niv et al., 1997; Sato et al., 2009). Chronic pancreatitis dramatically increases the risk of pancreatic ductal adenocarcinoma. Therefore, the observed similarities between pancreatitis and *Pdx1-Cre; Kif3a$^{D/D}$* pancreata could provide an insight into the early stages of the progression to ductal neoplasia. In addition, chronic pancreatitis also increases the risk of islet cell damage and subsequent *Diabetes mellitus*. Cilia are not only present on ductal but on islet cells of the pancreas, yet *Pdx1-Cre; Kif3a$^{D/D}$* mice have normal glucose tolerance at three months of age (Cano et al., 2006), suggesting that β-cell ciliary dysfunction alone does not impair glucose metabolism. A comparison between the different *Pdx1-Cre* mouse strains (inducible and constitutively active) could suggest that the main phenotypes observed are due to ciliary dysfunction of ductal cells.

The gene mutated in Von Hippel Lindau Disease, *VHL*, encodes a tumour suppressor and *bona fide* ciliary protein which targets Hypoxia Inducible Factor 1A (HIF1A) for proteasomal degradation (Maxwell et al., 1999). Vhl has subsequently been shown to regulate ciliogenesis independently of the Hif1α signalling pathway (Schermer et al., 2006; Lutz and Burk, 2006). Deletion of *Vhl* from β-cells (βVhlKO) or pancreatic cells (PVhlKO) does not produce overt morphological defects in pancreatic tissue organization. However, probably due to the activation of Cre-Lox recombination in a small population of insulin expressing cells in the brain and subsequent impairment of growth hormone release from the pituitary gland, βVhlKO mice are significantly smaller than control littermates or PVhlKO mice (Cantley et al., 2009). While the exocrine portion of the organ seems to be unaffected in PVhlKO mice, both βVhlKO and PVhlKO mice are glucose intolerant at 12 weeks of age. In *ad libitum* fed animals, blood glucose levels are elevated while plasma insulin levels are reduced compared to those of controls. *Ex vivo* experiments showed that isolated βVhlKO islets secrete basal insulin levels that are similar to those in controls, however, when stimulated with 20 mM D-glucose, they secrete significantly less insulin. On the other hand, α-Ketoisocaproic acid or potassium chloride, two compounds that stimulate insulin secretion independently of glucose, yielded comparable responses from both βVhlKO and control animals. This observation suggests that the effect is not directly mitochondrial, but proximal to mitochondrial metabolism (Cantley et al., 2009). Ensuing experiments suggest that glucose uptake is impaired, possibly via a downregulation of *Solute carrier 2a2/glucose transporter2* (*Slc2a2/GLUT2*), the main glucose transporter in murine islets. Knocking out both *Vhl* and *Hif1α* in β-cells rescued the phenotype, including the dwarfism observed in βVhlKO mice. PVhlKO mice show increased perinatal lethality compared to controls and this was linked to severe hypoglycaemia partially due to α-cell dysfunction (Puri et al., 2013). Although it is not clear to which degree the effects are linked to ciliogenesis and ciliary function, these lines of evidence strongly support a role for the Vhl/Hif1α signalling pathway in pancreas and islet function.

Transcriptional Regulation of Ciliary Components in the Pancreas

The regulatory factor X family

The term "neuroendocrine cell" has been coined to emphasise the parallels between endocrine islet cells and neuronal cells. Both cell types are excitable and secrete small molecules and peptides in response to specific stimuli; their function is tightly linked to Ca^{2+} dynamics and some neurotransmitters such as acetylcholine and γ-aminobutyric acid (GABA) are even secreted by islet cells (Caicedo, 2013). Interestingly, a sweet taste receptor, T1R3, is expressed in islet cells and its expression is regulated by the nutritional state (Medina et al., 2014), suggesting functional similarities between endocrine islet cells and sensory neurons. In *Caenorhabditis elegans*, cilia are present in sensory neurons and the reproductive tract. The dauer formation 19 gene (*daf-19*) is specific to the ciliated sensory neuron and encodes a *regulatory factor* (Rfx)-type transcription factor. Daf-19 has been shown to specifically regulate the expression of targets that are functional in all ciliated sensory neurons via the x-box binding motif (Swoboda et al., 2000). In mammals, the *Rfx*-family of transcription factors is orthologous to *daf-19*. Several of these are important for ciliogenesis depending on tissue type (Choksi et al., 2014). *Rfx3* is expressed in a number of tissues including early pancreatic endocrine progenitors and major endocrine cell lineages (Baas et al., 2006; Ait-Lounis et al., 2007). The main phenotypes of *Rfx3$^{-/-}$* mice are left-right patterning defects leading to embryonic lethality, situs inversus or hydrocephalus. Just before birth, *Rfx3$^{-/-}$* mice have a reduced number of ciliated islet cells and the remaining cilia are stunted (Ait-Lounis et al., 2007). In addition, *Ift88* expression is reduced in islets of these animals and the promoter region of *Ift88* contains an x-box binding motif. *Ift172* and *Kif3a* expression, a subunit of the cilia-specific motor protein kinesin-II, is not significantly altered and there are no conserved x-box motifs upstream of these genes. Islets appeared smaller and disorganized in the pancreata of *Rfx3$^{-/-}$* mice. In addition, pancreatic insulin and glucagon contents were significantly reduced in *Rfx3$^{-/-}$* mice, while somatostatin was unchanged and pancreatic polypeptide (PP) was significantly upregulated (Ait-Lounis et al., 2007). Consequently, both global and pancreas-specific *Rfx3$^{-/-}$* mice are glucose intolerant (Ait-Lounis et al., 2007; Ait-Lounis et al., 2010). Importantly, early pancreatic development is unaffected in *Rfx3$^{-/-}$* mice and the number of cells committed to the β-cell lineage is unchanged. In pancreas-specific *Rfx3* knockout mice, β-cell mass is significantly reduced as early as E15.5 (Figure 1). In adult animals, *Rfx3* is necessary for the expression of β-cell maturation markers such as *insulin*, β-*Glucokinase* (β*Gck*) and *glucose transporter 2* (Glut-2) (Ait-Lounis et al., 2010). β-cells develop from PP expressing precursor cells (Herrera, 2000) and thus, *Rfx3* is likely important for the maturation of β-cell progenitors into functional β-cells. It is still unclear, however, whether Rfx3 is acting through the formation of functional cilia or if there are cilia independent mechanisms through which Rfx3 affects β-cell maturation.

Another Rfx transcription factor, Rfx6, has been implicated in islet formation in both mice and humans (Smith et al., 2010). In zebrafish, loss of *rfx6* blocks differentiation of α-, δ- and ε-cells, the number of β-cells, however, is only slightly reduced (Soyer et al., 2010). *Rfx6*-expression is dependent on Ngn3 activity and, similar to *Rfx3$^{-/-}$* mice, *Rfx6$^{-/-}$* animals show increased PP cell mass compared to other islet cells, such as α- or β-cells. Mutations in

RFX6 cause Mitchell-Riley-Syndrome (MTCHRS) that presents with neonatal diabetes, pancreatic hypoplasia, and intestinal and gallbladder malformations (OMIM #615710). In human pancreatic islets, *RFX6* is expressed in both α- and β-cells. *Rfx6* maintains β-cell maturity via transcriptional activation of mature β-cell markers including *βGck*, *Glut-2* and voltage-gated calcium channels (VDCC) and transcriptional repression of *lactate dehydrogenase (Ldha)* and *solute carrier family 16 (Slc16a1)* among others (Piccand et al., 2014). The latter two genes belong to a family of genes which are specifically repressed in mature β-cells, referred to as the "disallowed" genes (Pullen et al., 2010). Chromatin-immunoprecipitation (ChIP) experiments revealed specific Rfx6 binding to x-boxes in the *β-cell specific Glucokinase (βGck)* promoter region and in intron 10 of *Sulfonylurea Receptor 1/ATP-binding cassette subfamily C 8 (Sur1/Abcc8)* that are key players in insulin secretion. Moreover, the human *insulin* promoter contains several x-box motifs that recruit RFX6 but not p.V506G RFX6, a novel mutation discovered in a case of neonatal diabetes, duodenal stenosis and jejunal atresia (Chandra et al., 2014). Both phases of insulin secretion are reduced in *Rfx6$^{-/-}$* mice and in an *Rfx6*-depleted human β-cell line (Piccand et al., 2014; Chandra et al., 2014) and these phenotypes are likely connected to perturbed Ca^{2+}dynamics. ChIP-analysis also helped discover x-box binding sites in subunits of L-type and P/Q-type VDCCs (Piccand et al., 2014; Chandra et al., 2014). L-type Ca^{2+}-channel inhibitor nifedipine phenocopied the signalling defect, suggesting a central role for Rfx6 in transcriptional regulation of these ion channels. However, because all Rfx transcription factors recognize the x-box motif, it is not clear whether Rfx3 or Rfx6 is the main factor involved in transcriptional regulation of *Ins, βGck* or *VDCCs*. Whether ciliogenesis or ciliary functions in pancreatic islet cells of Rfx6$^{-/-}$ animals are perturbed, remains unknown. Taking the human data together with those from *Rfx3* and *Rfx6* knockout mice, ciliary transcription factors are involved in islet formation, β-cell maturation and function, although it remains unclear whether these roles are directly linked to ciliary function or if ciliogenesis is an event that impacts on, but is not directly connected to, the transcriptional regulation of β-cell maturation markers.

The hepatocyte nuclear factor family

Several hereditary diseases present with pancreatic cysts that are the result of abnormal tubulogenesis. Pancreas formation is in part regulated by a complex transcriptional program that includes hepatocyte nuclear factors 1 β and 6 (Hnf1β and Hnf6). During mouse development, *Hnf6* is first expressed throughout most of the pancreatic epithelium (E13.5), before it becomes restricted to differentiating ductal cells (E15.5). In older embryos (E17.5), *Hnf6* expression is found in epithelial linings of intra- and inter-lobular ducts as well as the intercalating ducts that form the transition to acini (Pierreux et al., 2006). Between E13 and 18, ductal progenitors give rise to neurogenin 3 (Ngn3) positive islet progenitor cells, placing *Hnf6* upstream of *Ngn3* transcriptional activity (Maestro et al., 2003). In addition, Hnf6 suppresses *v-maf avian musculaponeutrotic fibrosarcoma oncogene homolog A (MafA)*, a β-cell maturation marker, and *Hnf6* expression is upregulated in islets of leptin receptor mutant *db/db* mice (Yamamoto et al., 2013). MafA binds in the same region and competitively to the Forkhead box A2 (FoxA2) transcription factor, which plays a key role in endoderm development. As early as E11.5, *Hnf6*$^{-/-}$ knockout embryos show enlarged pancreatic lumen and the ductal marker Mucin-1 (Muc-1) is reduced (Figure 1). Later, Muc-1 staining reveals irregularly organized ducts that give rise to cystic ducts connected to acini. In contrast, the intercalating epithelial or centroacinar cells appear normal. Cyst formation is not linked to increased proliferation, but epithelia lining the cysts show polarization defects (Pierreux et al., 2006). Ciliogenesis is perturbed during the early stages of ductal development in *Hnf6*$^{-/-}$ pancreata; epithelial cells lining the pancreatic cysts do not form cilia, but smaller ducts show normal ciliation. These observations support a role for Hnf6 in development of intra- and interlobular ductal epithelia including polarization and cell-cell adhesion (Pierreux et al., 2006).

In developing *Hnf6*$^{-/-}$ pancreata, the expression of ciliary disease genes *Pkhd1* and *Cys1*, but not *Pkd1*, is reduced, suggesting that Hnf6 acts upstream of a subset of ciliary genes. At the same time, *Hnf1β* expression is reduced between E12.5 and E16.5 and is lost completely within the epithelial lining of the pancreatic cysts (Pierreux et al., 2006). HNF1β is also known as transcription factor 2 (TCF2) or maturity-onset diabetes of the young 5 (MODY5), because mutations in this gene lead to Renal Cysts and Diabetes Syndrome or MODY5 (RCAD, OMIM #137920) (Horikawa et al., 1997). Affected individuals have non-diabetic renal disease resulting from abnormal kidney development, impaired glucose-stimulated insulin secretion and diabetes, which can occur earlier than 25 years of age. Different frameshift mutations in *HNF1β/TCF2/MODY5*, resulting from an 8 bp deletion in exon 2 and an 8 bp insertion in exon 7 respectively, led to pancreatic hypoplasia in addition to bilateral cystic kidneys in two human foetuses (Haumaitre et al., 2006). In humans, *HNF1β* is strongly expressed in the metanephric kidney and pancreatic, biliary, duodenal and stomach epithelium during development; in adult tissues, *HNF1β* expression is strong in the kidney and pancreas and lower levels are found in the gut, testes, liver, lung and ovary. In *HNF1β* mutant foetuses, acini were underdeveloped in addition to defective arborisation of the ductal epithelium. Insulin, glucagon and somatostatin expression appeared normal, but β-cell mass was reduced in *HNF1β* mutants. Constitutive *Hnf1β*$^{-/-}$ mice are embryonic lethal, therefore analysis of *Hnf1β* deficiency in whole body glucose metabolism has been hampered. β-cell specific *Hnf1β* depletion impairs glucose tolerance with reduced glucose-stimulated insulin secretion (Wang et al., 2004). Interestingly, depolarizing agents such as L-arginine, still stimulated insulin secretion in these mutants. This is a phenomenon often observed in (pre-)diabetic animals and individuals (Trent et al., 1984) and indicates a role for *Hnf1β* in glucose metabolism, insulin signalling and/or glucose uptake. Changes in mRNA levels of the transcription factors *Pdx-1, hepatocyte nuclear factors 1α* and *4* implicate *Hnf1β* as part of a β-cell specific transcriptional network and as a necessary factor for glucose sensing or metabolism (Wang et al., 2004). In murine kidneys, *Hnf1β* regulates transcription of cystic disease genes that also localize to the cilium (Gresh et al., 2004), suggesting that ciliary function might be involved in proper β-cell maturation and function.

The above observations suggest that in the context of the developing pancreas Hnf6 can act upstream of Hnf1β, which in turn is upstream of ciliary cystic disease genes. The transcriptional machinery seems to have some overlap with the one regulating the development of renal epithelia and might help explain the high co-morbidity of renal and pancreatic cysts. Importantly, the *Hnf6-Hnf1β* transcriptional network is partially acting through cilia, as suggested by the phenotypic overlap with animal models of polycystic kidney disease. However, because cilia are present in the small ducts of *Hnf6$^{-/-}$* pancreata and *Tg737/polaris* pancreata do not develop cysts until after birth, it is likely that the *Hnf6-Hnf1β* transcriptional activity is involved in other cellular processes as well.

TISSUE HOMEOSTASIS

Signalling Pathways

Hedgehog signalling

One of the best characterized ciliary signalling pathways is Sonic Hedgehog (Hh) signalling; IFT is required for Shh signalling (Huangfu et al., 2003; Huangfu and Anderson, 2005) and Smoothened (Smo), a Hh receptor, was observed in primary cilia (Corbit et al., 2005). At the early stages of pancreatic development, gain of Shh signalling inhibits the specialization of pancreatic mesenchyme and drives differentiation into intestinal tissue (Apelqvist et al., 1997). During the later stages of development and in adult pancreata, Hh signalling becomes activated: loss of pancreatic Hh signalling impairs expansion of pancreatic epithelial cells and dysregulates the balance between mesenchymal and epithelial tissue differentiation (Kawahira et al., 2005). In human embryonic pancreata, an accumulation of Hh signalling components Smo and GLI-Kruppel family member 2 (Gli2) is observed in primary cilia, while Gli3 levels are reduced in the pancreatic ductal epithelium at weeks 14 and 18 of embryonic development, supporting a role for Hh signalling in pancreatic epithelia (Nielsen et al., 2008).

In adult pancreata, Hh signalling is tightly regulated. Mice with pancreas-specific overexpression of a constitutively active Gli2 variant, GLI2ΔN, fail to activate Hh signalling within the pancreatic epithelium (Cervantes et al., 2010). Additional deletion of *Kif3a* in the pancreata of these mice, however, results in ectopic activation of Hh signalling and the expression of downstream effector genes, such as Patched1 (Ptch1). This is in line with the observation that cilia repress Hh signalling mediated by GLI2ΔN in medulloblastoma (Han et al., 2009) and basal cell carcinoma (BCC) (Wong et al., 2009). Mice with pancreas-specific deletion of *Kif3a* and overexpression of constitutively active Gli2, referred to as triple transgenics from here on, are born at near Mendelian ratios and have no overt morphological phenotypes compared to littermate controls. However, at 2–3 weeks of age, these mice are significantly smaller, fail to thrive and show postnatal lethality, preventing metabolic phenotyping (Cervantes et al., 2010). Surrounded by fibrous stroma, solid epithelial nests form within or adjacent to ducts of triple transgenic pancreata, containing cells that do not express markers of any of the three pancreatic lineages (acini, endocrine and ductal tissue). This implies an inability of GLI2ΔN-overexpressing cells to maintain their differentiated state. Instead, they revert to a progenitor-like state marked by *Forkhead box A2 (FoxA2)* and *(sex determining region Y)-box 9 (Sox9)* expression. Unlike the β-cell specific ciliary mutant or GLI2ΔN-overexpressing mice, triple transgenic mice have defects in the endocrine compartment of the pancreas, manifested by a reduction by 50% of islet cells that results from increased apoptosis and impaired expansion of the pancreatic epithelium as a source of progenitor cells for embryonic neogenesis of islets cells in triple transgenic mice (Cervantes et al., 2010). The functional characterization of the remaining β-cells was hampered by the early lethality of the triple transgenic mice. Repeating the experiment with an inducible triple transgenic mouse line rendered transgenic mice glucose intolerant as little as one week after tamoxifen-mediated induction at 8–10 weeks of age (Landsman et al., 2011). Recombination was efficient as indicated by elevated *Ptch1* and *Gli1* expression, two Hh reporter genes; expression levels of *Ins1* and *Ins2*, *Glut2* and the two subunits of the ATP-dependent potassium (K_{ATP}-)channel and other markers of mature β-cells were reduced in these islets. Insulin release in response to elevated glucose was reduced in β-cells, with activated Hh signalling observed both *in vivo* and *ex vivo*. Concomitantly, transcription factors characteristic of pancreatic progenitor cells, such as *Sox9* and *Hes1* were elevated. *Ins1-Cre; GLI2ΔN; Kif3a^{loxP/loxP}* mice showed a similar β-cell defect, implicating increased Hh signalling activity and β-cell de-differentiation. Animals with inducible as well as constitutive activation of Hh signalling reverted to normal blood glucose levels roughly 10 weeks after tamoxifen treatment, probably by post-translational silencing of GLI2ΔN. Seemingly, additional, non-ciliary mechanisms regulate Hh activation in the pancreas. In the constitutively active β-cell specific GLI2ΔN-overexpressing and ciliary mutant mouse line, some islet cells maintained GLI2ΔN overexpression and gave rise to pancreatic tumors that lacked *Ins* expression, but had high levels of *Sox9* and *Hes1* transcripts, demonstrating the importance of tight regulation of Hh activity in pancreatic tissue (Landsman et al., 2011).

mTOR signalling

The mammalian target of rapamycin (mTOR) signalling pathway plays a central role in energy metabolism, nutrient sensing and cell size regulation. Cilia have been implicated both in the regulation of mTOR complex 1 (mTORC1) (Boehlke et al., 2010) and mTORC2 (Cardenas-Rodriguez, Osborn, et al., 2013; Cardenas-Rodriguez et al., 2013) activity, two multiprotein complexes that regulate mTOR signalling. In the kidney, cilia are required for flow sensing and regulate cell size via the mTOR signalling pathway. A role for the primary cilium in cell size control is supported by the observation that cells lining the cysts are larger than normal tubular cells in polycystic kidneys (Grantham et al., 1987). Cultured renal epithelial cells (Madin-Darby-Canine-Kidney Epithelial (MDCK)) grown under flow conditions are smaller than those grown in the absence of flow or depleted of *Kif3a*- or *Ift88*-expression (Boehlke et al., 2010). Growth inhibition is independent of Ca^{2+}-influx or serine/threonine kinase Akt signalling but rather relies on tumor suppressor liver kinase B1 (Lkb1), also known as serine/threonine kinase 11 (Stk11), and the subsequent activation of metabolic sensor AMP-activated kinase (AMPK). Lkb1 localizes to primary cilia and phosphorylated AMPK was observed at the basal body of MDCK cells (Boehlke et al., 2010). Cilia-specific proximity labelling in another renal cell line confirmed Lkb1 and AMPK as *bona fide* ciliary proteins (Mick et al., 2015). Depleting Lkb1 in adult β-cells did not overtly affect fasted or fed blood glucose (Granot et al., 2009). In these mice, glucose tolerance is improved and β-cells mount an exaggerated insulin response to a glucose stimulus. Peripheral insulin sensitivity remains unchanged. Detailed morphological studies reveal abnormal polarization of Lkb1-deficient β-cells that are arranged in a rosette-like structure around islet capillaries: the position of nuclei changed from distal to proximal with respect to capillaries and the positioning of primary cilia changed to the distal membrane, reminiscent of apical-basal epithelial polarity. Depletion of Lkb1 from β-cells resulted in larger cells and this effect was mediated via loss of AMPK phosphorylation (Granot et al., 2009). Insulin hypersecretion was only partially linked to increased β-cell size or mass and did not affect basal insulin secretion, secretion kinetics or the biphasic nature of insulin release. Although it is not clear whether these effects are mediated through cilia directly, it is possible that cilia or Lkb1 itself act as a brake on insulin secretion or that loss of cilia or Lkb1 function attenuates signalling activities of an antagonist of insulin secretion.

Insulin receptor signalling

While insulin receptor (IR) signalling does not play a central role in pancreatic development, mature α- and β-cells require IR signalling for proper function. α-cell specific IR knockout mice (αIRKO) manifest mild glucose intolerance, hyperglycemia and hyperglucagonemia in the fed state. In addition, αIRKO mice developed hyperinsulinemia in the fed state over time (Kawamori et al., 2009). Because both elevated blood glucose levels and insulin (secreted from β-cells) are antagonists of glucagon-secretion from α-cells, the observed phenotypes could be an effect of either hypoinsulinemia or hyperglycemia. Specific ablation of β-cells did not yield an additional effect, suggesting that the phenotype is directly linked to loss of α-cell IR signalling (Kawamori et al., 2009). β-cell specific ablation of IR (βIRKO) in mice blunts 1st phase insulin secretion in response to glucose that leads to glucose intolerance and diabetes in 25% of βIRKO mice (Otani et al., 2004; Kulkarni et al., 1999). At 6 months of age, both male and female mice have normal blood glucose levels and mildly elevated plasma insulin levels. In response to glucose but not L-arginine, 1st phase insulin secretion is blunted in both sexes at three to four months of age (Kulkarni et al., 1999). In seven month old βIRKO mice, total pancreatic insulin content was reduced, likely due to a significant reduction in β-cell mass (Otani et al., 2004). These lines of evidence support a role for insulin signalling both in insulin secretion and β-cell proliferation. At 4 weeks of age, pancreata of *Bardet-Biedl-Syndrome 4* (*Bbs4*) knockout mice (Kulaga et al., 2004), a ciliopathy model, show no overt morphological abnormalities: islet architecture and size distribution, serum insulin levels and pancreatic insulin content are similar to age- and sex-matched wild type controls. However, as early as seven to nine weeks of age, male *Bbs4$^{-/-}$* mice show defective glucose handling which precedes obesity, suggesting a direct role for primary cilia in glucose metabolism (Gerdes et al., 2014). Concomitant monitoring of serum insulin levels in parallel with blood glucose sampling revealed that 1st phase insulin release was blunted in *Bbs4$^{-/-}$* mice. This phenotype was also observed *ex vivo* in murine islets that were transduced with short hairpin RNA (shRNA) targeting ciliary genes and in an insulinoma cell line, indicating that this is an islet- and even a β-cell specific phenotype; still, effects on α- and δ-cells cannot be excluded, since they are ciliated as well (Munger, 1958). Cellular glucose metabolism was not affected by loss of two different basal body proteins, Bbs4 and Ofd1 (Gerdes et al., 2014). The observed insulin secretion defect is reminiscent of that of βIRKO mice, suggesting that the β-cell cilium is required for IR dependent signalling. In insulin stimulated β-cells, IR is recruited to the primary cilium and insulin-mediated IR activation is required and sufficient for ciliary localization. Disrupting ciliary function with shRNAs targeting different ciliary genes in an insulinoma line and in primary murine islets reduces phospho-activation of the downstream IR signalling nodes, phosphoinositide-3-kinase regulatory subunit p85/p55 and Akt. In addition, Forkhead box O1 (FoxO1)-

dependent transcription of exocytotic *Soluble N-ethylmaleimide-sensitive factor Attachment protein Receptor* (*SNARE*) components *synaptosomal-associated protein 25* (*SNAP25*) and *syntaxin1A (Kaneko et al., 2010)* is reduced in these cells as well as in *Bbs4−/−* islets (Gerdes et al., 2014). Because the acute or 1st phase of insulin release largely relies on pre-assembled SNARE-complexes, while the 2nd more sustained phase of insulin release relies on transiently assembled SNARE complexes, a reduction in *Snap25* or *Syntaxin1a* would over-proportionately affect the equilibrium of 1st phase rather than the steady state of 2nd phase insulin release (Takahashi et al., 2010). In summary, ciliary disruption leads to loss of ciliary IR-dependent signalling and subsequently to a reduction in SNARE components, which in turn reduces 1st phase insulin release. In a Type 2 *Diabetes Mellitus* (T2DM) rat model, Goto-Kakizaki (GK) rats (Ostenson and Efendic, 2007; Ling et al., 1998), the number of β-cell cilia was significantly reduced and several ciliary, basal body and/or centrosomal genes and proteins were misregulated, suggesting that the insulin secretion defects observed in GK rats might be a result of ciliary impairment.

Other pathways

Ciliary disruption in pancreata leads to regions of fibrosis around the ducts and this accumulation of extracellular matrix proteins resembles human chronic pancreatitis (Cano et al., 2006). Members of the Transforming growth factor Tgf-β superfamily play an important role in the biosynthesis and turnover of the extracellular matrix; indeed, the expression pattern of Tgf-β2 and Tgf-β3 is changed in *Pdx1-Cre*; *Kif3a$^{D/D}$* pancreata, while TGF-β1 levels are stable. The expression levels of the Tgf-β target gene *Connective tissue growth factor* (*Ctgf*) are elevated in the absence of *Kif3a*, implicating elevated Tgf-β signalling in *Pdx1-Cre*; *Kif3a$^{D/D}$* pancreata. Primary cilia impairment has been linked to reduced Tgf-β signalling in murine orpk fibroblasts and a stem cell line during cardiomyogenesis (Clement et al., 2013). This work was primarily done *ex vivo* and *in vitro*. In pancreatic tissue, cilia seem to inhibit Tgf-β signalling rather than promote it. Further work is necessary to address this seeming contradiction; obvious disparities such the differences between cultured cells and primary cells in living organisms are only one of the several possible explanations. Tissue specificity is another possible explanation and certainly the potential for non-ciliary roles of Kif3a and Ift88 should not be neglected. In addition, matrix metalloproteinases are involved in reversible pancreatic fibrosis in both humans and animal models. *Matrix metalloproteinase 7* (*Mmp-7*) expression levels are significantly elevated in *Kif3a$^{D/D}$* pancreata along with two members of the tissue inhibitors of metalloproteinases (Timp) superfamily, *Timp-1* and *Timp-4*. While little is known about the link between Mmps and ciliary function, one report links Mmp-7 activity to cell differentiation and the inhibition of ciliogenesis in murine airway epithelia (Gharib et al., 2013). The effect of loss of ciliary function on Mmp-7 signaling, however, is still largely unexplored.

Finally, *Somatostatin receptor 3* (*Sstr3*), a gold standard of receptors observed in primary cilia, localises at the cilia of murine β- and a subset of α-cells (Iwanaga et al., 2011). Somatostatin (Sst) is known to inhibit cAMP-dependent signalling; at the same time, cAMP-dependent signalling has been shown to amplify glucose-stimulated insulin secretion. The role of Sstr3 in islet function is unknown as is, indeed, if ciliary localization is required for proper Sstr3 signalling. Unlike cilia in the adenohypophysis that are double-positive for Sstr3 and adenylyl cyclase III (AcIII), AcIII immune reactivity is not observed in the pancreas, indicating that the Sst signalling pathway in the adenohypophysis is different from that in pancreata. Of note, global *Sst* knockout mice manifest hyperglucagonemia and hyperinsulinemia without affecting islet morphology or total pancreatic glucagon or insulin content (Hauge-Evans et al., 2009). The function of Sstr3 in islets is still unclear, but there might be a role for cilia in the regulation of islet peptide hormone secretion involving other hormones than insulin.

Metabolic Disease

A subset of ciliopathies present with metabolic disease, including but not limited to obesity (Hildebrandt et al., 2011). Expansion of fat cell mass requires IR dependent signalling: an adipose tissue specific IR knockout (FIRKO) mouse does not gain weight on High Fat Diet (HFD) (Bluher et al., 2002). Elevated insulin levels are characteristic of obesity though they might not always be causal. Insulin resistance is defined as the inability of the body to respond to insulin, i.e., transport glucose across cell membranes of insulin target tissues such as muscle, liver or fat. In this context, increased insulin levels in the circulating blood can be a sign of insulin resistance. Two ciliopathies are characterized by early onset childhood obesity and metabolic complications: Alström Syndrome (ALMS) (Alstrom et al., 1959) and Bardet Biedl Syndrome (BBS). While ALMS has been linked to a single gene, *ALMS1* (Collin et al., 2002), BBS has been linked to mutations in 19 different genes (M'Hamdi et al., 2014; Aldahmesh et al., 2014). Many BBS gene products form a multiprotein complex at the basal body (Nachury et al., 2007). In *D. rerio*, suppression of *alms1*, *bbs1* or *bbs4* with splice-blocking (alms1) or translation-blocking morpholino antisense nucleotides revealed that β-cell proliferation was differentially affected in *alms1* versus *bbs* morphants (Lodh et al., 2015). β-cells of *bbs1/4* morphants showed higher proliferation levels compared to those of *alms1* morphants or even controls. When exposed to 20 mM glucose, β-cell expansion in *alms1* morphants increased similar to that in controls, but *bbs1* and *bbs4* morphants showed no additional increase. After prolonged glucose exposure, β-cells continued to expand in controls, but not in morphants. *Alms1* morphants showed reduced β-cell proliferation, while *bbs* morphants showed both increased β-cell proliferation and apoptosis. The differential effects on β-cell proliferation could explain the difference in age of onset of T2DM in ALMS and BBS patients, although transient *alms1* and *bbs1/ bbs4* suppression prevents a far reaching conclusion about overall disease progression. More extensive studies will have to address disease progression in *Alms1* or *Bbs1/4* depleted pancreata including β-cell mass, total pancreatic insulin content and islet function at various time points during adolescence and adult tissue homeostasis. Moreover, the signalling pathways that are perturbed and result in the observed difference in β-cell proliferation remain unclear.

A study in ALMS patients found an inverse correlation between Body Mass Index (BMI) and age in male patients, indicating that the obesity phenotype is most pronounced early on and then decreases slightly as judged by a decrease in BMI (Bettini et al., 2012). Of note, this is in contrast to what is reported for the general population, where weight and BMI commonly increase with age. Importantly, none of the available ALMS mouse models recapitulate this phenotype. At the same time, oral glucose tolerance tests (OGTT) revealed no significant difference between ALMS patients and age- and weight-matched control patients. Yet, insulin response was higher in younger ALMS patients than controls and the parameters of β-cell function decreased with age. In the oldest age group (older than 18 years), 67% of ALMS patients manifest impaired glucose metabolism which can be attributed to both decreased systemic insulin resistance and β-cell function (Bettini et al., 2012). An excess of 22% of ALMS patients have T2DM, which constitutes a dramatic increase compared to the general population. According to data from the United States Centers for Disease Control and Prevention, the median age at diagnosis of T2DM was 54.2 years in 2011 (Control, 2015). In BBS patients, T2DM is also more prevalent than in the general population (Badano et al., 2006), although the average age of onset is not as young as that of ALMS patients (Beales et al., 1999). The degree of obesity is comparable among ALMS and BBS patients, while the age of onset of T2DM is not. This could imply that the incidence of T2DM is not a secondary feature of obesity in these syndromes, but it might be linked directly to ciliary function.

Interestingly, ADPKD patients have decreased 1st phase insulin secretion and β-cell function (Pietrzak-Nowacka et al., 2010). In β-cells, Ca^{2+} signalling is an essential step preceding insulin secretion (Yang and Berggren, 2006). Polycystin 1 and 2, the gene products mutated in ADPKD, encode an heterodimeric Ca^{2+}-channel that interacts with ryanodine receptor (RyR)-mediated Ca^{2+}-signaling pathways (Nauli et al., 2003). In β-cells, RyR release of Ca^{2+}-channels partially controls insulin release independently of glucose, implicating a role for polycystin 1 and 2 in the RyR-dependent insulin secretory pathway (Johnson et al., 2004). Further studies are required to characterize the role of ciliary function in β-cell Ca^{2+}-dynamics and subsequent insulin release with or without a glucose stimulus.

In *Bbs4*$^{-/-}$ mice, 1st phase insulin secretion is blunted and glucose handling is delayed as early as 7–9 weeks of age (see above and in ref. Gerdes et al.,

2014). Impaired 1st phase insulin secretion is one of the earliest detectable anomalies in individuals at risk of T2DM (Haeften et al., 2000), although family history remains one of the best predictors of T2DM disease risk. Children with one parent diagnosed with T2DM before the age of 50 have a 14% risk of developing T2DM, while those with both parents suffering from T2DM have a 50% chance. A clinical study investigated the family history of T2DM (FH), β-cell function and tissue insulin sensitivity of 283 healthy volunteers (Haeften et al., 2000). The participants were subjected to oral glucose tolerance tests and subsequent hyperglycemic clamps, i.e., their blood glucose levels were maintained at 10 mM with variable infusion of a 20% glucose solution. 185 individuals had normal glucose tolerance (NGT), while 98 individuals had impaired glucose tolerance (IGT) according to the World Health Organization criteria (these involve normal glucose levels between 7.8 and 11.1 mM over 2 hours and a fasting level below 7 mM glucose). During the hyperglycemic clamp, IGT individuals secreted less insulin than those with NGT, but taking into account FH, serum insulin levels were not significantly different between NGT and FH and IGT individuals without FH of T2DM. 1st phase insulin secretion, however, was a better predictor: acute insulin secretion was highest in NGT without FH individuals and lowest in IGT individuals with or without FH. Importantly, NGT individuals with FH (i.e., subjects at risk of T2DM who manifest no symptoms yet) showed reduced insulin secretion but not to the same degree as in IGT. Consequently, reduced 1st phase insulin release is a sign of T2DM risk even before glucose tolerance is impaired (Haeften et al., 2000). Ciliary dysfunction might confer T2DM risk by impairing 1st phase insulin secretion. Additional insults to glucose and potentially lipid metabolism could then lead to T2DM.

In islets, α-, β- and δ-cells are ciliated (Munger, 1958). The effect of ciliary impairment on α- and δ-cells is still unknown. Given that an α-cell specific IR knockout mouse displays hyperglucagonemia and hyperglycemia (Kawamori et al., 2009), it is possible that the hyperglycemia phenotype in *Bbs4$^{-/-}$* mice is caused by a combination of α- and β-cell dysfunction, in addition to defects in peripheral target tissues of insulin action. This hypothesis gains further support when taking into account that glucagon levels are elevated in T2DM (Larsson et al., 2000) and both glucose and insulin are antagonists of glucagon secretion.

Discussion

Several morphogenetic signalling pathways have been linked to the primary cilium and total ciliary mouse knockouts die as early as E9.5 to E10.5, which is the stage at which the pancreatic bud forms from the gut endoderm (Gittes, 2009). The majority of endocrine cells are specified during the secondary transition between E13 and E14. Disrupting ciliary function early during pancreatic development results in dilated ducts or ductal hyperplasia, as well as acinar degeneration at various ages of onset. Although Tgf-β and Mmp signalling pathways have been shown to be mis-regulated in pancreas-specific *Kif3a* mutant mice, this is more likely linked to fibrosis and the deposition of extracellular matrix proteins. Although some lines of evidence implicate mis-regulated Wnt signalling, it has not yet been conclusively shown which signalling pathways lead to ductal hyperplasia. Consequently, the mechanism by which cilia control the balance between the growth of ductal epithelium and acini is unclear. Interestingly, cilia are present in the pancreatic ducts and islet cells but not on acinar cells, yet the main defects in ciliary mutants are observed in the acini. This could implicate cilia as a tissue organizer around birth, when acinar degeneration sets in.

Whole body metabolism is the result of a complex interplay of autocrine, paracrine, exo- and endocrine hormones. Even a slight imbalance can result in global dysregulation of the finely tuned network over time, i.e., metabolic disease. Both Orpk and $Bbs4^{-/-}$ mice show impaired glucose handling. On the other hand, pancreas specific ciliary mutants ($Pdx1$-Cre; $Kif3a^{D/D}$) show no such defect when tested at the age of three months. These observations support the hypothesis that T2DM is not a single-organ disease. Instead, impairment of tissue homeostasis in several tissues will eventually manifest in disease. The metabolically active cell types in liver and fat tissue, two targets of insulin action, are not ciliated. On the other hand, cilia are found in skeletal muscle, islet cells and the brain, which has recently emerged as a target tissue of insulin action. Concerted dysregulation of ciliary signalling in these tissues might be sufficient to manifest T2DM.

Recently, genetic fine mapping revealed a role for FOXA2-dependent transcriptional regulation in T2DM (Gaulton et al., 2015). Cross-referencing of enhancer sites in T2DM susceptibility genes with Chromatin-immunoprecipitation (ChIP) data of FOXA2 binding sites showed more pronounced association in islet than in liver cells, but was significant in both cell types. Importantly, significant enrichment was only observed for binding sites shared with at least one other transcription factor, while FoxA2 binding sites that were not in the proximity of another transcription factor binding site were not enriched. Consequently, FOXA2-binding sites emerged as a genomic marker for T2DM susceptibility genes such as *Melatonin Receptor 1B* (MTNR1B; please refer to Figure 2). In islets, the risk variant increases the FOXA2-bound enhancer activity upstream of *MTNR1B*, affecting both insulin secretion and insulin signalling. FoxA2 plays an important role during pancreas development and serves as a marker of pancreatic epithelium, separating it from surrounding tissue such as the mesenchyme, endothelial cells or neurons (Willmann et al.). Because FoxA2 acts upstream of several important ciliogenic transcription factors including the mouse homeobox gene *Noto* (Abdelkhalek et al., 2004; Gerdes et al., 2009), this enrichment might suggest a role for cilia in T2DM susceptibility. In addition, FoxA2 is part of a transcriptional network consisting of, but not limited to, MODY5/Hnf1β

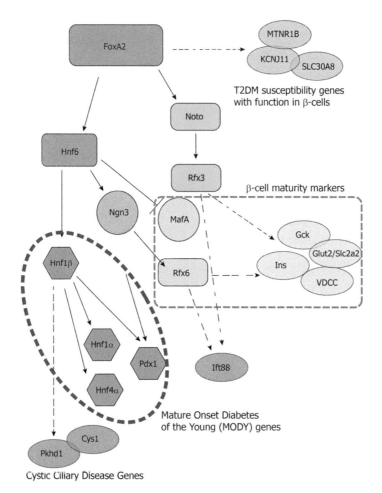

Figure 2. FoxA2, Hnf6 and Noto-dependent transcriptional regulation defines an islet-cell/ciliary network that might play a role in metabolic disease. Ciliary transcription factors *Rfx3* and *Rfx6* can bind to x-box motifs upstream of β-cell maturity markers including β-*Glucokinase* and *Insulin* as well as core ciliary genes, such as *Ift88*. At the same time, a subset of MODY genes is regulated by Hnf1β (MODY5) activity along with genes mutated in ciliary cystic disease. In addition, FoxA2 has been reported to be upstream of the MODY transcriptional network and the ciliary Rfx network and β-cell maturity markers. Recently, FoxA2 binding was identified as a genomic marker for T2DM susceptibility genes.

which regulates Hnf4α, Hnf1α and Pdx-1. In addition, Hnf6 acts upstream of Hnf1β and both regulate a subset of ciliary cystic disease genes. On the other hand, genes characteristic of functional β-cells such as β*Gck*, *Ins*, *Glut2/Slc2a2* and *VDCC*s contain X-box binding motifs in their promoter regions, defining an interconnected ciliary-β-cell transcriptional network that might inform us further about the role of ciliary function in metabolic disease and β-cell maturation and function (Figure 2). Further studies are needed to answer the question as to whether these transcription factors indeed act through ciliogenesis, ciliary maintenance or function or whether they control parallel processes that are independent of cilia.

Acknowledgments

My heartfelt thanks go to all group members and colleagues at our institute for the lively discussion. Especially, I am grateful to Dr. Ingo Burtscher for his critical evaluation of this manuscript.

References

Abdelkhalek, H. B., Beckers, A., Schuster-Gossler, K. et al. 2004. The mouse homeobox gene Not is required for caudal notochord development and affected by the truncate mutation. Genes Dev. 18(14): 1725–36.

Ait-Lounis, Aouatef, Baas, Dominique, Barras, Emmanuèle et al. 2007. Novel function of the ciliogenic transcription factor RFX3 in development of the endocrine pancreas. Diabetes 56(4): 950–959.

Ait-Lounis, Aouatef, Bonal, Claire, Seguín-Estévez, Queralt et al. 2010. The transcription factor Rfx3 regulates β-cell differentiation, function, and glucokinase expression. Diabetes 59(7): 1674–1685.

Al-Hasani, Keith, Pfeifer, Anja, Courtney, Monica et al. 2013. Adult duct-lining cells can reprogram into β-like cells able to counter repeated cycles of toxin-induced diabetes. Dev. Cell 26(1): 86–100.

Aldahmesh, Mohammed A., Yuanyuan, Li, Amal, Alhashem et al. 2014. IFT27, encoding a small GTPase component of IFT particles, is mutated in a consanguineous family with Bardet-Biedl syndrome. Hum. Mol. Genet. 23(12): 3307–3315.

Alstrom, C. H., Hallgren, B., Nilsson, L. B. and Asander, H. 1959. Retinal degeneration combined with obesity, diabetes mellitus and neurogenous deafness: a specific syndrome (not hitherto described) distinct from the Laurence-Moon-Biedl syndrome. A clinical endocrinological and genetic examination based on a large pedigree. Acta Psychiat. Neurol. Scand. 34: 1–35.

Apelqvist, Åsa, Ahlgren, Ulf and Edlund, Helena. 1997. Sonic hedgehog directs specialised mesoderm differentiation in the intestine and pancreas. Curr. Biol. 7(10): 801–804.

Baas, D., Meiniel, A., Benadiba, C. et al. 2006. A deficiency in RFX3 causes hydrocephalus associated with abnormal differentiation of ependymal cells. Eur. J. Neurosci. 24(4): 1020–30.

Badano, J. L., Mitsuma, N., Beales, P. L. and Katsanis, N. 2006. The ciliopathies: an emerging class of human genetic disorders. Annu. Rev. Genomics Hum. Genet. 7: 125–48.

Beales, P. L., Elcioglu, N., Woolf, A. S., Parker, D. and Flinter, F. A. 1999. New criteria for improved diagnosis of Bardet-Biedl syndrome: results of a population survey. J. Med. Genet. 36: 437–446.

Bettini, V., Maffei, P., Pagano, C. et al. 2012. The progression from obesity to type 2 diabetes in Alstrom syndrome. Pediatr Diabetes 13(1): 59–67.

Bluher, M., Michael, M. D., Peroni, O. D. et al. 2002. Adipose tissue selective insulin receptor knockout protects against obesity and obesity-related glucose intolerance. Dev. Cell 3(1): 25–38.

Boehlke, C., Kotsis, F., Patel, V. et al. 2010. Primary cilia regulate mTORC1 activity and cell size through Lkb1. Nat. Cell Biol. 12(11): 1115–22.

Caicedo, Alejandro. 2013. Paracrine and autocrine interactions in the human islet: More than meets the eye. Seminars in Cell & Developmental Biology 24(1): 11–21.

Cano, D. A., Murcia, N. S., Pazour, G. J. and Hebrok, M. 2004. Orpk mouse model of polycystic kidney disease reveals essential role of primary cilia in pancreatic tissue organization. Development 131: 3457–3467.

Cano, D. A., Sekine, S. and Hebrok, M. 2006. Primary cilia deletion in pancreatic epithelial cells results in cyst formation and pancreatitis. Gastroenterology 131: 1856–1869.

Cantley, James, Colin Selman, Deepa Shukla et al. 2009. Deletion of the von Hippel-Lindau gene in pancreatic β cells impairs glucose homeostasis in mice. J. Clin. Invest. 119(1): 125–135.

Cardenas-Rodriguez, M., Irigoín, F., Osborn, D. P. et al. 2013. The bardet-biedl syndrome-related protein CCDC28B modulates mTORC2 function and interacts with SIN1 to control cilia length independently of the mTOR complex. Hum. Mol. Genet. 22(20): 4031–42.

Cardenas-Rodriguez, M., Osborn, D. P., Irigoín, F. et al. 2013. Characterization of CCDC28B reveals its role in ciliogenesis and provides insight to understand its modifier effect on Bardet-Biedl syndrome. Hum. Genet. 132(1): 91–105.

Cervantes, S., Lau, J., Cano, D. A., Borromeo-Austin, C. and Hebrok, M. 2010. Primary cilia regulate Gli/hedgehog activation in pancreas. Proc. Natl. Acad. Sci. USA 107(22): 10109–10114.

Chandra, Vikash, Albagli-Curiel, Olivier, Hastoy, Benoit et al. 2014. RFX6 regulates insulin secretion by modulating Ca2+ homeostasis in human β cells. Cell Reports 9(6): 2206–2218.

Choksi, Semil P., Gilbert Lauter, Peter Swoboda and Sudipto Roy. 2014. Switching on cilia: transcriptional networks regulating ciliogenesis. Development 141(7): 1427–1441.

Clement, Christian Alexandro, Katrine Dalsgaard Ajbro, Karen Koefoed et al. 2013. TGF-β signaling is associated with endocytosis at the pocket region of the primary cilium. Cell Reports 3(6): 1806–1814.

Collin, G. B., Marshall, J. D., Ikeda, A. et al. 2002. Mutations in *ALMS1* cause obesity, type 2 diabetes and neurosensory degeneration in Alstrom syndrome. Nat. Genet. 31: 74–78.

Control, Centers for Disease. Diabetes Public Health Resource 2015. Available from http: // www.cdc.gov/diabetes/statistics/age/detailtable4.htm.

Corbit, K. C., Aanstad, P., Singla, V., Norman, A. R., Stainier, D. Y. and Reiter, J. F. 2005. Vertebrate smoothened functions at the primary cilium. Nature 437: 1018–1021.

Fliegauf, M., Benzing, T. and Omran, H. 2007. When cilia go bad: cilia defects and ciliopathies. Nat. Rev. Mol. Cell Biol. 8(11): 880–93.

Gaulton, Kyle J., Ferreira, Teresa, Lee, Yeji et al. 2015. Genetic fine mapping and genomic annotation defines causal mechanisms at type 2 diabetes susceptibility loci. Nat. Genet. 47(12): 1415–1425.

Gerdes, J. M., Davis, E. E. and Katsanis, N. 2009. The vertebrate primary cilium in development, homeostasis, and disease. Cell 137(1): 32–45.

Gerdes, J. M., Christou-Safina, S., Xiong, Y. et al. 2014. Ciliary dysfunction impairs beta-cell insulin secretion and promotes development of type 2 diabetes in rodents. Nat. Commun. 5: 5308.

Gharib, Sina A., William A. Altemeier, Laura S. Van Winkle et al. 2013. Matrix metalloproteinase-7 coordinates airway epithelial injury response and differentiation of ciliated cells. American Journal of Respiratory Cell and Molecular Biology 48(3): 390–396.

Gittes, George K. 2009. Developmental biology of the pancreas: a comprehensive review. Developmental Biology 326(1): 4–35.

Granot, Z., Swisa, A., Magenheim, J. et al. 2009. LKB1 regulates pancreatic beta cell size, polarity, and function. Cell Metab. 10(4): 296–308.

Grantham, J. J., Geiser, J. L. and Evan, A. P. 1987. Cyst formation and growth in autosomal dominant polycystic kidney disease. Kidney Int. 31(5): 1145–52.

Gresh, Lionel, Fischer, Evelyne, Reimann, Andreas et al. 2004. A transcriptional network in polycystic kidney disease. The EMBO Journal 23(7): 1657–1668.

Gu, G., Dubauskaite, J. and Melton, D. A. 2002. Direct evidence for the pancreatic lineage: Ngn3+ cells are islet progenitors and are distinct from duct progenitors. Development 129(10): 2447–2457.

Haeften, T. W. van, Pimenta, W., Mitrakou, A. et al. 2000. Relative contributions of beta-cell function and tissue insulin sensitivity to fasting and postglucose-load glycemia. Metabolism 49(10): 1318–1325.

Han, Young-Goo, Hong Joo Kim, Andrzej A. Dlugosz, David W. Ellison, Richard J. Gilbertson and Arturo Alvarez-Buylla. 2009. Dual and opposing roles of primary cilia in medulloblastoma development. Nat. Med. 15(9): 1062–1065.

Hauge-Evans, A. C., King, A. J., Carmignac, D. et al. 2009. Somatostatin secreted by islet delta-cells fulfills multiple roles as a paracrine regulator of islet function. Diabetes 58(2): 403–411.

Haumaitre, Cécile, Fabre, Mélanie, Cormier, Sarah, Baumann, Clarisse, Delezoide, Anne-Lise and Cereghini, Silvia. 2006. Severe pancreas hypoplasia and multicystic renal dysplasia in two human fetuses carrying novel HNF1β/MODY5 mutations. Human Molecular Genetics 15(15): 2363–2375.

Herrera, P. L. 2000. Adult insulin- and glucagon-producing cells differentiate from two independent cell lineages. Development 127(11): 2317–2322.

Hildebrandt, Friedhelm, Thomas Benzing and Nicholas Katsanis. 2011. Ciliopathies. New England Journal of Medicine 364(16): 1533–1543.

Horikawa, Yukio, Naoko Iwasaki, Manami Hara et al. 1997. Mutation in hepatocyte nuclear factor-1[beta] gene (TCF2) associated with MODY. Nat. Genet. 17(4): 384–385.

Huangfu, D., Liu, A., Rakeman, A. S., Murcia, N. S., Niswander, L. and Anderson, K. V. 2003. Hedgehog signalling in the mouse requires intraflagellar transport proteins. Nature 426: 83–87.

Huangfu, D. and Anderson, K. V. 2005. Cilia and Hedgehog responsiveness in the mouse. Proc. Natl. Acad. Sci. USA 102: 11325–11330.

Iwanaga, T., Miki, T. and Takahashi-Iwanaga, H. 2011. Restricted expression of somatostatin receptor 3 to primary cilia in the pancreatic islets and adenohypophysis of mice. Biomed. Res. 32(1): 73–81.

Johnson, James D., Shihuan Kuang, Stanley Misler and Kenneth S. Polonsky. 2004. Ryanodine receptors in human pancreatic β cells: localization and effects on insulin secretion. The FASEB Journal.

Kaneko, K., Ueki, K., Takahashi, N. et al. 2010. Class IA phosphatidylinositol 3-kinase in pancreatic beta cells controls insulin secretion by multiple mechanisms. Cell Metab. 12(6): 619–32.

Kawahira, Hiroshi, David W. Scheel, Stuart B. Smith, Michael S. German and Matthias Hebrok. 2005. Hedgehog signaling regulates expansion of pancreatic epithelial cells. Developmental Biology 280(1): 111–121.

Kawamori, D., Kurpad, A. J., Hu, J. et al. 2009. Insulin signaling in alpha cells modulates glucagon secretion *in vivo*. Cell Metab. 9(4): 350–61.

Kulaga, H. M., Leitch, C. C., Eichers, E. R. et al. 2004. Loss of BBS proteins causes anosmia in humans and defects in olfactory cilia structure and function in the mouse. Nat. Genet. 36(9): 994–8.

Kulkarni, R. N., Bruning, J. C., Winnay, J. N., Postic, C., Magnuson, M. A. and Kahn, C. R. 1999. Tissue-specific knockout of the insulin receptor in pancreatic beta cells creates an insulin secretory defect similar to that in type 2 diabetes. Cell 96(3): 329–39.

Landsman, Limor, Audrey Parent and Matthias Hebrok. 2011. Elevated hedgehog/gli signaling causes β-cell dedifferentiation in mice. Proceedings of the National Academy of Sciences 108(41): 17010–17015.

Larsson, H., Ahren, B. and Iannaccone, A. 2000. Islet dysfunction in insulin resistance involves impaired insulin secretion and increased glucagon secretion in postmenopausal women with impaired glucose tolerance. Diabetes Care 23: 650–657.

Ling, Z. C., Efendic, S., Wibom, R. et al. 1998. Glucose metabolism in Goto-Kakizaki rat islets. Endocrinology 139(6): 2670–2675.

Lodh, Sukanya, Timothy L. Hostelley, Carmen C. Leitch, Elizabeth A. O'Hare and Norann A. Zaghloul. 2015. Differential effects on β-cell mass by disruption of Bardet-Biedl syndrome or Alstrom syndrome genes. Human Molecular Genetics 25(1): 57–68.

Lutz, Mallory S. and Robert D. Burk. 2006. Primary cilium formation requires von hippel-lindau gene function in renal-derived cells. Cancer Research 66(14): 6903–6907.

M'Hamdi, O., Ouertani, I. and Chaabouni-Bouhamed, H. 2014. Update on the genetics of Bardet-Biedl syndrome. Molecular Syndromology 5(2): 51–56.

Maestro, Miguel A., Sylvia F. Boj, Reini F. Luco et al. 2003. Hnf6 and Tcf2 (MODY5) are linked in a gene network operating in a precursor cell domain of the embryonic pancreas. Human Molecular Genetics 12(24): 3307–3314.

Maxwell, Patrick H., Michael S. Wiesener, Gin-Wen Chang et al. 1999. The tumour suppressor protein VHL targets hypoxia-inducible factors for oxygen-dependent proteolysis. Nature 399(6733): 271–275.

Medina, Anya, Yuko Nakagawa, Jinhui Ma et al. 2014. Expression of the glucose-sensing receptor T1R3 in pancreatic islet: changes in the expression levels in various nutritional and metabolic states. Endocrine Journal 61(8): 797–805.

Mick, David U., Rachel B. Rodrigues, Ryan D. Leib et al. 2015. Proteomics of primary cilia by proximity labeling. Developmental Cell 35(4): 497–512.

Moyer, J. H., Lee-Tischler, M. J., Kwon, H. Y. et al. 1994. Candidate gene associated with a mutation causing recessive polycystic kidney disease in mice. Science 264(5163): 1329–1333.

Munger, B. L. 1958. A light and electron microscopic study of cellular differentiation in the pancreatic islets of the mouse. Am. J. Anat. 103(2): 275–311.

Nachury, M. V., Loktev, A. V., Zhang, Q. et al. 2007. A core complex of BBS proteins cooperates with the GTPase Rab8 to promote ciliary membrane biogenesis. Cell 129(6): 1201–13.

Nauli, S. M., Alenghat, F. J., Luo, Y. et al. 2003. Polycystins 1 and 2 mediate mechanosensation in the primary cilium of kidney cells. Nat. Genet. 33(2): 129–37.

Nielsen, S. K., Mollgard, K., Clement, C. A. et al. 2008. Characterization of primary cilia and hedgehog signaling during development of the human pancreas and in human pancreatic duct cancer cell lines. Dev. Dyn. 237(8): 2039–52.

Niv, Y., Turani, C., Kahan, E. and Fraser, G. M. 1997. Association between pancreatic cystadenocarcinoma, malignant liver cysts, and polycystic disease of the kidney. Gastroenterology 112(6): 2104–2107.

Nurnberger, J., Bacallao, R. L. and Phillips, C. L. 2002. Inversin forms a complex with catenins and N-cadherin in polarized epithelial cells. Mol. Biol. Cell 13(9): 3096–106.

Ohlsson, H., Karlsson, K. and Edlund, T. 1993. IPF1, a homeodomain-containing transactivator of the insulin gene. The EMBO Journal 12(11): 4251–4259.

Ostenson, C. G. and Efendic, S. 2007. Islet gene expression and function in type 2 diabetes; studies in the Goto-Kakizaki rat and humans. Diabetes Obes. Metab. 9 Suppl 2 (Suppl 2): 180–6.

Otani, K., Kulkarni, R. N., Baldwin, A. C. et al. 2004. Reduced beta-cell mass and altered glucose sensing impair insulin-secretory function in betaIRKO mice. Am. J. Physiol. Endocrinol. Metab. 286(1): E41–9.

Pan, F. C. and Brissova, M. 2014. Pancreas development in humans. Curr. Opin. Endocrinol. Diabetes Obes. 21: 77–82.

Pan, Fong Cheng and Chris Wright. 2011. Pancreas organogenesis: from bud to plexus to gland. Developmental Dynamics 240(3): 530–565.

Piccand, Julie, Perrine Strasser, David J. Hodson et al. 2014. Rfx6 maintains the functional identity of adult pancreatic β cells. Cell Reports 9(6): 2219–2232.

Pierreux, Christophe E., Aurélie V. Poll, Caroline R. Kemp et al. 2006. The transcription factor hepatocyte nuclear factor-6 controls the development of pancreatic ducts in the mouse. Gastroenterology 130(2): 532–541.

Pietrzak-Nowacka, M., Safranow, K., Byra, E., Nowosiad, M., Marchelek-Mysliwiec, M. and Ciechanowski, K. 2010. Glucose metabolism parameters during an oral glucose tolerance test in patients with autosomal dominant polycystic kidney disease. Scand J. Clin. Lab. Invest. 70(8): 561–7.

Pullen, Timothy J., Arshad M. Khan, Geraint Barton, Sarah A. Butcher, Gao Sun and Guy A. Rutter. 2010. Identification of genes selectively disallowed in the pancreatic islet. Islets 2(2): 89–95.

Puri, Sapna, Alejandro García-Núñez, Matthias Hebrok and David A. Cano. 2013. Elimination of Von Hippel-Lindau function perturbs pancreas endocrine homeostasis in mice. PLoS ONE 8(8): e72213.

Sato, Yasunori, Mukai, Munenori, Sasaki, Motoko et al. 2009. Intraductal papillary-mucinous neoplasm of the pancreas associated with polycystic liver and kidney disease. Pathology International 59(3): 201–204.

Schermer, B., Ghenoiu, C., Bartram, M. et al. 2006. The von Hippel-Lindau tumor suppressor protein controls ciliogenesis by orienting microtubule growth. J. Cell Biol. 175(4): 547–54.

Smith, Stuart, B., Qu, Hui-Qi, Taleb, Nadine et al. 2010. Rfx6 directs islet formation and insulin production in mice and humans. Nature 463(7282): 775–780.

Soyer, Josselin, Flasse, Lydie, Raffelsberger, Wolfgang et al. 2010. Rfx6 is an Ngn3-dependent winged helix transcription factor required for pancreatic islet cell development. Development (Cambridge, England) 137(2): 203–212.

Swoboda, P., Adler, H. T. and Thomas, J. H. 2000. The RFX-type transcription factor DAF-19 regulates sensory neuron cilium formation in *C. elegans*. Mol. Cell 5(3): 411–21.

Takahashi, N., Hatakeyama, H., Okado, H., Noguchi, J., Ohno, M. and Kasai, H. 2010. SNARE conformational changes that prepare vesicles for exocytosis. Cell Metab. 12(1): 19–29.

Taulman, P. D., Haycraft, C. J., Balkovetz, D. F. and Yoder, B. K. 2001. Polaris, a protein involved in left-right axis patterning, localizes to basal bodies and cilia. Mol. Biol. Cell 12(3): 589–99.

Trent, D. F., Fletcher, D. J., May, J. M., Bonner-Weir, S. and Weir, G. C. 1984. Abnormal islet and adipocyte function in young B-cell-deficient rats with near-normoglycemia. Diabetes 33(2): 170–5.

Wang, L., Coffinier, C., Thomas, M. K. et al. 2004. Selective deletion of the Hnf1 beta (MODY5) gene in beta-cells leads to altered gene expression and defective insulin release. Endocrinology 145(8): 3941–3949.

Wells, J. M. and Melton, D. A. 1999. Vertebrate endoderm development. Ann. Rev. Cell Dev. Biol. 15: 393–410.

Willmann, Stefanie, J., Nikola S. Mueller, Silvia Engert et al. The global gene expression profile of the secondary transition during pancreatic development. Mechanisms of Development.

Wong, Sunny Y., Allen D. Seol, Po-Lin So et al. 2009. Primary cilia can both mediate and suppress Hedgehog pathway-dependent tumorigenesis. Nat. Med. 15(9): 1055–1061.

Yamamoto, Kaoru, Matsuoka, Taka-aki, Kawashima, Satoshi et al. 2013. A novel function of onecut1 protein as a negative regulator of MafA gene expression. Journal of Biological Chemistry 288(30): 21648–21658.

Yang, S. N. and Berggren, P. O. 2006. The role of voltage-gated calcium channels in pancreatic beta-cell physiology and pathophysiology. Endocr. Rev. 27(6): 621–76.

Zhang, Q., Davenport, J. R., Croyle, M. J., Haycraft, C. J. and Yoder, B. K. 2005. Disruption of IFT results in both exocrine and endocrine abnormalities in the pancreas of Tg737(orpk) mutant mice. Lab. Invest. 85(1): 45–64.

CHAPTER 6

The Role of Cilia in Skeletal Development and Disease

Miriam Schmidts

INTRODUCTION

Inherited conditions resulting from ciliary malfunction affecting skeletal development in human and other mammals are often summarized under the term "ciliary chondrodysplasias", as the primary site of disease onset seems to be the cartilage including chondrocytes, rather than the bony skeleton. The exact disease frequencies are difficult to predict, as all of these conditions occur very rarely so that larger patient studies cannot be performed. For example, Jeune Syndrome (JATD) is thought to occur in about 1 in 200,000 persons (Oberklaid et al., 1977), while Ellis-van-Creveld-Syndrome (EVC) is about 5 times less common in western populations (Baujat and Le Merrer, 2007). Due to the largely autosomal recessive inheritance pattern, however, frequencies can be much higher in genetically isolated populations (Kovacs et al., 2006).

As described in more detail in the first chapter of this book, cilia are small hair-like organelles projecting from the surface of most cells in mammalian organisms. One can distinguish 2 main subgroups: motile cilia, occurring in bundles of hundreds and single, non-motile (primary) cilia. A third type, motile monocilia, is only found in the embryonic node. In contrast to the immotile single so called "primary" cilia, bundles of hundreds of motile cilia are only found in certain organs and specialised tissues such as the respiratory tract, the ependyma of brain ventricles, oviducts in females (bundles of hundreds of cilia) and the embryonic node in mammals as mentioned above.

Human genetics Department, Genome Research Division, Radboud University Medical Center and Radboud Institute for Molecular Life Sciences, Geert Grooteplein Zuid 10, 6525GA Nijmegen, The Nederlands.

The tail of male sperm in fact has a highly similar protein composition as motile cilia, however, the movement pattern is more closely related to the one observed in the flagella of prokaryotes. Each cilium contains 9 pairs of microtubules (9 + 0 structure) and motile cilia possess a 10th pair, the so-called central pair (9 + 2 structure), except for the motile cilia of the embryonic node, which also displays a 9 + 0 structure. Motile cilia also contain additional proteins required to enable microtubule sliding and therefore cilia motility that requires the presence of dynein arms and radial spokes is not found in primary cilia. Motile cilia are essential for mucociliary clearance of the airways and are thought to move fluid and particles in the brain ependymal, the female reproductive tract and the embryonic node. In the latter, this seems essential to establish a left-right body axis. If genes encoding for components of the ciliary motile apparatus are defective, this leads to a cystic fibrosis (CF)-like phenotype in humans termed Primary Ciliary Dyskinesia (PCD, OMIM244400) with main symptoms including recurrent upper and lower airways infection and laterality defects, probably due to the randomization of the left-right body axis (Fliegauf et al., 2007), in about half of the patients. Laterality defects not only occur secondarily to impaired cilia movement at the embryonic node, but also as a result of non-motile cilia dysfunction. While the exact mechanism has remained elusive, it is generally believed that disturbed cellular signalling pathways as a result of primary cilia malfunction and/or direct influence of ciliary proteins on those signalling pathways (ciliary protein functions outside of the cilium) may be the cause. In contrast to PCD, laterality defects observed in individuals with non-motile ciliopathies often occur in combination with other complex phenotypes affecting the kidneys, liver, eyes, brain and/or the skeleton. Lately, an ever-increasing number of severe human developmental genetic conditions can be linked to malfunction of primary cilia, with polycystic kidney disease (PKD, OMIM173900, OMIM613095, OMIM236200) being one of the first followed by nephronophthisis (NPHP, OMIM256100), Joubert syndrome (JS, OMIM213300), Bardet-Biedl syndrome (BBS, OMIM# 209900), Alström syndrome (ALMS, OMIM203800) and the ciliary chondrodysplasias, which will be described in this chapter (Baker and Beales, 2009).

Ciliary chondrodysplasias, also often referred to as skeletal ciliopathies, are mainly autosomal-recessive phenotypes caused by a large number of genes. While there is significant phenotypic and genetic overlap between the different conditions encompassed in this term, important clinical and radiological differences are also observed and they have been historically used to distinguish the different conditions from each other. Further, progress in gene identification over the last 5 or so years has revealed some consistent genetic associations predicting certain clinical features in the patients. Ciliary chondrodysplasias include short rib-polydactyly syndromes/short-rib thoracic dystrophy (SRPS, OMIM611263, 613091, 263520, 269860, 614091), Jeune asphyxiating thoracic dystrophy or Jeune syndrome (JATD, OMIM208500), Axial Spondylometaphyseal Dysplasia (Axial SMD, OMIM602271) Mainzer-

Saldino syndrome (MZSDS, OMIM266920), Sensenbrenner syndrome or cranio-ectodermal dysplasia (CED, OMIM218330), Orofaciodigital syndromes (OFD, OMIM31120, OMIM252100, OMIM248850, OMIM258860, OMIM174300, OMIM277170, OMIM608518, OMIM300484, OMIM258865, OMIM165590, OMIM6129, OMIM615944) and Ellis-van Creveld syndrome (EVC, OMIM225500). Exceptions from this autosomal-recessive inheritance pattern are Weyers acrodental dysostosis (OMIM193530), which is caused by heterozygous mutations in the EVC genes, as well as X-chromosomal OFD-forms. Relatively recently, ciliary chondrodysplasias with narrow thorax and polydactyly such as SRPS and JATD, have also been summarized under the term short-rib thoracic dystrophy (SRTD) in OMIM. However, this novel classification is rather unfortunate as it only takes into account the underlying genetic cause, instead of the clinical and radiological hallmarks. This results in different phenotypes caused by mutations in the same gene summarised under the same OMIM number, while the former SRPS and JATD classification was based on phenotypic features.

The ciliary chondrodysplasia group shares shortened limbs and ribs, polydactyly and sometimes craniofacial malformations such as craniosynostosis as its main clinical hallmarks. Additional extraskeletal findings include renal, liver, eye, heart and other organ involvement, however, phenotype severity varies significantly between the different conditions as well as between patients with the same condition, sometimes even amongst affected family members (Huber and Cormier-Daire, 2012). If thoracic constriction is used to define phenotype severity, SRPS can be considered the leading condition followed by JATD and axial SMD. This is reflected by the finding that the thoracic phenotype in SRPS is never compatible with life beyond the early neonatal age due to cardiorespiratory failure, while ~ 40% of affected individuals with JATD survive into adulthood, although often only after a period of intensive care and the threat of cardiopulmonary arrest during infancy (Schmidts et al., 2013a); thorax expansion and/or correction surgery is sometimes performed in JATD and axial SMD. Compared to these phenotypes, MZSDS- or CED-affected individuals usually display a significantly milder rib phenotype, not normally requiring mechanical ventilation or surgery and not associated with respiratory mortality. However, unfortunately, extraskeletal disease such as childhood onset end-stage renal disease requiring dialysis and renal transplant, is observed in nearly all MZSDS and CE patients. Further, some affected individuals also display liver disease including cysts and fibrosis and MZSDS patients also nearly always develop childhood-onset retinal degeneration (Perrault et al., 2012). Similarly, OFD patients often present with a complex phenotype involving multiple organ systems. CED and EVC further lead to additional ectodermal defects such as dysplastic fingers and toenails, and hair and teeth abnormalities (Levin et al., 1977). Finally, EVC is frequently associated with congenital heart anomalies, such as primary atrial septation defects (Blackburn and Belliveau, 1971; Hills et al., 2011). A complete overview of ciliary chondrodysplasias is shown in Table 1.

How do genetic defects in ciliary genes cause these complex phenotypes and why can defects in the same genes cause different symptom combinations in different people?

In contrast to PCD where lack of ciliary motility leads to the impaired fluid movement, disease mechanisms for non-motile ciliopathies are far less well understood. However, the skeletal phenotype including polydactyly observed in the main ciliopathies in humans and mice strongly indicates that defective hedgehog signalling could be responsible at least for some of the features observed. This chapter will describe the clinical phenotype, the genetic basis, current understanding of the underlying disease mechanisms as well as future prospects for ciliary chondrodysplasias.

Clinical Features of Ciliary Chondrodysplasias

While ciliary chondrodysplasias can be distinguished from other chondrodysplasias, especially thanks to some specific radiological skeletal signs and extraskeletal features discussed in more detail below, an exact clinical diagnosis can be difficult to make, especially in fetal cases, due to the extensive phenotypic overlap (Elcioglu and Hall, 2002). The precise clinical classification may appear arbitrary or of academic interest only, however, making a precise diagnosis often helps to predict the clinical course of the disease and is therefore of great significance to patients and their families as well as the doctors involved in clinical management. Further, identification of the causative genes not only improves the opportunities for genetic counseling and prenatal diagnosis but is also of value for the prediction of the clinical course of the disease, as certain genes seem to be associated with certain phenotypic features regardless of the clinical disease classification, e.g., IFT genes seem to be associated with a high rate of renal and retinal disease in contrast to dynein genes, which rather cause severe rib shortening in JATD, MZSDS and CED.

One easy way of clinical classification would be the recognition that SRPS are always lethal (Huber and Cormier-Daire, 2012) while generally, JATD is compatible with life and so are MZSDS, CED, OFD and EVC. However, in comparison to JATD, MZSDS and CED have a milder rib phenotype but a higher rate of renal, retinal and liver disease (Halbritter et al., 2013; Lin et al., 2013; Perrault et al., 2012; Schmidts et al., 2013b) while in EVC, severe heart defects are most frequent (Hills et al., 2011; O'Connor and Collins, 2012).

The phenotypic spectrum of perinatal lethal short rib-polydactyly syndromes

The phenotypic SRPS disease group consists of SRPS I-V, but generally JATD is considered as the mild end of the same spectrum, with reduced lethality

Table 1. Ciliary chondrodysplasia—overview.

Condition	Inhertitance	Skeletal phenotype	Renal phenotype	Retinopathy	Liver phenotype	Obesity	Developmental delay	Situs inversus	Other	Gene
Short-Rib-Polydactyly Syndromes (SRPS)	AR	Most often polydactyly, short ribs, shortened long bones, brachydactyly, abnormal pelvis configuration, sometimes orofacial clefts	Often, NPHP-like or polycystic	not evident in utero/at birth	Cysts and/or fibrosis may occur	-	na (early lethality)	rarely	Always lethal perinatal due to cardiorespiratory insufficiency resulting from severe thoracic constriction. Heart defects, gastro-intestinal, cardiac defects occur	*DYNC2H1, NEK1, WDR60, TCTN3, WDR35, DYNC2L1, KIAA0586, INTU, C2CD3, ICK, IFT52*
OFD4	AR	Variable rib shortening often polydactyly, micromelia, orofacial clefts, disproportional small and oval tibia	Cystic-dysplastic kidneys		Intrahepatic cysts, hepatic fibrosis	-	brain malformations observed	-	Lobulated tongue Coloboma, ambiguous genitalia, anal atresia, deafness have been observed	*TCTN3*
Other OFD subtypes	AR. X-chr.	Clefts, poly-, brachy dactyly	Cystic-dysplastic	Usually not	Rarely	-	yes	Usually not	Lobulated tongue, tongues hamartoma, brain malformations	*OFD1, C5ORF42, TMEM231, TMEM216, TMEM107, TBC1D32, SCLT1, C2CD3, WDPCP, DDX59*
ECO syndrome	AR	Short limbs, Poly(syn)dactyly, brachydactyly, clefts	Cystic	-	-	-	possible	-	Brain malformations, facial dysmorhism	*ICK*

Jeune Asphyxiating Thoracic Dystrophy (JATD)	AR	Short ribs, shortened long bones, polydactyly, abnormal pelvis configuration, scoliosis	< 30%, mainly NPHP-like, rarely cystic	rarely	Frequently elevated liver enzymes but rarely progression into liver failure	Single cases	-	Not described	Often severe cardiorespiratory distress with ~ 30% lethality	DYNC2H1, WDR34, WDR60, DYNC2LI1, TCTEX1D2, IFT80, IFT172, IFT140, IFT144 (WDR19), IFT139 (TTC21B), CEP120, CSPP1, KIAA0586, C21orf2
Mainzer-Saldino-Syndrome (MZSDS)	AR	(mildly) shortened ribs, cone shaped epiphyses	Always, mainly NPHP-like, rarely cystic	always	Sometimes cholestasis and hepatic fibrosis	-	Single cases	Not described	Mild thorax phenotype, usually no cardiorespiratory lethality	IFT140, IFT172
Axial SMD	AR	Short ribs, thorax deformities, hip dysplasia, scoliosis	-	-	-	-	-	-		C21orf2
Sensenbrenner syndrome (CED)	AR	(mildly) shortened ribs, brachydactyly, craniosynostosis leading to dolichocephalus	Very often, mainly NPHP-like	sometimes	Inconsistent hepatic cysts and fibrosis/ hepatic ductal plate malformation	-	Sometimes	Usually not	Facial dysmorphism with telecanthus/ epicanthus, high forehead, broad nasal bridge, low-set prominent ears, Thin and sparse growing hair, nail dysplasia, teeth abnormalities, heart defects. Usually mild thorax phenotype, usually no cardiorespiratory lethality	IFT122, IFT144 (WDR19), IFT43, WDR35

Table 1 contd....

...Table 1 contd.

Condition	Inheritance	Skeletal phenotype	Renal phenotype	Retinopathy	Liver phenotype	Obesity	Developmental delay	Situs inversus	Other	Gene
Ellis-van Creveld Syndrome (EVC)	AR	Short ribs and long bones, abnormal pelvis configuration, polydactyly of the hands	-	-	-	-	-	-	Hypoplastic nails, teeth abnormalities, heart defects	*EVC1, EVC2*
Weyer's acrofacial dysostosis	AD	Short stature, short extremities, Polydactyly of the hands	-	-	-	-	-	-	Hypoplastic nails, teeth abnormalities	*EVC1, EVC2*
Hydrolethalus Syndrome	AR	Polydactyly, clefts, microganthia	-	-	-	-	na		Midline defects, complex brain malformations	*KIF7*
Acrocallosal Syndrome	AR	Polydactyly	-	-	-	-	yes		Brain malformations, often absent corpus callosum, midline defects	*KIF7*

AR: autosomal recessive; AD: autosomal-dominant; X-chr:: X-chromosomal; na: not applicable.

due to milder rib shortening and rare polydactyly, which in contrast is a consistent feature of SRPS. Also extraskeletal organ involvement such as malformations of the brain, heart, kidneys, liver and pancreas frequently observed in SRPS in prenatal ultrasound examinations, are less common in JATD. Five subtypes of SRPS are historically phenotypically distinguished: SRPS-I (Saldino-Noonan type, OMIM613091) (Saldino and Noonan, 1972); SRPS-II (Majewski type; OMIM263520) (Majewski et al., 1971), SRPS-III (Verma-Naumoff type, OMIM613091) (Verma et al., 1975), SRPS-IV (Beemer-Langer type, OMIM269860) (Beemer et al., 1983) and SRPS-V (OMIM614091) (Mill et al., 2011).

SRPS-I and SRPS-III

SRPS-I and SRPS-III are often grouped together due to their overlapping clinical features. However, historically these two phenotypes have been described as different entities. Differences include the more pronounced limb shortening in SRPS I with "flipper" like extremities and similar to SRPS-II and -IV, polydactyly, hydropic appearance and a small thorax with short horizontal ribs causing cardiorespiratory lethality, is observed in SRPS-I (Saldino and Noonan, 1972). Ossification defects of the vertebrae,

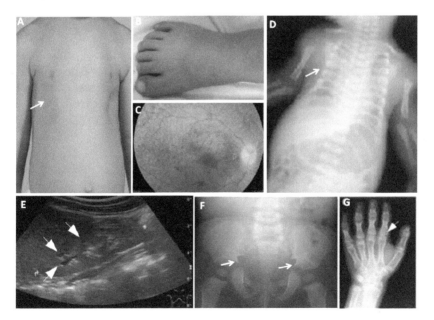

Figure 1. The most common clinical and radiological findings in skeletal dysplasias. (A) Narrow ribcage due to shortened ribs and thorax deformity. (B) Polydactyly. (C) Fundus of a patient with retinal degeneration. (D) Shortened horizontal ribs and thorax narrowing, as well as handlebar clavicles visualised by radiography. (E) Increased renal echogeneity as a hallmark of nephronophthisis and renal cysts visualised by ultrasound. (F) Dysplastic pelvis with acetabular spurs. (G) Cone shaped epiphyses of the hands.

pelvis, as well as of the bones of the hands and feet are also frequently observed. Pelvis anomalies include small ilia with flattened acetabular roofs with ossified spurs, resembling the pelvis configuration in EVC and JATD. Polycystic kidneys, transposition of the great vessels as well as atretic lesions of the gastrointestinal and genitourinary systems are extraskeletal features described with both subtypes (Le Marec et al., 1973; Spranger et al., 1974a; Spranger et al., 1974b).

In addition to the classic SRPS phenotypic features such as hydropic appearance at birth, short long bones, short horizontal ribs, narrow thorax and protuberant abdomen, SRPS-III is additionally characterised by a short cranial base, bulging forehead, depressed nasal bridge and flat occiput (Verma et al., 1975). Radiologically, the long tubular bones show a corticomedullary demarcation, slightly widened metaphyses and marked longitudinal spurs. Similar to SRPS-I, the pelvis radiographs in SRPS-III resemble radiographs in EVC and JATD with spurs present. Polydactyly often occurs but is not found in all cases of SRPS-I and III. Cleft lip and palate are reported as well as malformations of the larynx and epiglottis (Bernstein et al., 1985; Naumoff et al., 1977). In contrast to SRPS-I, where developmental urogenital defects seem common, these are rarely observed in SRPS-III (Yang et al., 1987).

SRPS-II, SRPS-IV and OFD-IV

In contrast to other SRPS forms, SRPS-IV does not usually result in polydactyly, while the remaining clinical picture in SRPS-II and SRPS-IV is quite similar with hydropic appearance at birth, short long bones, small and narrow thorax with horizontal ribs and protuberant abdomen (Majewski et al., 1971). SRPS II and IV are mainly distinguished by the disproportionally shortened tibia observed in SRPS-II (the tibia is shorter than the fibula) but not SRPS-IV (Spranger et al., 1974b). However, dysplastic tibia can also be found in OFD4 (Mohr-Majewski syndrome, Baraitser-Burn Syndrome, OMIM258860) and cause problems in distinguishing these two conditions, especially prenatally. OFD4 affected individuals also present with variable thoracic constriction, pre- and post-axial polydactyly of the hands and feet, cleft and pseudo-cleft of the lip and/or palate, lobulated tongue, cystic dysplastic kidney and liver involvement, brain malformations (Baraitser, 1986), severe bilateral deafness (Nevin and Thomas, 1989) and coloboma of the eye (Ades et al., 1994). Orofacial clefts, cerebral malformations and renal involvement are also observed in SRPS-II (Chen et al., 1980; Spranger et al., 1974b) and SRPS-IV (Beemer et al., 1983; Kovacs et al., 2006; Passarge, 1983; Yang et al., 1991).

SRPS-V

SRPS-V has only been described relatively recently, including the identification of the underlying genetic defect in two foetuses from a Maori family from

New Zealand. The affected pregnancies presented with hydrops fetalis, narrow chest and severely shortened and bowed long bones displaying lack of ossification, hypoplastic scapulae and peritoneal calcifications, postaxial poly-syndactyly and cleft palate. Extraskeletal organ involvement such as urogenital malformations (bilateral cystic hygroma, hypospadia, glomerular kidney cysts) and intestinal malrotation were also noted. Acromesomelic hypomineralisation and campomelia distinguish SRPS-V from SRPS subtypes I-III (Kannu et al., 2007; Mill et al., 2011).

Other perinatally lethal OFD subtypes

To date, at least 14 OFD subtypes caused by 12 genes have been distinguished based on different clinical and/or genetic findings, with OFD-I being most frequently diagnosed. In contrast to most other conditions described in this chapter, OFD-I is inherited in an X-linked fashion and is lethal in males (Jorgenson, 1971; Segni et al., 1970; Yeamans, 1973). While apart from OFD-IV, rib shortening is usually not a feature in other OFD forms. Limb shortening, brachydactyly, orofacial clefts and fibular agenesis have been reported for OFD-X and polydactyly can occur in several OFD-subtypes (Franco and Thauvin-Robinet, 2016). Therefore, despite not usually been classified as ciliary chondrodysplasias, OFDs are briefly mentioned in this chapter. However, apart from *OFD-1* mutation carrying males and some individuals with *C2CD3* mutations, OFD is not usually lethal perinatally, but rather leads to complex developmental disorders that become evident in early childhood (Cortes et al., 2016; Franco and Thauvin-Robinet, 2016; Ruess et al., 1965).

Hydrolethalus syndrome

Hydrolethalus Syndrome 2 (HLS2, OMIM# 200990) is allelic to Acrocallosal Syndrome (ACLS), as well as Joubert Syndrome caused by KIF7 defects and may therefore be considered the most severe end of this spectrum of conditions. The syndrome is characterized by polydactyly with or without hallux duplication, sometimes cleft palate and micrognathia, in combination with very severe brain malformations such as anencephaly, hydrocephalus and cerebellar vermis hypoplasia, incompatible with post-natal live (Putoux et al., 2011).

Endocrine-cerebro-osteodysplasia syndrome (ECO syndrome, OMIM#612651)

This rare condition was first described by Lahiry et al. within the Amish community (Lahiry et al., 2009) and manifests with the main hallmarks of endocrine, cerebral (mainly hydrocephalus but also holoprosencephaly, absent corpus callosum and cortical malformations) as well as skeletal abnormalities (facial clefting, short extremities, bowed forearms, ulnar deviation, poly(syn)dactyly, brachydactyly). Facial dysmorphism can also be observed and includes braid nasal bridge, dysplastic ears and fused eyelids. Extraskeletal findings include renal cysts. The ECO syndrome seems somewhat allelic to SRPS as in both conditions, *Intestinal Cell Kinase* (ICK) mutations have been identified (Oud et al., 2016; Paige Taylor et al., 2016).

The Phenotypic Spectrum of Childhood Chondrodysplasias

A number of ciliary chondrodysplasias display some degree of rib shortening and show significant overlap with SRPS but are less severe and therefore result in the survival of affected individuals during childhood. Recently, these conditions together with SRPS were summarized under the term "short-rib thoracic dystrophy" (SRTD), including Jeune asphyxiating thoracic dystrophy (JATD, OMIM208500), Mainzer-Saldino syndrome (MZSDS, OMIM266920), Sensenbrenner syndrome (CED, OMIM218330), Ellis-van Creveld syndrome (EVC, OMIM225500). Weyers acrodental dysostosis (OMIM193530) and orofaciodigital syndromes in contrast do not usually affect the thorax.

Jeune asphyxiating thoracic dystrophy (JATD)

The main characteristics of JATD include a small or narrow, sometimes bell-shaped, thorax with shortened, more horizontally orientated shortening ribs. It is thought that ribcage abnormalities restrict pulmonary development and that this causes the severe respiratory distress sometimes observed during the first 2 years of life. However, there might also be a certain element of pulmonary dysplasia contributing independently of the rib shortening. Compared to SRPS, the thorax phenotype is less severe and compatible with life in the majority of the cases. Nevertheless, respiratory problems account for most of the mortality in JATD, ranging from 20–60% depending on the study (Baujat et al., 2013; Oberklaid et al., 1977; Schmidts et al., 2013a). Nevertheless, patients seem to somewhat "grow out" of the respiratory phenotype, meaning it becomes less pronounced from childhood to adolescence (Baujat et al., 2013; de Vries et al., 2010; Schmidts et al., 2013a). In accordance with this, fatalities are mainly observed during the first 2 years of life and are associated with respiratory decomposition during acute airway infections. In addition to shortened ribs and extremities, other skeletal hallmarks include pelvis abnormalities similar to those observed in SRPS-I, -III and EVC with trident acetabulum with spurs (so called "handlebar clavicles") and cone-shaped epiphyses of fingers and toes, and occasionally polydactyly. Importantly, while the pelvic changes are usually only observed during the first year of life, cone-shaped epiphyses are noticed later on. NPHP-like (and rarely cystic) kidney involvement is observed in about 1/5th of all individuals with JATD and usually results in end-stage renal disease requiring dialysis and/or renal transplant. This phenotype seems to depend mainly on the underlying genetic defects, with mutations in *IFT140* and *IFT172* predicting this phenotype with close to 100% probability (Halbritter et al., 2013; Perrault et al., 2012; Schmidts et al., 2013b). Patients with mutations in one of these two genes also usually develop retinal degeneration resulting in blindness, a phenotype which otherwise has been described as "rare" in literature (Allen et al., 1979; Bard et al., 1978). However, Baujat et al. observed electroretinogram (ERG) abnormalities in up to 50% of JATD patients with *DYNC2H1* mutations (Baujat et al., 2013) with no vision impairment present at the time of examination. It is not clear, however, how many of these patients will develop clinically relevant retinal disease in the future. Pancreatic lesions (Hopper et al., 1979), elevated liver enzymes (Halbritter et al., 2013; Yerian et al., 2003) and brain malformations (Halbritter et al., 2013; Lehman et al., 2010; Tuz et al., 2014) have also been described in JATD. The latter are rare in JATD, but common in Joubert syndrome and can be imaged as a so-called molar tooth sign on MRI, due to cerebellar vermis hypoplasia.

Axial spondylometaphyseal dysplasia (SMD, OMIM# 602271)

This extremely rare condition was first described in 3 Asian children by Ehara et al., 1997 (Ehara et al., 1997) and for a long time, it was not recognised as a ciliopathy. The phenotypic features resemble those observed in Jeune Syndrome and include a small thorax, short extremities, short stature and pelvis dysplasia. However, the anterior ends of the ribs appear cupped and there are irregular proximal femur epiphyses as well as malformed vertebrae (platyspondyly), which are usually not noted in Jeune Syndrome. Further, although the pelvis' configuration appears dysplastic, in contrast to Jeune Syndrome where a trident configuration with acetabular spurs is very common, in SMD the iliac crest appears lacy. Interestingly, affected individuals develop retinal degeneration (retinitis pigmentosa, cone-rod dystrophy) similar to what is observed in some Jeune Syndrome cases, but renal or hepatic involvement has not been observed. Additional facultative symptoms include frontal bossing and craniofacial dysmorphism (Isidor et al., 2010).

It took 19 years from the description of the first syndrome to the elucidation of the genetic basis of SMD, which was recently identified as dysfunctional *C21orf2* by Wang et al., 2016 (Wang et al., 2016). 6 of the 9 investigated families carried biallelic *C21orf2* mutations and based on the localisation of the protein to the connecting cilium of photoreceptors, the authors suggested for the first time that SMD is in fact a ciliopathy. Interestingly, shortly before their publication, *C21orf2* was identified as the causative gene in a subset of Jeune Syndrome patients with rather mild thorax phenotype, but early onset retinal degeneration without renal function impairment (Wheway et al., 2015).

Mainzer-saldino syndrome (MZSDS)

Together with JATD, MZSDS falls under the disease term "cono-renal syndrome", arising from the fact that radiographs show cone-shaped epiphyses of the fingers and toes. By definition, MZSDS is a combination of skeletal symptoms and impaired renal function as well as retinal disease. In addition to the typical epiphyseal changes, patients exhibit a narrower than normal ribcage, however, unlike many JATD patients, they do not usually require ventilation assistance (Mainzer et al., 1970; Popovic-Rolovic et al., 1976; Robins et al., 1976; Spranger et al., 1974b). MZSDS and JATD not only display overlapping phenotypic features, but also result from mutations in the same genes. Therefore these two conditions can be considered truly allelic, making these two disorders truly allelic (Halbritter et al., 2013; Perrault et al., 2012; Schmidts et al., 2013b). While all MZSDS individuals progress to end-stage renal disease and manifest vision loss by late adolescence, the onset and speed of progression of renal and retinal disease is extremely variable and therefore hard to predict for individual cases. In addition, the diagnosis of retinal involvement is based on different tests (Electroretinogram (ERG), fundoscopy) for different patients and these tests display variable sensitivity: fundoscopy may appear normal in the case of retinal degeneration without pigmentary deposits and it is also difficult to obtain reliable ERG results in non-cooperative young children. So although accurate conclusions cannot be drawn at the moment, the childhood onset of renal and retinal disease seems common in both JATD and MZSDS patients (Halbritter et al., 2013; Perrault et al., 2012; Schmidts et al., 2013b). Further, liver disease, sometimes with cholestasis as well as liver fibrosis, has been described for some MZSDS cases (Halbritter et al., 2013; Perrault et al., 2012). Lastly, some of the originally described MZSDS cases were found to display ataxia (Giedion, 1979; Mainzer et al., 1970), however, subsequent studies failed to confirm that this is a common disease presentation (Halbritter et al., 2013; Perrault et al., 2012).

Sensenbrenner syndrome (Chondroectodermal dysplasia, CED)

CED can be distinguished from other ciliary chondrodysplasias such as JATD and MZSDS by the presence of characteristic facial dysmorphic features (dolichocephalus resulting from craniosynostosis of the sutura sagittalis (Amar et al., 1997; Arts and Knoers, 1993), downwards pointing palpebral fissures, small mouth, lowset ears, to name a few) and ectodermal defects such as nail, hair and teeth abnormalities as well as skin laxity, sometimes resulting in the development of hernias. The skeletal phenotype involves shortening of the extremities (rhizomelic micromelia and brachydactyly) and variable narrowing of the ribcage, similar to what is seen in MZSDS and therefore milder than what is observed in JATD cases (Levin et al., 1977; Lin et al., 2013). Extraskeletal organ involvement includes progressive renal failure due to a nephronophthisis-like renal phenotype with small hyperechogenic kidneys, the histological picture of tubulointerstitial nephritis and microscopic glomerular and tubular cysts (Eke et al., 1996; Lang and Young, 1991), liver cysts, liver fibrosis and hepatic ductal plate malformations (Konstantinidou et al., 2009; Walczak-Sztulpa et al., 2010; Zaffanello et al., 2006) and less frequently congenital heart defects or retinal degeneration (Lin et al., 2013).

Ellis-van creveld syndrome (EVC)

EVC was first described about 75 years ago by Ellis and Creveld (Ellis and van Creveld, 1940), and found to be fairly common (in relation to the overall rarity of the condition) in the consanguineous Amish population in Pennsylvania in 1964 (McKusick et al., 1964). EVC is mainly characterized by acromelic dwarfism, polydactyly of the hands (but not feet), dysplastic fingers and toenails and teeth abnormalities (prenatal eruption of teeth, hypodontia and malformed teeth) and in many cases cardiac defects, such as primarily atrial septation defects (Blackburn and Belliveau, 1971; McKusick et al., 1964), affecting approximately half of all patients. Hydrocephalus due to a Dandy-Walker malformation has also been described (Zangwill et al., 1988), but its overall frequency is probably low. Radiologically, EVC resembles JATD and SRPS I and III especially in fetal and neonatal cases, with the pelvis displaying a trident acetabulum with spurs in combination with brachydactyly and a narrow thorax. The diagnosis of the condition can be based on cardiac defects as well as ectodermal defects which are not a feature of JATD, while older JATD patients may rather present with additional renal-liver and retinal disease that is not usually observed in EVC (Baujat and Le Merrer, 2007).

Weyers acrodental dysostosis

Weyers acrofacial dysostosis (also referred to as Curry-Hall) is allelic to EVC, although inherited in an autosomal dominant manner (Weyers, 1952). Affected individuals display polydactyly of the hands, teeth abnormalities such as abnormal shape and numbers of teeth, dystrophic fingers and toe nails as well as short stature with short extremities. In contrast to EVC, there is usually no ribcage or visceral organ involvement observed (Curry and Hall, 1979; Roubicek and Spranger, 1984, 1985; Weyers, 1952).

Acrocallosal syndrome (ACLS)

ACLS (OMIM# 200990) is not usually listed as a ciliary chondrodysplasia and dysplastic cartilage may not play a role in the phenotype expression in the proper sense, however, as the clinical hallmarks include craniofacial dysmorphism with a prominent forehead and hypertelorism as well as polydactyly/hallux duplication and further, *KIF7* that is important for the organisation of microtubules at the ciliary tip is found to be defective in affected individuals, the disease definitely fulfils the criteria of the ciliary skeletal syndromes (Putoux et al., 2011). Its main extraskeletal features include developmental delay and brain malformations, such as severe hypoplasia or agenesis of the corpus callosum with or without Dandy-Walker malformation (Schinzel and Schmid, 1980). ACLS is allelic to Joubert Syndrome, as well as Hydrolethalus Syndrome (HLS-2).

The Genetic Basis of Ciliary Chondrodysplasias/Skeletal Ciliopathies

Ciliary chondrodysplasias are largely inherited in an autosomal recessive manner apart from Weyers acrodental dysostosis, which is autosomal dominant and the X-linked OFD subtypes (OFD-I OFD-VII and OFD-VIII (Franco and Thauvin-Robinet, 2016)). Like in other ciliopathies, amazing genetic heterogeneity has been noted in ciliary chondrodysplasias. This is especially true for JATD and CED and genetic overlap is observed between the different SRPS types as well as between SRPS, JATD, MZSDS, OFD and CED (Table 1). This has hampered genetic diagnosis and gene discovery in the past, however, the introduction of next-generation sequencing (NGS) has revolutionised patient genotyping and has led to huge advances in research, in the form of gene panel sequencing and whole exome and genome sequencing. As a result, over the last 5 years, a large number of novel human disease causing genes have been identified as causative of skeletal ciliopathies, many of which encode intraflagellar transport (IFT) proteins (*IFT172* (Halbritter et al., 2013), *IFT52* (Zhang et al., 2016) *WDR19* (Bredrup et al., 2011), *TTC21B* (Davis et al., 2011a), *IFT140* (Perrault et al., 2012; Schmidts et al., 2013b), *IFT122* (Walczak-Sztulpa et al., 2010), *IFT80* (Beales et al., 2007; Cavalcanti et al., 2011), *IFT43* (Arts et al., 2011), and *WDR35* (Gilissen et al., 2010; Mill et al., 2011) or components of the retrograde IFT motor complex dynein-2 (*DYNC2H1*) (Dagoneau et al., 2009; El Hokayem et al., 2012; Merrill et al., 2009), *WDR34* (Huber et al., 2013; Krock et al., 2009; Rompolas et al., 2007; Schmidts et al., 2013c), *WDR60* (McInerney-Leo et al., 2013; Patel-King et al., 2013), *DYNC2LI1* (Kessler et al., 2015; Taylor et al., 2015) and *TCTEX1D21* (Schmidts et al., 2015)). Further, proteins localising to the base of the cilium (basal body, peri-basal body region or peri-centriolar matrix) seem to play a role in proper cilium function, as indicated by mutations in genes encoding for NEK1 (El Hokayem et al., 2012; Thiel et al., 2011), *CSPP1* (Tuz et al., 2014), *WDPCP*, *INTU* (Aguilar, 2016; Saari et al., 2015), *C2CD3* (Cortes et al., 2016; Thauvin-Robinet et al., 2014), *TBC1D32* (Adly et al., 2014; Ko et al., 2010), *EVC/EVC2* (Ruiz-Perez et al., 2000; Tompson et al., 2007), *KIAA0586* (Alby et al., 2015; Malicdan et al., 2015; Stephen et al., 2015; Yin et al., 2009) and *TMEM216* (Valente et al., 2010). *TCTN3*, *C5orf42*, *TMEM231* and *TMEM107* were found to be mutated in OFD and localise to the ciliary transition zone (Thomas et al., 2012; Lopez et al., 2014; Romani et al., 2015; Roberson et al., 2015; Shylo et al., 2016), while DDX59, also causing OFD when defective, is found in the cytoplasm and nucleus and therefore its relation to cilia is unclear. However, OFD is considered to represent a ciliopathy as proteins encoded by other OFD genes have a clear relationship to the cilium (Shamseldin et al., 2013). ICK, found to be causative of ECO and SRPS syndromes, localizes to the ciliary axoneme and the ciliary base (Chaya et al., 2014; Oud et al., 2016; Paige Taylor et al., 2016). The proportion of human ciliary chondrodysplasias caused by mutations in known disease-causing genes varies greatly between the different conditions: while the vast majority of EVC cases occurs due to mutations in *EVC* and *EVC2* (D'Asdia et al., 2013) and causative mutations in *NEK1* and *DYNC2H1* are found in over 2/3 of all SRPS-II cases (El Hokayem et al., 2012), the majority of all MZSDS cases are caused by IFT140 (IFT-A subcomplex) and *IFT172* (IFT-B subcomplex) mutations (Halbritter et al., 2013; Perrault et al., 2012). Interestingly, the main cause of axial SMD seems to be a single defective gene, *C21orf2* (Wang et al., 2016), and mutations in this gene have also been identified in rare cases of JATD. 50% of OFD4 cases appears to result from mutations in *TCTN3* (Thomas et al., 2012) and 50% of JATD cases are due to *DYNC2H1* mutations (Baujat et al., 2013; Schmidts et al., 2013a). However, JATD is genetically more heterogeneous, with 14 known genes accounting for more than 70% of all cases. In addition to *DYNC2H1* mutations, mutations in other dynein-2 complex genes such as *WDR34* (approximately 10%) (Huber et al., 2013; Schmidts et al., 2013c), *WDR60*, *DYNC2LI1* and *TCTEX1D2* (all rare) cause JATD and SRPS. Similarly, multiple genes encoding for IFT-A complex components have been found to cause CED and the frequency of *WDR19*, *IFT122* and *WDR35* mutations seems fairly equally distributed, whereas mutations in *IFT43* are less commonly identified. The total number of cases reported to date, however, is too small to draw final conclusions in this respect (Lin et al., 2013). Likewise, a fairly vast number of genes have been identified to cause OFD when mutated, most of which encode proteins localising to the base of the cilium, with OFD1 representing the most commonly mutated gene. For several rare OFD subtypes, the genetic cause remains elusive to date.

This mutational data suggests that JATD and MZSDS, JATD and Axial Spondylometaphyseal Dysplasia as well as JATD and SRPS-II/III and are indeed allelic diseases. Further, genetic overlap has been observed between JATD and CED (*IFT144/WDR19*) as well as JATD and OFD (*CEP120*). However, due to the low number of reported JATD cases carrying mutations in these genes, the degree to which these genes cause distinct phenotypes or whether this could rather be an issue of clinical-phenotypical categorisation remains unclear.

Genotype-phenotype correlations in JATD, MZSDS and CED further suggest that mutations in IFT complex A components lead to a rather mild rib phenotype compared with dynein-2 complex mutations, but often cause significant extraskeletal symptoms such as renal, hepatic and retinal disease (Halbritter et al., 2013; Lin et al., 2013; Perrault et al., 2012; Schmidts et al., 2013b). In contrast, *DYNC2H1* mutations (and other dynein-2 complex mutations, except mutations in *TCTEX1D2*) are associated with a very severe thorax phenotype and an infrequent clinically relevant renal or eye involvement in affected children (Baujat et al., 2013; Schmidts et al., 2013a).

Interestingly, *C21orf2* mutations have been recently identified in both JATD with a mild rib phenotype, scoliosis and retinal degeneration (but in contrast to IFT-A mutations, no renal disease manifests) (Wheway et al., 2015) and axial SMD (Wang et al., 2016), where vertebral defects are observed in addition to mild JATD-like rib shortening combined with retinal degeneration. This raises the question of the degree to which these are clinically different conditions or whether axial SMD could be considered to be a specific JATD phenotype. Likewise, mutations in *CEP120* have been reported to cause a JATD phenotype with oral clefts (Shaheen et al., 2015) as well as OFD-like phenotypes and Joubert Syndrome (Roosing et al., 2016). One could thus conclude that the reported JATD cases may in fact represent a specific sub-category of the OFD spectrum, rather than that *CEP120* mutations cause classical JATD and OFD.

While many ciliopathies are primarily autosomal recessive (or more rarely autosomal dominant) disorders, the influence of additional "third" alleles on phenotype expression has been contemplated for many years. This was first suggested for Bardet-Biedl-Syndrome (BBS) by Katsanis et al. (Davis et al., 2011b; Katsanis et al., 2001) who termed it "triallelic inheritance" and has also been suggested for retinal disease (Khanna et al., 2009), as well as polycystic kidney disease (Bergmann et al., 2011). Without doubt, the genetic background will influence the phenotype in humans, similar to what has been demonstrated in other mammals such as mice. However, proving that a certain allele has a specific effect is difficult to achieve, as large scale modifier studies would be needed which in contrast to mice, cannot be performed in humans. For ciliary chondrodysplasias, the so-called "triallelic inheritance" has not been described and digenic inheritance of heterozygous mutations in two different genes is most likely very rare, since this has only been shown in a single case of SRPS-II with a heterozygous mutation in *NEK1* and *DYNC2H1* (Thiel et al., 2011). However, there are indications that the "mutational load" (the total number of mutations in genes known to cause the disease), could potentially play a role in ciliary dysplasias (Schmidts et al., 2013b). Further, it seems possible that protective modifier alleles may exist (at least regarding dynein-2 light chain mutations), as for example a nonsense *TCTEX1D2* mutation exhibits incomplete penetrance possibly due to functional redundancy from other dynein-2 light chains (Schmidts et al., 2015).

Molecular Mechanisms

Ciliogenesis and IFT

Despite long-lasting efforts over several decades now, how defects in genes encoding for ciliary proteins result in the complex phenotypes of the so-called ciliopathies has remained astonishingly unclear. Especially, we still don't understand why mutations in certain genes give rise to only certain phenotypes, when nearly all the cells in the human body exhibit a cilium and therefore ciliary protein expression is rather ubiquitous. This is also the case for the ciliary chondrodysplasias, where many of the genes implicated encode for proteins involved in intraflagellar transport (IFT), the process essential for the building and maintenance of cilia and for many ciliary and cellular signalling pathways. Many genes found to be defective in ciliary chondrodysplasias encode for components of the IFT complex A (retrograde transport from the ciliary tip back to the base) (Arts et al., 2011; Bredrup et al., 2011; Davis et al., 2011a; Gilissen et al., 2010; Mill et al., 2011; Perrault et al., 2012; Schmidts et al., 2013b; Walczak-Sztulpa et al., 2010), but likewise, components of IFT complex B (anterograde transport form the ciliary base to the tip) (Beales et al., 2007; Halbritter et al., 2013) as well as of the retrograde motor complex cytoplasmic dynein-2 powering retrograde IFT (Dagoneau et al., 2009; Huber et al., 2013; Kessler et al., 2015; McInerney-Leo et al., 2013; Merrill et al., 2009; Schmidts et al., 2015; Schmidts et al., 2013c; Taylor et al., 2015) have been shown to play a role.

In addition to defects in genes encoding for IFT components or the retrograde dynein-2 motor complex, a large number of genes encoding for proteins localising to the base of the cilium and/or to centrioles have been found to be defective in skeletal ciliopathies. These include *EVC* and *EVC2* (Ruiz-Perez et al., 2007), *NEK1* (Thiel et al., 2011), a serine-threonine kinase thought to be involved in cell-cycle-dependent control of ciliogenesis, the NEK1 interactor *C21orf2* which could potentially be involved in the regulation of NEK1 phosphorylation (Wheway et al., 2015), *CSPP1* (Akizu et al., 2014; Shaheen et al., 2014; Tuz et al., 2014) and *CEP120* (Mahjoub et al., 2010; Roosing et al., 2016; Shaheen et al., 2015). KIAA0586/TALPID3 also localises to the distal ends of centrioles and is thought to be involved in the initiation of ciliogenesis (Stephen et al., 2015). In contrast, KIF7 is found at the ciliary tip, where it is thought to control ciliary length by binding to the microtubule plus ends and reducing the microtubule growth rate (He et al., 2014), while TCTN3 is found in a complex with TCTN1 and TCTN2 at the ciliary transition zone (Thomas et al., 2012) (Figure 2).

Building and maintaining a cilium requires protein synthesis within the cytosol of cells and subsequent transport of those proteins to and within the cilium to the exact site where they are needed. Likewise, components have to be transported back from the ciliary axoneme to the base of the cilium

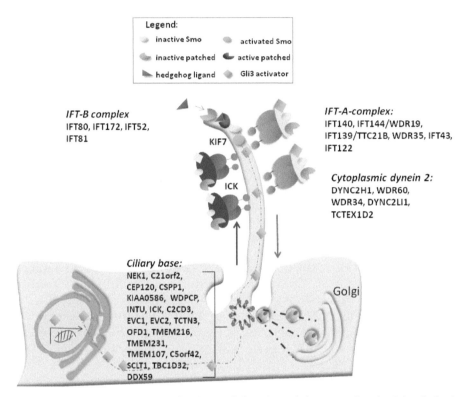

Figure 2. Subcellular/ciliary localisation and function of the genes involved in skeletal ciliopathies and a simplified schematic of ciliary regulation of hedgehog signalling. Vesicular transport brings ciliary proteins to the cilium and vesicle membranes probably merge with the ciliary membrane in the ciliary pocket area. Anterograde intraflagellar transport (IFT) from the ciliary base to the tip is facilitated by the IFT complex B, while retrograde transport back to the base is enabled by the IFT complex A. However, IFT-complex A components are also transported in the anterograde direction and IFT-B components in the retrograde direction. While many genes defective in skeletal ciliopathies encode for the proteins of complex A and the associated motor complex, cytoplasmic dynein-2 or localise to the ciliary base, a minority also localise to the transition zone or the ciliary axoneme. No human mutations have been identified in the motor for complex B, kinesin-2 so far. Hedgehog (Hh) signalling is essential for proper skeletal development by maintaining a balance between proliferation and cellular differentiation at the growth plates and creating a gradient at the limb bud. Several pathway components such as Smoothened, Patched and Gli proteins localise to the cilium. Activated Gli proteins finally translocate from the cilium to the cell body and to the nucleus where they influence gene expression. Adapted from Schmidts 2014 (Schmidts, 2014).

and the cytosol if they are no longer required or in the case of certain cell signalling pathways, if signalling components require translocation, e.g., Gli translocation to the nucleus. To achieve this, proteins are loaded onto IFT particles for transport along the ciliary axoneme towards the ciliary tip (anterograde IFT, complex B, powered by kinesin motor complexes) as well from the ciliary tip back to the base (retrograde IFT, complex A, powered by the dynein-2 complex) (Rosenbaum and Witman, 2002; Cole and Snell, 2009; Hao et al., 2011). Dysfunctional IFT therefore results in defects of the ciliary architecture with absent or very short cilia in extreme cases (Fliegauf et al., 2007; Ocbina et al., 2011). Human patients are considered to represent "hypomorphs" rather than true "nulls" and likewise, mild forms of such ciliary defects have been described in patients' fibroblasts with mutations in *DYNC2H1, NEK1, WDR19* and *WDR60* (Merrill et al., 2009; Bredrup et al., 2011; McInerney-Leo et al., 2013; Thiel et al., 2011). However, in other patients with mutations in genes encoding for retrograde IFT proteins or the retrograde motor protein DYNC2H1, cilia seem to be present on skin fibroblasts in normal numbers and reach normal length, but disturbance of retrograde IFT with accumulation of IFT-B components at the ciliary tip can be observed (Arts et al., 2011; Schmidts et al., 2013a; Rix et al., 2011; Jonassen et al., 2012; Moore et al., 2013). Further, mutations in *IFT172* impairing anterograde transport, result in reduced incorporation of ciliary proteins, such as adenylate cyclase III and IFT molecules, into the ciliary axoneme (Halbritter et al., 2013).

Retinal degeneration in skeletal ciliopathies

Retinal photoreceptor cells do not exhibit a typical cilium but instead possess a structure named the "connecting cilium" bridging two parts crucial for light processing: the inner and outer segment of the actual cell. The connecting cilium hereby serves as a bridge and although somewhat different in architecture compared to a classic cilium, the protein composition is similar to the transition zone composition in monocilia. IFT-dependent transport via the connecting cilium enables protein exchange between the inner and outer segment, especially the transport of rhodopsin and so it is not surprising that mutations in IFT genes very often result in the accumulation of rhodopsin, causing photoreceptor death (Krock et al., 2009; Marszalek et al., 2000; Pazour et al., 2002; Pazour and Rosenbaum, 2002). This associated retinal phenotype can therefore be considered degenerative rather than developmental in its origin, in contrast to the primarily developmental skeletal phenotype resulting from the same mutations. Interestingly, similar to the renal phenotype observed in humans with IFT gene mutations (Halbritter et al., 2013; Lin et al., 2013; Perrault et al., 2012; Schmidts et al., 2013b; Zhang et al., 2016), retinal degeneration is not usually associated with *DYNC2H1* mutations, despite the fact that DYNC2H1 represents an IFT motor component (Baujat et al., 2013; Schmidts et al., 2013a). The reason for this discrepancy has remained elusive to date, but is a likely cause of additional IFT gene function unconnected to the cilium, such as cell signalling pathway influence or may be due to the loss of specific cargo, compared to slower but intact IFT in the case of dynein mutations.

Disturbances in cell signalling pathways in skeletal ciliopathies

It is widely accepted that the complex developmental phenotypes observed within the skeletal ciliopathy spectrum are a result of defective cellular signalling. The cilium can be imagined as an "antenna" transmitting signals from outside the cell to the inside but at the same time, it is also critically involved in processing signals coming from the cell body or nucleus (Ishikawa and Marshall, 2011; Christensen et al., 2007; Veland et al., 2009; Satir et al., 2010). Disturbed ciliary architecture and dysfunctional IFT cause the accumulation or lack of signal transduction particles at the site of the cilium and within the cell, interrupting the flow of cellular signalling information (Moore et al., 2013; Goetz and Anderson, 2010; Gerdes et al., 2009). Many fundamental cell signalling pathways are known to be required for proper cartilage and bone development and growth. One pathway in particular, the Hedgehog signalling pathway, has been the focus of attention, as it was found to be mis-regulated in many human and animal models of ciliopathies and it has been identified as crucially influencing chondrogenic (and subsequently osteogenic) proliferation and differentiation (Kronenberg, 2003).

Imbalanced hedgehog signalling as a cause of skeletal ciliopathy symptoms

Most of our ideas regarding the pathophysiology behind rib and extremity shortening in skeletal ciliopathies comes from the mouse models of *Evc*, *Dync2h1* and *IFT80*, suggesting that IFT defects result in imbalances in the hedgehog signalling pathway, causing a premature stop of chondrogenic proliferation and the induction of chondrogenic differentiation at the growth plates (Ruiz-Perez et al., 2007; Rix et al., 2011). Further, IFT-A mouse mutants have been shown to display ligand-independent expansion of hedgehog signalling, resulting in abnormal digit numbers and oro-facial clefts (Ocbina et al., 2011).

It is well-accepted that the canonical hedgehog signaling pathway is dependent on cilia, as its signaling components use ciliary trafficking. Anterograde IFT transports smoothened (Smo) and Patched to the ciliary tip, where Smo becomes activated and releases GLI3 activator. The latter then travels back to the ciliary base using retrograde IFT and finally continues its journey to the nucleus where, with regards to skeletal development, it activates genes regulating chondrogenic and osteogenic differentiation and proliferation (simplified schematic in Figure 2). A hedgehog gradient within the limb bud is also required for the proper development of digits and feet and hence, in the case of disturbed signaling as found in skeletal ciliopathies, poly(syn)dactyly may occur.

Apart from altered skeletal development, impaired hedgehog signalling also results in complex developmental defects in mammalian organisms including heart defects and midline defects such as clefting and holoprosencephaly, to name a few. Potentially, distrubed hedgehog signalling could also play a role in kidney involvement in ciliopathies: renal abnormalities, mainly ectopic kidneys but also cystic-dysplastic renal changes, have been observed in mice (Hu et al., 2006) and human subjects affected by Smith-Lemli-Opitz syndrome, a condition thought to result from altered hedgehog signalling as a consequence of a cholesterol biosynthesis defect (Kelley and Hennekam, 2000). However, regarding skeletal ciliopathies, neither human subjects with *IFT80* or *EVC1/EVC2* mutations nor the corresponding mouse models exhibit a renal phenotype, despite findings of impaired hedgehog signaling in the developing skeleton. On the other hand, human subjects with mutations in other *IFT* genes such as *WDR19/IFT144*, *TTC21B/IFT139*, *IFT140*, *IFT43*, *WDR35* or *IFT172* are affected by childhood-onset cystic or nephronophthisis-like renal disease (Arts et al., 2011; Bredrup et al., 2011; Davis et al., 2011a; Gilissen et al., 2010; Halbritter et al., 2013; Perrault et al., 2012; Schmidts et al., 2013b) and corresponding knockout mouse models, such as *Ift140* and *Ift172 Null* mice display early onset cystic (dysplastic) kidney disease (Friedland-Little et al., 2011; Jonassen et al., 2012). While the skeletal phenotype observed in these human subjects likely results from disturbed hedgehog signaling, no such pathway disturbances have been found in the kidneys of *Ift140* knockout mice (Jonassen et al., 2012).

Pathophysiology underlying the renal phenotype associated with many skeletal ciliopathies

Numerous ciliopathies combine skeletal defects with renal as well as retinal disease (Huber and Cormier-Daire, 2012). This is particularly true for IFT gene mutations (Lin et al., 2013; Perrault et al., 2012; Schmidts et al., 2013b; Halbritter et al., 2013; Zhang et al., 2016). The precise underlying pathophysiological mechanism remains to be defined. However, renal symptoms such as cysts were one of the first clinical ciliopathy hallmarks noted and therefore, much research effort has been targeted to this field over the last 2 decades. Like most mammalian tissues, renal tubule cells exhibit primary cilia and both the loss of the physical structure of the cilium as well as the lack of or defective ciliary proteins result in kidney cyst formation in mice (Davenport et al., 2007; Lin et al., 2003; Moyer et al., 1994). Disruption of *Ift140* or *Ift172* in mice produces an early onset kidney phenotype (Friedland-Little et al., 2011; Jonassen et al., 2012), resembling the human phenotype. In contrast, hypomorphic *Ift80* knockout mice do not exhibit any renal phenotype (Rix et al., 2011) and neither do human subjects with *IFT80* mutations. Loss of planar cell polarity, potentially as a result of mis-orientated mitotic spindles has been suggested as a pathomechanism of renal cyst formation in other models, however, *Ift140* knockout mice show normally orientated mitotic spindle axis, although hyper-proliferation of renal tubule cells was noted in the pre-cystic epithelium. Also, increased canonical Wnt and hedgehog signalling was only observed in cystic tissue of *Ift140* mice (Jonassen et al., 2012). This is of note as imbalances between the canonical and non-canonical Wnt pathways with the reduction of non-canonical and augmented canonical signalling have been previously detected in nephronophthisis models, especially Inversin (INV) (Lienkamp et al., 2012; Simons et al., 2005). In summary, neither disturbances in hedgehog nor Wnt signalling seem to initiate cyst formation in *Ift140* knockout mice but might contribute to cyst progression. No studies have been performed using renal tissue from human *IFT140* cases, so it remains unclear if either pathway may play a role for nephronophthisis development in JATD or MZSDS cases.

Future perspectives

Unfortunately, no therapy is available to date for skeletal ciliopathies and their treatment has remained purely symptomatic. Therapeutically influencing the skeletal phenotype will be very challenging; not only because treatment would have to be commenced *in utero* or latest at birth but also because the developing skeleton is difficult to access by gene therapy, in comparison to other organs and tissues, such as the airway system. The huge size of some of the genes involved (e.g., *DYNC2H1* and *IFT172*) represents an additional hurdle. Lastly, many of the mutations identified to date in skeletal ciliopathies are mainly found in single families and often represent missense changes. This renders the development of mutation-specific therapies, such as CRISPR/Cas9 based gene editing, very time- and labour-intensive and economically difficult to perform, as each family would require their specific therapy. The relative lack of premature stop mutations further implies that read-through drugs as PTC124/Ataluren which overcome premature stop codons and are currently being tested to improve other genetic conditions such as cystic fibrosis, will only be applicable to a minority of patients. The most promising approach may therefore be the pharmacological treatment of potentially affected cell signalling pathways such as hedgehog signalling and others and hopefully, progress will be made within this area over the coming years.

Acknowledgement

Sincere apologies to all colleagues whose findings could not be cited due to space limits. I am grateful for funding to the Radboud university Nijmegen (Excellence Fellowship), Radboud University Medical Center Nijmegen (Hypatia Tenure track fellowship), the German Research Foundation (Collaborative research Center KIDGEM CRC-1140) and the ERC (ERC-StG-TREATCilia, grant agreement 716344).

References

Ades, L. C., Clapton, W. K., Morphett, A., Morris, L. L. and Haan, E. A. 1994. Polydactyly, campomelia, ambiguous genitalia, cystic dysplastic kidneys, and cerebral malformation in a fetus of consanguineous parents: a new multiple malformation syndrome, or a severe form of oral-facial-digital syndrome type IV? Am. J. Med. Genet. 49: 211–217.

Adly, N., Alhashem, A., Ammari, A. and Alkuraya, F. S. 2014. Ciliary genes TBC1D32/C6orf170 and SCLT1 are mutated in patients with OFD type IX. Human Mutation 35: 36–40.

Aguilar, A. 2016. Ciliopathies: CPLANE regulates intraflagellar transport. Nature Reviews Nephrology 12: 376.

Akizu, N., Silhavy, J. L., Rosti, R. O., Scott, E., Fenstermaker, A. G., Schroth, J., Zaki, M. S., Sanchez, H., Gupta, N., Kabra, M. et al. 2014. Mutations in CSPP1 lead to classical Joubert syndrome. Am. J. Hum. Genet. 94: 80–86.

Alby, C., Piquand, K., Huber, C., Megarbane, A., Ichkou, A., Legendre, M., Pelluard, F., Encha-Ravazi, F., Abi-Tayeh, G., Bessieres, B. et al. 2015. Mutations in KIAA0586 cause lethal ciliopathies ranging from a hydrolethalus phenotype to short-Rib polydactyly syndrome. Am. J. Hum. Genet. 97: 311–318.

Allen, A. W., Jr., Moon, J. B., Hovland, K. R. and Minckler, D. S. 1979. Ocular findings in thoracic-pelvic-phalangeal dystrophy. Arch. Ophthalmol. 97: 489–492.

Amar, M. J., Sutphen, R. and Kousseff, B. G. 1997. Expanded phenotype of cranioectodermal dysplasia (Sensenbrenner syndrome). Am. J. Med. Genet. 70: 349–352.

Arts, H. and Knoers, N. 1993. Cranioectodermal dysplasia. In: Pagon, R. A., Adam, M. P., Ardinger, H. H., Bird, T. D., Dolan, C. R., Fong, C. T., Smith, R. J. H. and Stephens, K. (eds.). GeneReviews(R), (Seattle (WA)).

Arts, H. H., Bongers, E. M., Mans, D. A., van Beersum, S. E., Oud, M. M., Bolat, E., Spruijt, L., Cornelissen, E. A., Schuurs-Hoeijmakers, J. H., de Leeuw, N. et al. 2011. C14ORF179 encoding IFT43 is mutated in Sensenbrenner syndrome. J. Med. Genet. 48: 390–395.

Baker, K. and Beales, P. L. 2009. Making sense of cilia in disease: the human ciliopathies. Am. J. Med. Genet. C. Semin. Med. Genet. 151C: 281–295.

Baraitser, M. 1986. The orofaciodigital (OFD) syndromes. J. Med. Genet. 23: 116–119.

Bard, L. A., Bard, P. A., Owens, G. W. and Hall, B. D. 1978. Retinal involvement in thoracic-pelvic-phalangeal dystrophy. Arch. Ophthalmol. 96: 278–281.

Baujat, G. and Le Merrer, M. 2007. Ellis-van creveld syndrome. Orphanet. J. Rare. Dis. 2: 27.

Baujat, G., Huber, C., El Hokayem, J., Caumes, R., Do Ngoc Thanh, C., David, A., Delezoide, A. L., Dieux-Coeslier, A., Estournet, B., Francannet, C. et al. 2013. Asphyxiating thoracic dysplasia: clinical and molecular review of 39 families. J. Med. Genet. 50: 91–98.

Beales, P. L., Bland, E., Tobin, J. L., Bacchelli, C., Tuysuz, B., Hill, J., Rix, S., Pearson, C. G., Kai, M., Hartley, J. et al. 2007. IFT80, which encodes a conserved intraflagellar transport protein, is mutated in Jeune asphyxiating thoracic dystrophy. Nat. Genet. 39: 727–729.

Beemer, F. A., Langer, L. O., Jr., Klep-de Pater, J. M., Hemmes, A. M., Bylsma, J. B., Pauli, R. M., Myers, T. L. and Haws, C. C., 3rd. 1983. A new short rib syndrome: report of two cases. Am. J. Med. Genet. 14: 115–123.

Bergmann, C., von Bothmer, J., Ortiz Bruchle, N., Venghaus, A., Frank, V., Fehrenbach, H., Hampel, T., Pape, L., Buske, A., Jonsson, J. et al. 2011. Mutations in multiple PKD genes may explain early and severe polycystic kidney disease. Journal of the American Society of Nephrology: JASN 22: 2047–2056.

Bernstein, R., Isdale, J., Pinto, M., Du Toit Zaaijman, J. and Jenkins, T. 1985. Short rib-polydactyly syndrome: a single or heterogeneous entity? A re-evaluation prompted by four new cases. J. Med. Genet. 22: 46–53.

Blackburn, M. G. and Belliveau, R. E. 1971. Ellis-van Creveld syndrome. A report of previously undescribed anomalies in two siblings. Am. J. Dis. Child. 122: 267–270.

Bredrup, C., Saunier, S., Oud, M. M., Fiskerstrand, T., Hoischen, A., Brackman, D., Leh, S. M., Midtbo, M., Filhol, E., Bole-Feysot, C. et al. 2011. Ciliopathies with skeletal anomalies and renal insufficiency due to mutations in the IFT-A gene WDR19. Am. J. Hum. Genet. 89: 634–643.

Cavalcanti, D. P., Huber, C., Sang, K. H., Baujat, G., Collins, F., Delezoide, A. L., Dagoneau, N., Le Merrer, M., Martinovic, J., Mello, M. F. et al. 2011. Mutation in IFT80 in a fetus with the phenotype of Verma-Naumoff provides molecular evidence for Jeune-Verma-Naumoff dysplasia spectrum. J. Med. Genet. 48: 88–92.

Chaya, T., Omori, Y., Kuwahara, R. and Furukawa, T. 2014. ICK is essential for cell type-specific ciliogenesis and the regulation of ciliary transport. The EMBO Journal 33: 1227–1242.

Chen, H., Yang, S. S., Gonzalez, E., Fowler, M. and Al Saadi, A. 1980. Short rib-polydactyly syndrome, Majewski type. Am. J. Med. Genet. 7: 215–222.

Christensen, S. T., Pedersen, L. B., Schneider, L. and Satir, P. 2007. Sensory cilia and integration of signal transduction in human health and disease. Traffic. 8: 97–109.

Cole, D. G. and Snell, W. J. 2009. SnapShot: Intraflagellar transport. Cell 137: 784–784 e781.

Cortes, C. R., McInerney-Leo, A. M., Vogel, I., Rondon Galeano, M. C., Leo, P. J., Harris, J. E., Anderson, L. K., Keith, P. A., Brown, M. A., Ramsing, M. et al. 2016. Mutations in human C2CD3 cause skeletal dysplasia and provide new insights into phenotypic and cellular consequences of altered C2CD3 function. Scientific Reports 6: 24083.

Curry, C. J. and Hall, B. D. 1979. Polydactyly, conical teeth, nail dysplasia, and short limbs: a new autosomal dominant malformation syndrome. Birth Defects Orig. Artic. Ser. 15: 253–263.

D'Asdia, M. C., Torrente, I., Consoli, F., Ferese, R., Magliozzi, M., Bernardini, L., Guida, V., Digilio, M. C., Marino, B., Dallapiccola, B. et al. 2013. Novel and recurrent EVC and EVC2 mutations in Ellis-van Creveld syndrome and Weyers acrofacial dyostosis. Eur. J. Med. Genet. 56: 80–87.

Dagoneau, N., Goulet, M., Genevieve, D., Sznajer, Y., Martinovic, J., Smithson, S., Huber, C., Baujat, G., Flori, E., Tecco, L. et al. 2009. DYNC2H1 mutations cause asphyxiating thoracic dystrophy and short rib-polydactyly syndrome, type III. Am. J. Hum. Genet. 84: 706–711.

Davenport, J. R., Watts, A. J., Roper, V. C., Croyle, M. J., van Groen, T., Wyss, J. M., Nagy, T. R., Kesterson, R. A. and Yoder, B. K. 2007. Disruption of intraflagellar transport in adult mice leads to obesity and slow-onset cystic kidney disease. Curr. Biol. 17: 1586–1594.

Davis, E. E., Zhang, Q., Liu, Q., Diplas, B. H., Davey, L. M., Hartley, J., Stoetzel, C., Szymanska, K., Ramaswami, G., Logan, C. V. et al. 2011a. TTC21B contributes both causal and modifying alleles across the ciliopathy spectrum. Nat. Genet. 43: 189–196.

Davis, E. E., Zhang, Q., Liu, Q., Diplas, B. H., Davey, L. M., Hartley, J., Stoetzel, C., Szymanska, K., Ramaswami, G., Logan, C. V. et al. 2011b. TTC21B contributes both causal and modifying alleles across the ciliopathy spectrum. Nat. Genet. 43: 189–196.

de Vries, J., Yntema, J. L., van Die, C. E., Crama, N., Cornelissen, E. A. and Hamel, B. C. 2010. Jeune syndrome: description of 13 cases and a proposal for follow-up protocol. Eur. J. Pediatr. 169: 77–88.

Ehara, S., Kim, O. H., Maisawa, S., Takasago, Y. and Nishimura, G. 1997. Axial spondylometaphyseal dysplasia. Eur. J. Pediatr. 156: 627–630.

Eke, T., Woodruff, G. and Young, I. D. 1996. A new oculorenal syndrome: retinal dystrophy and tubulointerstitial nephropathy in cranioectodermal dysplasia. Br. J. Ophthalmol. 80: 490–491.

El Hokayem, J., Huber, C., Couve, A., Aziza, J., Baujat, G., Bouvier, R., Cavalcanti, D. P., Collins, F. A., Cordier, M. P., Delezoide, A. L. et al. 2012. NEK1 and DYNC2H1 are both involved in short rib polydactyly Majewski type but not in Beemer Langer cases. J. Med. Genet. 49: 227–233.

Elcioglu, N. H. and Hall, C. M. 2002. Diagnostic dilemmas in the short rib-polydactyly syndrome group. Am. J. Med. Genet. 111: 392–400.

Ellis, R. W. and van Creveld, S. 1940. A syndrome characterized by ectodermal dysplasia, polydactyly, chondro-dysplasia and congenital morbus cordis: report of three cases. Arch. Dis. Child 15: 65–84.

Fliegauf, M., Benzing, T. and Omran, H. 2007. When cilia go bad: cilia defects and ciliopathies. Nat. Rev. Mol. Cell Biol. 8: 880–893.

Franco, B. and Thauvin-Robinet, C. 2016. Update on oral-facial-digital syndromes (OFDS). Cilia 5: 12.

Friedland-Little, J. M., Hoffmann, A. D., Ocbina, P. J., Peterson, M. A., Bosman, J. D., Chen, Y., Cheng, S. Y., Anderson, K. V. and Moskowitz, I. P. 2011. A novel murine allele of intraflagellar transport protein 172 causes a syndrome including VACTERL-like features with hydrocephalus. Hum. Mol. Genet. 20: 3725–3737.

Gerdes, J. M., Davis, E. E. and Katsanis, N. 2009. The vertebrate primary cilium in development, homeostasis, and disease. Cell 137: 32–45.

Giedion, A. 1979. Phalangeal cone shaped epiphysis of the hands (PhCSEH) and chronic renal disease—the conorenal syndromes. Pediatr. Radiol. 8: 32–38.

Gilissen, C., Arts, H. H., Hoischen, A., Spruijt, L., Mans, D. A., Arts, P., van Lier, B., Steehouwer, M., van Reeuwijk, J., Kant, S. G. et al. 2010. Exome sequencing identifies WDR35 variants involved in sensenbrenner syndrome. Am. J. Hum. Genet. 87: 418–423.

Goetz, S. C. and Anderson, K. V. 2010. The primary cilium: a signalling centre during vertebrate development. Nat. Rev. Genet. 11: 331–344.

Halbritter, J., Bizet, A. A., Schmidts, M., Porath, J. D., Braun, D. A., Gee, H. Y., McInerney-Leo, A. M., Krug, P., Filhol, E., Davis, E. E. et al. 2013. Defects in the IFT-B component IFT172 cause Jeune and Mainzer-Saldino syndromes in humans. Am. J. Hum. Genet. 93: 915–925.

Hao, L., Efimenko, E., Swoboda, P. and Scholey, J. M. 2011. The retrograde IFT machinery of *C. elegans* cilia: two IFT dynein complexes? PLoS One 6: e20995.

He, M., Subramanian, R., Bangs, F., Omelchenko, T., Liem, K. F., Jr., Kapoor, T. M. and Anderson, K. V. 2014. The kinesin-4 protein Kif7 regulates mammalian Hedgehog signalling by organizing the cilium tip compartment. Nature Cell Biology 16: 663–672.

Hills, C. B., Kochilas, L., Schimmenti, L. A. and Moller, J. H. 2011. Ellis-van creveld syndrome and congenital heart defects: presentation of an additional 32 cases. Pediatr. Cardiol. 32: 977–982.

Hopper, M. S., Boultbee, J. E. and Watson, A. R. 1979. Polyhydramnios associated with congenital pancreatic cysts and asphyxiating thoracic dysplasia. A case report. S. Afr. Med. J. 56: 32–33.

Hu, M. C., Mo, R., Bhella, S., Wilson, C. W., Chuang, P. T., Hui, C. C. and Rosenblum, N. D. 2006. GLI3-dependent transcriptional repression of Gli1, Gli2 and kidney patterning genes disrupts renal morphogenesis. Development 133: 569–578.

Huber, C. and Cormier-Daire, V. 2012. Ciliary disorder of the skeleton. Am. J. Med. Genet. C. Semin. Med. Genet. 160C: 165–174.

Huber, C., Wu, S., Kim, A. S., Sigaudy, S., Sarukhanov, A., Serre, V., Baujat, G., Le Quan Sang, K. H., Rimoin, D. L., Cohn, D. H. et al. 2013. WDR34 mutations that cause short-rib polydactyly syndrome type III/severe asphyxiating thoracic dysplasia reveal a role for the NF-kappaB pathway in cilia. Am. J. Hum. Genet. 93: 926–931.

Ishikawa, H. and Marshall, W. F. 2011. Ciliogenesis: building the cell's antenna. Nat. Rev. Mol. Cell Biol. 12: 222–234.

Isidor, B., Baron, S., Khau van Kien, P., Bertrand, A. M., David, A. and Le Merrer, M. 2010. Axial spondylometaphyseal dysplasia: Confirmation and further delineation of a new SMD with retinal dystrophy. Am. J. Med. Genet. A. 152A: 1550–1554.

Jonassen, J. A., SanAgustin, J., Baker, S. P. and Pazour, G. J. 2012. Disruption of IFT complex A causes cystic kidneys without mitotic spindle misorientation. J. Am. Soc. Nephrol. 23: 641–651.

Jorgenson, R.J. 1971. Orofaciodigital syndrome I (OFD I). Birth Defects Orig. Artic. Ser. 7: 270.

Kannu, P., McFarlane, J. H., Savarirayan, R. and Aftimos, S. 2007. An unclassifiable short rib-polydactyly syndrome with acromesomelic hypomineralization and campomelia in siblings. Am. J. Med. Genet. A. 143A: 2607–2611.

Katsanis, N., Ansley, S. J., Badano, J. L., Eichers, E. R., Lewis, R. A., Hoskins, B. E., Scambler, P. J., Davidson, W. S., Beales, P. L. and Lupski, J. R. 2001. Triallelic inheritance in Bardet-Biedl syndrome, a Mendelian recessive disorder. Science 293: 2256–2259.

Kelley, R. I. and Hennekam, R. C. 2000. The Smith-Lemli-Opitz syndrome. J. Med. Genet. 37: 321–335.

Kessler, K., Wunderlich, I., Uebe, S., Falk, N. S., Giessl, A., Brandstatter, J. H., Popp, B., Klinger, P., Ekici, A. B., Sticht, H. et al. 2015. DYNC2LI1 mutations broaden the clinical spectrum of dynein-2 defects. Scientific Reports 5: 11649.

Khanna, H., Davis, E. E., Murga-Zamalloa, C. A., Estrada-Cuzcano, A., Lopez, I., den Hollander, A. I., Zonneveld, M. N., Othman, M. I., Waseem, N., Chakarova, C. F. et al. 2009. A common allele in RPGRIP1L is a modifier of retinal degeneration in ciliopathies. Nat. Genet. 41: 739–745.

Ko, H. W., Norman, R. X., Tran, J., Fuller, K. P., Fukuda, M. and Eggenschwiler, J. T. 2010. Broad-minded links cell cycle-related kinase to cilia assembly and hedgehog signal transduction. Developmental Cell 18: 237–247.

Konstantinidou, A. E., Fryssira, H., Sifakis, S., Karadimas, C., Kaminopetros, P., Agrogiannis, G., Velonis, S., Nikkels, P. G. and Patsouris, E. 2009. Cranioectodermal dysplasia: a probable ciliopathy. Am. J. Med. Genet. A. 149A: 2206–2211.

Kovacs, N., Sarkany, I., Mohay, G., Adamovich, K., Ertl, T., Kosztolanyi, G. and Kellermayer, R. 2006. High incidence of short rib-polydactyly syndrome type IV in a Hungarian Roma subpopulation. Am. J. Med. Genet. A. 140: 2816–2818.

Krock, B. L., Mills-Henry, I. and Perkins, B. D. 2009. Retrograde intraflagellar transport by cytoplasmic dynein-2 is required for outer segment extension in vertebrate photoreceptors but not arrestin translocation. Invest. Ophthalmol. Vis. Sci. 50: 5463–5471.

Kronenberg, H. M. 2003. Developmental regulation of the growth plate. Nature 423: 332–336.

Lahiry, P., Wang, J., Robinson, J. F., Turowec, J. P., Litchfield, D. W., Lanktree, M. B., Gloor, G. B., Puffenberger, E. G., Strauss, K. A., Martens, M. B. et al. 2009. A multiplex human syndrome implicates a key role for intestinal cell kinase in development of central nervous, skeletal, and endocrine systems. Am. J. Hum. Genet. 84: 134–147.

Lang, G. D. and Young, I. D. 1991. Cranioectodermal dysplasia in sibs. J. Med. Genet. 28: 424.

Le Marec, B., Passarge, E., Dellenbach, P., Kerisit, J., Signargout, J., Ferrand, B. and Senecal, J. 1973. [Lethal neonatal forms of chondroectodermal dysplasia. Apropos of 5 cases]. Ann. Radiol. (Paris) 16: 19–26.

Lehman, A. M., Eydoux, P., Doherty, D., Glass, I. A., Chitayat, D., Chung, B. Y., Langlois, S., Yong, S. L., Lowry, R. B., Hildebrandt, F. et al. 2010. Co-occurrence of Joubert syndrome and Jeune asphyxiating thoracic dystrophy. Am. J. Med. Genet. A. 152A: 1411–1419.

Levin, L. S., Perrin, J. C., Ose, L., Dorst, J. P., Miller, J. D. and McKusick, V. A. 1977. A heritable syndrome of craniosynostosis, short thin hair, dental abnormalities, and short limbs: cranioectodermal dysplasia. J. Pediatr. 90: 55–61.

Lienkamp, S., Ganner, A. and Walz, G. 2012. Inversin, Wnt signaling and primary cilia. Differentiation 83: S49–55.

Lin, A. E., Traum, A. Z., Sahai, I., Keppler-Noreuil, K., Kukolich, M. K., Adam, M. P., Westra, S. J. and Arts, H. H. 2013. Sensenbrenner syndrome (Cranioectodermal dysplasia): clinical and molecular analyses of 39 patients including two new patients. Am. J. Med. Genet. A. 161A: 2762–2776.

Lin, F., Hiesberger, T., Cordes, K., Sinclair, A. M., Goldstein, L. S., Somlo, S. and Igarashi, P. 2003. Kidney-specific inactivation of the KIF3A subunit of kinesin-II inhibits renal ciliogenesis and produces polycystic kidney disease. Proc. Natl. Acad. Sci. USA 100: 5286–5291.

Lopez, E., Thauvin-Robinet, C., Reversade, B., Khartoufi, N. E., Devisme, L., Holder, M., Ansart-Franquet, H., Avila, M., Lacombe, D., Kleinfinger, P. et al. 2014. C5orf42 is the major gene responsible for OFD syndrome type VI. Hum. Genet. 133: 367–377.

Mahjoub, M. R., Xie, Z. and Stearns, T. 2010. Cep120 is asymmetrically localized to the daughter centriole and is essential for centriole assembly. J. Cell Biol. 191: 331–346.

Mainzer, F., Saldino, R. M., Ozonoff, M. B. and Minagi, H. 1970. Familial nephropathy associated with retinitis pigmentosa, cerebellar ataxia and skeletal abnormalities. Am. J. Med. 49: 556–562.

Majewski, F., Pfeiffer, R. A., Lenz, W., Muller, R., Feil, G. and Seiler, R. 1971. [Polysyndactyly, short limbs, and genital malformations—a new syndrome?]. Z. Kinderheilkd 111: 118–138.

Malicdan, M. C., Vilboux, T., Stephen, J., Maglic, D., Mian, L., Konzman, D., Guo, J., Yildirimli, D., Bryant, J., Fischer, R. et al. 2015. Mutations in human homologue of chicken talpid3 gene (KIAA0586) cause a hybrid ciliopathy with overlapping features of Jeune and Joubert syndromes. J. Med. Genet. 52: 830–839.

Marszalek, J. R., Liu, X., Roberts, E. A., Chui, D., Marth, J. D., Williams, D. S. and Goldstein, L. S. 2000. Genetic evidence for selective transport of opsin and arrestin by kinesin-II in mammalian photoreceptors. Cell 102: 175–187.

McInerney-Leo, A. M., Schmidts, M., Cortes, C. R., Leo, P. J., Gener, B., Courtney, A. D., Gardiner, B., Harris, J. A., Lu, Y., Marshall, M. et al. 2013. Short-rib polydactyly and Jeune syndromes are caused by mutations in WDR60. Am. J. Hum. Genet. 93: 515–523.

McKusick, V. A., Eldridge, R., Hostetler, J. A. and Egeland, J. A. 1964. Dwarfism in the Amish. Trans. Assoc. Am. Physicians. 77: 151–168.

Merrill, A. E., Merriman, B., Farrington-Rock, C., Camacho, N., Sebald, E. T., Funari, V. A., Schibler, M. J., Firestein, M. H., Cohn, Z. A., Priore, M. A. et al. 2009. Ciliary abnormalities due to defects in the retrograde transport protein DYNC2H1 in short-rib polydactyly syndrome. Am. J. Hum. Genet. 84: 542–549.

Mill, P., Lockhart, P. J., Fitzpatrick, E., Mountford, H. S., Hall, E. A., Reijns, M. A., Keighren, M., Bahlo, M., Bromhead, C.J., Budd, P. et al. 2011. Human and mouse mutations in WDR35 cause short-rib polydactyly syndromes due to abnormal ciliogenesis. Am. J. Hum. Genet. 88: 508–515.

Moore, D. J., Onoufriadis, A., Shoemark, A., Simpson, M. A., zur Lage, P. I., de Castro, S. C., Bartoloni, L., Gallone, G., Petridi, S., Woollard, W. J. et al. 2013. Mutations in ZMYND10, a gene essential for proper axonemal assembly of inner and outer dynein arms in humans and flies, cause primary ciliary dyskinesia. Am. J. Hum. Genet. 93: 346–356.

Moyer, J. H., Lee-Tischler, M. J., Kwon, H. Y., Schrick, J. J., Avner, E. D., Sweeney, W. E., Godfrey, V. L., Cacheiro, N. L., Wilkinson, J. E. and Woychik, R. P. 1994. Candidate gene associated with a mutation causing recessive polycystic kidney disease in mice. Science 264: 1329–1333.

Naumoff, P., Young, L. W., Mazer, J. and Amortegui, A. J. 1977. Short rib-polydactyly syndrome type 3. Radiology 122: 443–447.

Nevin, N. C. and Thomas, P. S. 1989. Orofaciodigital syndrome type IV: report of a patient. Am. J. Med. Genet. 32: 151–154.

O'Connor, M. J. and Collins, R. T. 2nd 2012. Ellis-van Creveld syndrome and congenital heart defects: presentation of an additional 32 cases. Pediatr. Cardiol. 33: 491; discussion 491–492.

Oberklaid, F., Danks, D. M., Mayne, V. and Campbell, P. 1977. Asphyxiating thoracic dysplasia. Clinical, radiological, and pathological information on 10 patients. Arch. Dis. Child 52: 758–765.

Ocbina, P. J., Eggenschwiler, J. T., Moskowitz, I. and Anderson, K. V. 2011. Complex interactions between genes controlling trafficking in primary cilia. Nat. Genet. 43: 547–553.

Oud, M. M., Bonnard, C., Mans, D. A., Altunoglu, U., Tohari, S., Ng, A. Y., Eskin, A., Lee, H., Rupar, C. A., de Wagenaar, N. P. et al. 2016. A novel ICK mutation causes ciliary disruption and lethal endocrine-cerebro-osteodysplasia syndrome. Cilia 5: 8.

Paige Taylor, S., Kunova Bosakova, M., Varecha, M., Balek, L., Barta, T., Trantirek, L., Jelinkova, I., Duran, I., Vesela, I., Forlenza, K. N. et al. 2016. An inactivating mutation in intestinal cell kinase, ICK, impairs hedgehog signalling and causes short rib-polydactyly syndrome. Hum. Mol. Genet.

Passarge, E. 1983. Familial occurrence of a short rib syndrome with hydrops fetalis but without polydactyly. Am. J. Med. Genet. 14: 403–405.

Patel-King, R. S., Gilberti, R. M., Hom, E. F. and King, S. M. 2013. WD60/FAP163 is a dynein intermediate chain required for retrograde intraflagellar transport in cilia. Mol. Biol. Cell 24: 2668–2677.

Pazour, G. J., Baker, S. A., Deane, J. A., Cole, D. G., Dickert, B. L., Rosenbaum, J. L., Witman, G. B. and Besharse, J. C. 2002. The intraflagellar transport protein, IFT88, is essential for vertebrate photoreceptor assembly and maintenance. J. Cell Biol. 157: 103–113.

Pazour, G. J. and Rosenbaum, J. L. 2002. Intraflagellar transport and cilia-dependent diseases. Trends Cell Biol. 12: 551–555.

Perrault, I., Saunier, S., Hanein, S., Filhol, E., Bizet, A. A., Collins, F., Salih, M. A., Gerber, S., Delphin, N., Bigot, K. et al. 2012. Mainzer-Saldino syndrome is a ciliopathy caused by IFT140 mutations. Am. J. Hum. Genet. 90: 864–870.

Popovic-Rolovic, M., Calic-Perisic, N., Bunjevacki, G. and Negovanovic, D. 1976. Juvenile nephronophthisis associated with retinal pigmentary dystrophy, cerebellar ataxia, and skeletal abnormalities. Arch. Dis. Child 51: 801–803.

Putoux, A., Thomas, S., Coene, K. L., Davis, E. E., Alanay, Y., Ogur, G., Uz, E., Buzas, D., Gomes, C., Patrier, S. et al. 2011. KIF7 mutations cause fetal hydrolethalus and acrocallosal syndromes. Nat. Genet. 43: 601–606.

Rix, S., Calmont, A., Scambler, P. J. and Beales, P. L. 2011. An Ift80 mouse model of short rib polydactyly syndromes shows defects in hedgehog signalling without loss or malformation of cilia. Hum. Mol. Genet. 20: 1306–1314.

Roberson, E. C., Dowdle, W. E., Ozanturk, A., Garcia-Gonzalo, F. R., Li, C., Halbritter, J., Elkhartoufi, N., Porath, J. D., Cope, H., Ashley-Koch, A. et al. 2015. TMEM231, mutated in orofaciodigital and Meckel syndromes, organizes the ciliary transition zone. J. Cell Biol. 209: 129–142.

Robins, D. G., French, T. A. and Chakera, T. M. 1976. Juvenile nephronophthisis associated with skeletal abnormalities and hepatic fibrosis. Arch. Dis. Child 51: 799–801.

Romani, M., Mancini, F., Micalizzi, A., Poretti, A., Miccinilli, E., Accorsi, P., Avola, E., Bertini, E., Borgatti, R., Romaniello, R. et al. 2015. Oral-facial-digital syndrome type VI: is C5orf42 really the major gene? Hum. Genet. 134: 123–126.

Rompolas, P., Pedersen, L. B., Patel-King, R. S. and King, S. M. 2007. Chlamydomonas FAP133 is a dynein intermediate chain associated with the retrograde intraflagellar transport motor. J. Cell Sci. 120: 3653–3665.

Roosing, S., Romani, M., Isrie, M., Rosti, R. O., Micalizzi, A., Musaev, D., Mazza, T., Al-Gazali, L., Altunoglu, U., Boltshauser, E. et al. 2016. Mutations in CEP120 cause Joubert syndrome as well as complex ciliopathy phenotypes. J. Med. Genet. 53: 608–615.

Rosenbaum, J. L. and Witman, G. B. 2002. Intraflagellar transport. Nat. Rev. Mol. Cell Biol. 3: 813–825.

Roubicek, M. and Spranger, J. 1984. Weyers acrodental dysostosis in a family. Clin. Genet. 26: 587–590.

Roubicek, M. and Spranger, J. 1985. Syndrome of polydactyly, conical teeth and nail dysplasia. Am. J. Med. Genet. 20: 205–207.

Ruess, A. L., Pruzansky, S. and Lis, E. F. 1965. Intellectual development and the OFD syndrome: a review. The Cleft Palate Journal 2: 350–356.

Ruiz-Perez, V. L., Ide, S. E., Strom, T. M., Lorenz, B., Wilson, D., Woods, K., King, L., Francomano, C., Freisinger, P., Spranger, S. et al. 2000. Mutations in a new gene in Ellis-van Creveld syndrome and Weyers acrodental dysostosis. Nat. Genet. 24: 283–286.

Ruiz-Perez, V. L., Blair, H. J., Rodriguez-Andres, M. E., Blanco, M. J., Wilson, A., Liu, Y. N., Miles, C., Peters, H. and Goodship, J. A. 2007. Evc is a positive mediator of Ihh-regulated bone growth that localises at the base of chondrocyte cilia. Development 134: 2903–2912.

Saari, J., Lovell, M. A., Yu, H. C. and Bellus, G. A. 2015. Compound heterozygosity for a frame shift mutation and a likely pathogenic sequence variant in the planar cell polarity-ciliogenesis gene WDPCP in a girl with polysyndactyly, coarctation of the aorta, and tongue hamartomas. Am. J. Med. Genet. A. 167A: 421–427.

Saldino, R. M. and Noonan, C. D. 1972. Severe thoracic dystrophy with striking micromelia, abnormal osseous development, including the spine, and multiple visceral anomalies. Am. J. Roentgenol. Radium. Ther. Nucl. Med. 114: 257–263.

Satir, P., Pedersen, L. B. and Christensen, S. T. 2010. The primary cilium at a glance. J. Cell Sci. 123: 499–503.

Schinzel, A. and Schmid, W. 1980. Hallux duplication, postaxial polydactyly, absence of the corpus callosum, severe mental retardation, and additional anomalies in two unrelated patients: a new syndrome. Am. J. Med. Genet. 6: 241–249.

Schmidts, M., Arts, H. H., Bongers, E. M., Yap, Z., Oud, M. M., Antony, D., Duijkers, L., Emes, R. D., Stalker, J., Yntema, J. B. et al. 2013a. Exome sequencing identifies DYNC2H1 mutations as a common cause of asphyxiating thoracic dystrophy (Jeune syndrome) without major polydactyly, renal or retinal involvement. J. Med. Genet. 50: 309–323.

Schmidts, M., Frank, V., Eisenberger, T., Al Turki, S., Bizet, A. A., Antony, D., Rix, S., Decker, C., Bachmann, N., Bald, M. et al. 2013b. Combined NGS approaches identify mutations in the intraflagellar transport gene IFT140 in skeletal ciliopathies with early progressive kidney Disease. Hum. Mutat. 34: 714–724.

Schmidts, M., Vodopiutz, J., Christou-Savina, S., Cortes, C. R., McInerney-Leo, A. M., Emes, R. D., Arts, H. H., Tuysuz, B., D'Silva, J., Leo, P. J. et al. 2013c. Mutations in the gene encoding IFT dynein complex component WDR34 cause Jeune asphyxiating thoracic dystrophy. Am. J. Hum. Genet. 93: 932–944.

Schmidts, M. 2014. Clinical genetics and pathobiology of ciliary chondrodysplasias. Journal of Pediatric Genetics 3: 46–94.

Schmidts, M., Hou, Y., Cortes, C. R., Mans, D.A., Huber, C., Boldt, K., Patel, M., van Reeuwijk, J., Plaza, J.M., van Beersum, S. E. et al. 2015. TCTEX1D2 mutations underlie Jeune asphyxiating thoracic dystrophy with impaired retrograde intraflagellar transport. Nature Communications 6: 7074.

Segni, G., Serra, A., Mastrangelo, R., Polidori, G. and Massasso, J. 1970. [OFD syndrome in a male. Clinical-genetic analysis of 33 families]. Acta Geneticae Medicae et Gemellologiae 19: 546–566.

Shaheen, R., Shamseldin, H. E., Loucks, C. M., Seidahmed, M. Z., Ansari, S., Ibrahim Khalil, M., Al-Yacoub, N., Davis, E. E., Mola, N. A., Szymanska, K. et al. 2014. Mutations in CSPP1, encoding a core centrosomal protein, cause a range of ciliopathy phenotypes in humans. Am. J. Hum. Genet. 94: 73–79.

Shaheen, R., Schmidts, M., Faqeih, E., Hashem, A., Lausch, E., Holder, I., Superti-Furga, A., Consortium, U. K., Mitchison, H. M., Almoisheer, A. et al. 2015. A founder CEP120 mutation in Jeune asphyxiating thoracic dystrophy expands the role of centriolar proteins in skeletal ciliopathies. Hum. Mol. Genet. 24: 1410–1419.

Shamseldin, H. E., Rajab, A., Alhashem, A., Shaheen, R., Al-Shidi, T., Alamro, R., Al Harassi, S. and Alkuraya, F. S. 2013. Mutations in DDX59 implicate RNA helicase in the pathogenesis of orofaciodigital syndrome. Am. J. Hum. Genet. 93: 555–560.

Shylo, N. A., Christopher, K. J., Iglesias, A., Daluiski, A. and Weatherbee, S. D. 2016. TMEM107 is a critical regulator of ciliary protein composition and is mutated in orofaciodigital syndrome. Human Mutation 37: 155–159.

Simons, M., Gloy, J., Ganner, A., Bullerkotte, A., Bashkurov, M., Kronig, C., Schermer, B., Benzing, T., Cabello, O. A., Jenny, A. et al. 2005. Inversin, the gene product mutated in nephronophthisis type II, functions as a molecular switch between Wnt signaling pathways. Nat. Genet. 37: 537–543.

Spranger, J., Grimm, B., Weller, M., Weissenbacher, G., Herrmann, J., Gilbert, E. and Krepler, R. 1974a. Short rib-polydactyly (SRP) syndromes, types Majewski and Saldino-Noonan. Z. Kinderheilkd 116: 73–94.

Spranger, J., Langer, L. O., Weller, M. H. and Herrmann, J. 1974b. Short rib-polydactyly syndromes and related conditions. Birth Defects Orig. Artic. Ser. 10: 117–123.

Stephen, L. A., Tawamie, H., Davis, G. M., Tebbe, L., Nurnberg, P., Nurnberg, G., Thiele, H., Thoenes, M., Boltshauser, E., Uebe, S. et al. 2015. TALPID3 controls centrosome and cell polarity and the human ortholog KIAA0586 is mutated in Joubert syndrome (JBTS23). eLife 4.

Taylor, S. P., Dantas, T. J., Duran, I., Wu, S., Lachman, R. S., University of Washington Center for Mendelian Genomics, C., Nelson, S. F., Cohn, D. H., Vallee, R. B. and Krakow, D. 2015. Mutations in DYNC2LI1 disrupt cilia function and cause short rib polydactyly syndrome. Nature Communications 6: 7092.

Thauvin-Robinet, C., Lee, J. S., Lopez, E., Herranz-Perez, V., Shida, T., Franco, B., Jego, L., Ye, F., Pasquier, L., Loget, P. et al. 2014. The oral-facial-digital syndrome gene C2CD3 encodes a positive regulator of centriole elongation. Nat. Genet. 46: 905–911.

Thiel, C., Kessler, K., Giessl, A., Dimmler, A., Shalev, S. A., von der Haar, S., Zenker, M., Zahnleiter, D., Stoss, H., Beinder, E. et al. 2011. NEK1 mutations cause short-rib polydactyly syndrome type majewski. Am. J. Hum. Genet. 88: 106–114.

Thomas, S., Legendre, M., Saunier, S., Bessieres, B., Alby, C., Bonniere, M., Toutain, A., Loeuillet, L., Szymanska, K., Jossic, F. et al. 2012. TCTN3 mutations cause Mohr-Majewski syndrome. Am. J. Hum. Genet. 91: 372–378.

Tompson, S. W., Ruiz-Perez, V. L., Blair, H. J., Barton, S., Navarro, V., Robson, J. L., Wright, M. J. and Goodship, J. A. 2007. Sequencing EVC and EVC2 identifies mutations in two-thirds of Ellis-van Creveld syndrome patients. Hum. Genet. 120: 663–670.

Tuz, K., Bachmann-Gagescu, R., O'Day, D. R., Hua, K., Isabella, C. R., Phelps, I. G., Stolarski, A. E., O'Roak, B. J., Dempsey, J. C., Lourenco, C. et al. 2014. Mutations in CSPP1 cause primary cilia abnormalities and Joubert syndrome with or without Jeune asphyxiating thoracic dystrophy. Am. J. Hum. Genet. 94: 62–72.

Valente, E. M., Logan, C. V., Mougou-Zerelli, S., Lee, J. H., Silhavy, J. L., Brancati, F., Iannicelli, M., Travaglini, L., Romani, S., Illi, B. et al. 2010. Mutations in TMEM216 perturb ciliogenesis and cause Joubert, Meckel and related syndromes. Nat. Genet. 42: 619–625.

Veland, I. R., Awan, A., Pedersen, L. B., Yoder, B. K. and Christensen, S. T. 2009. Primary cilia and signaling pathways in mammalian development, health and disease. Nephron. Physiol. 111: 39–53.

Verma, I. C., Bhargava, S. and Agarwal, S. 1975. An autosomal recessive form of lethal chondrodystrophy with severe thoracic narrowing, rhizoacromelic type of micromelia, polydacytly and genital anomalies. Birth Defects Orig. Artic. Ser. 11: 167–174.

Walczak-Sztulpa, J., Eggenschwiler, J., Osborn, D., Brown, D. A., Emma, F., Klingenberg, C., Hennekam, R. C., Torre, G., Garshasbi, M., Tzschach, A. et al. 2010. Cranioectodermal Dysplasia, Sensenbrenner syndrome, is a ciliopathy caused by mutations in the IFT122 gene. Am. J. Hum. Genet. 86: 949–956.

Wang, Z., Iida, A., Miyake, N., Nishiguchi, K. M., Fujita, K., Nakazawa, T., Alswaid, A., Albalwi, M. A., Kim, O. H., Cho, T. J. et al. 2016. Axial spondylometaphyseal dysplasia is caused by C21orf2 mutations. PLoS One 11: e0150555.

Weyers, H. 1952. Ueber eine korrelierte Missbildung der Kiefer und Extremitatenakren (Dysostosis acro-facialis). Fortschr Roentgenstr 77: 5.

Wheway, G., Schmidts, M., Mans, D. A., Szymanska, K., Nguyen, T. M., Racher, H., Phelps, I. G., Toedt, G., Kennedy, J., Wunderlich, K. A. et al. 2015. An siRNA-based functional genomics screen for the identification of regulators of ciliogenesis and ciliopathy genes. Nature Cell Biology 17: 1074–1087.

Yang, S. S., Langer, L. O., Jr., Cacciarelli, A., Dahms, B. B., Unger, E. R., Roskamp, J., Dinno, N. D. and Chen, H. 1987. Three conditions in neonatal asphyxiating thoracic dysplasia (Jeune) and short rib-polydactyly syndrome spectrum: a clinicopathologic study. Am. J. Med. Genet. Suppl 3: 191–207.

Yang, S. S., Roth, J. A. and Langer, L. O., Jr. 1991. Short rib syndrome Beemer-Langer type with polydactyly: a multiple congenital anomalies syndrome. Am. J. Med. Genet. 39: 243–246.

Yeamans, E. H. 1973. OFD I syndrome and mental retardation. The Cleft Palate Journal 10: 84–91.

Yerian, L. M., Brady, L. and Hart, J. 2003. Hepatic manifestations of Jeune syndrome (asphyxiating thoracic dystrophy). Semin Liver Dis. 23: 195–200.

Yin, Y., Bangs, F., Paton, I. R., Prescott, A., James, J., Davey, M. G., Whitley, P., Genikhovich, G., Technau, U., Burt, D. W. et al. 2009. The Talpid3 gene (KIAA0586) encodes a centrosomal protein that is essential for primary cilia formation. Development 136: 655–664.

Zaffanello, M., Diomedi-Camassei, F., Melzi, M. L., Torre, G., Callea, F. and Emma, F. 2006. Sensenbrenner syndrome: a new member of the hepatorenal fibrocystic family. Am. J. Med. Genet. A. 140: 2336–2340.

Zangwill, K. M., Boal, D. K. and Ladda, R. L. 1988. Dandy-Walker malformation in Ellis-van Creveld syndrome. Am. J. Med. Genet. 31: 123–129.

Zhang, W., Taylor, S. P., Nevarez, L., Lachman, R. S., Nickerson, D. A., Bamshad, M., University of Washington Center for Mendelian Genomics, C., Krakow, D. and Cohn, D. H. 2016. IFT52 mutations destabilize anterograde complex assembly, disrupt ciliogenesis and result in short rib polydactyly syndrome. Hum. Mol. Genet.

CHAPTER 7

The Role of Cilia in Development and Disease of the Eye

S. Patnaik,[a] V. Kretschmer[b] and H. May-Simera*

INTRODUCTION

Cilia were once presumed to be vestigial organelles (Federman and Nichols, 1974; Webber and Lee, 1975), but have now emerged to be indispensable for various developmental and cellular processes. Although the existence of these structures was reported in the early 1980s, their functional role and involvement in development and disease has only become apparent in the last few decades. Primary cilia are microtubule based structures emerging from the surface of cells and predominantly act as signaling antennae, receiving external signals. Cilia function is involved in photosensation, osmosensation, mechanosensation, thermosensation, olfactory sensation, hormonal regulation and initiation of various signaling cascades (Fliegauf et al., 2007). Since cilia are ubiquitously present on all cell types, they play an important role in numerous functions in various tissues and organs. Our understanding of cilia function has been boosted since the advent of genetic testing, through which it was identified that ciliary defects are associated with various syndromic and non-syndromic diseases, collectively referred to as ciliopathies (Schwartz and Hildebrandt, 2011). Defects in ciliary genes have been associated with inherited disorders, most commonly including polycystic kidney disease (PKD), obesity syndromes, developmental disorders and retinal degeneration (Pazour and Rosenbaum, 2002; Pazour, 2005). Visual impairment is a common denominator observed in various

Institute of Molecular Physiology, Johannes-Gutenberg University, Mainz, 55128, Germany.
[a] E-mail: spatnaik@uni-mainz.de
[b] E-mail: vikretsc@uni-mainz.de
* Corresponding author: hmaysime@uni-mainz.de

syndromic and non-syndromic ciliopathies. The goal of this chapter is to discuss the role of cilia in the eye and how cilia defects affect visual function.

Anatomy of the Eye

Global eye anatomy

The eye is one of the most important sensory organs with a remarkable and complicated visual cycle, enabling us to visualize the external environment. The vertebrate eye is made up of three distinct layers, the outer layer, the middle layer and the inner layer. The outer layer consists of the anterior cornea and the posterior sclera. The cornea is a transparent layer that refracts the light and transmits it to the lens. The sclera is a white fibrous tissue comprised of extracellular matrix (ECM) and fibroblasts. It plays an important structural role by providing rigidity and elasticity, thus protecting the eye from intraocular pressure and external forces. It determines the shape and size of the eye and therefore influences the refractive state (Summers Rada et al., 2006).

The iris, ciliary body and choroid form the middle layer of the eye. The iris is conical shaped and rests on the lens (McCaa, 1982; Hughes, 2004). It consists of two sheets of smooth muscles that control the diameter and size of the pupil and thus determine the amount of light entering the retina. The melanocytes in the iris along with extracellular matrix components determine the different shades of eye color (Nischler et al., 2013). The ciliary body (so called because of the presence of ciliary processes, which are radial ridges, not because of its association with cilia) is positioned at the posterior surface of the iris and is attached to the lens via structures called zonules. The ciliary body shifts the focus by controlling the shape of the lens and also secretes the aqueous humor, which is essential for regulating the intraocular pressure within the eye (Tamm and Lütjen-Drecoll, 1996; Delamere, 2005). Finally, the choroid is a vascular structure that nourishes and provides oxygen to the outer retinal layers. It also participates in thermoregulation, light absorption and maintenance of intraocular pressure by controlling blood flow to the eye (Nickla and Wallman, 2010).

The inner layer of the eye is comprised of the retinal pigment epithelia and the multicellular retina, which is capable of transforming light energy into electric signals. This is arguably the most important component of the visual system as it contains and maintains the highly specialized photoreceptor neurons and all the other neuronal cell types required for signal transduction and visual processing.

Organization of the retina and retinal pigment epithelium

The inner layer of the eye, namely the retina and closely associated retinal pigment epithelium (RPE) line the inner surface of the eye, surrounding the vitreous cavity (Figure 1a). The retinal pigment epithelium is a monolayer epithelium whilst the retina is a multilayered neuroepithelium consisting of six major types of neurons arranged along parallel layers (Hildebrand and Fielder, 2011). Each will be discussed in more detail below.

The retinal pigment epithelium lies interposed between the photoreceptors and choroid and is comprised of a monolayer of cuboidal epithelial cells densely packed with melanin (Panda-Jonas et al., 1996). The retinal pigment epithelium's many roles include providing nutritional support to the photoreceptors and other neuronal cells and acting as a barrier between the blood supply of the choroid and the outer retina. It also participates in vitamin A retinal metabolism and is vital for the phagocytosis of the outer segment of photoreceptors (Katz and Gao, 1995). Photoreceptors (discussed below) undergo a daily renewal process, wherein 10% of the outer-segments are shed from the distal end and are subsequently phagocytosed by the retinal pigment epithelium (Strauss, 2005; Kevany and Palczewski, 2010). This process is highly metabolic and requires close interaction between the retinal pigment epithelium and the retina. From the apical surface of the retinal pigment epithelium extend numerous actin based apical processes

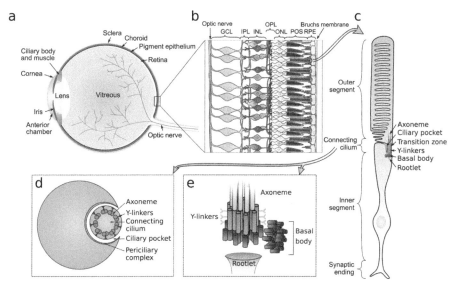

Figure 1. Schematic of the eye (**a**) focusing on the retina, highlighting the different layers of the retina (**b**). (**c**) Schematic of a rod photoreceptor showing specialized compartments. The inner segment (IS) contains the protein synthesizing machinery, the specialized outer segment (OS) contains the signal transduction machinery and the connecting cilium (CC)/transition zone connects the inner segment (IS) and outer segment (OS). (**d, e**) An enlarged view of the photoreceptor connecting cilium/transition zone showing the organization of the basal body and the axoneme. (**d**) is a whole mount view and (**e**) a cross section view. The photoreceptor rootlet projects from the basal body and extends into the IS. The Y-linkers connect the microtubule axoneme to the ciliary membrane. The periciliary complex invaginates and closely surrounds the CC axoneme, which acts as a docking site for intraflagellar transport (IFT) cargo.

that surround the photoreceptor outer segments and facilitate efficient phagocytosis.

The photoreceptor layer of the retina is the first cell type that directly associates with the retinal pigment epithelium (Figure 1b). Photoreceptor nuclei are arranged in the outer nuclear layer (ONL). Adjacent to the outer nuclear layer is the outer plexiform layer (OPL), a network of synapses between rod and cone axons, horizontal and bipolar cells (Willoughby et al., 2010; Hildebrand and Fielder, 2011). Then follows the inner nuclear layer (INL), which contains the cell bodies of amacrine cells, horizontal cells, bipolar cells and Müller glial cells. Next is the inner plexiform layer (IPL), where bipolar and amacrine cells synapse with retinal ganglion cells (RGCs) (Masland, 2001; Pignatelli and Strettoi, 2004; Willoughby et al., 2010). The retinal ganglion cell layer is the innermost layer of the retina, adjacent to the vitreous; it contains the cell bodies of ganglion and dispatched amacrine cells. Bipolar cells relay information from photoreceptor cells to ganglion cells and receive extensive synaptic feedback from amacrine cells and horizontal cells. Metabolic support is provided by Müller glial cells. The ganglion cells are the final output neurons of the retina. They transmit visual information transmitted vertically through bipolar cells processed laterally through horizontal and amacrine cells to deeper layers of the brain. This is followed by the retinal ganglion axons, the nerve fiber layer, which relays the processed signal from the retina to the brain (Rizzo III, 2005).

Photoreceptor structure

Photoreceptors are the photosensitive cells of the retina and are responsible for the process of phototransduction (Figure 1c). There are two types of photoreceptors, rods and cones, named after their differing appearance. They also differ in terms of their function, type of photopigment and pattern of synaptic connections (Ebrey and Koutalos, 2001). Rods are responsible for vision at low light levels (scotopic or night vision) and are very sensitive even at low light. Cones are responsible for color or day vision. They are much less sensitive than rods and generate signals at higher levels of light.

Photoreceptors consist of four distinct components: an outer segment (OS) and an inner segment (IS) connected by the so-called connecting cilium (CC) and lastly a synaptic terminal. The outer portion of the photoreceptor is a highly modified primary cilium that contains numerous light sensitive stacked discs. These membranous disc lamellae are the actual sites of phototransduction. Structurally, rods and cones share a similar architecture. However, there are differences in the way their outer segments are built. Rods have numerous membrane discs, stacked on top of each other and are not connected to the ciliary membrane (Palczewski, 2006), while its thought that the discs in cones are invaginations of the plasma membrane (Mustafi et al., 2009). The inner segment contains cellular organelles such as mitochondria, endoplasmic reticulum (ER) and Golgi, which do not interfere with biochemical reactions in the outer segment.

The ciliary nature of photoreceptors was first described by De Robertis in 1956 (Rods and Robertis, 1956). The photoreceptor outer segment is comprised of a highly modified primary cilium specialized to perform phototransduction. Similar to a primary cilium, the photoreceptor cilium is anchored via the basal body, derived from the mother centriole (Figure 1d, e). The basal body acts as a microtubule organizing centre (MTOC) and nucleates the ciliary axoneme (Besharse and Horst, 1990; Muresan et al., 1993; Troutt et al., 1990). The connecting cilium (CC) emerges directly from the basal body (Röhlich, 1975). Although termed the connecting cilium, the region connecting the inner and outer segments corresponds to the ciliary transition zone. In this region, fibrous links referred to as Y-linkers, connect the microtubule axoneme to the ciliary membrane. The function of the transition zone is most likely to regulate ciliary trafficking into and out of the ciliary axoneme (Omran, 2010; Szymanska and Johnson, 2012). Similar to other cilia, the ciliary rootlet extends away from the proximal end (non-axonemal end) of the basal body and is thought to be required for trafficking to the basal body, structural support for the cilium and long term maintenance of ciliated cells (Yang et al., 2005; Yang and Li, 2006). The ciliary rootlet is composed of longitudinally arranged microfilament bundles, of which the main component is Rootletin (Yang et al., 2002). A purely structural role for Rootletin is less likely, considering that loss of this protein only causes mild photoreceptor disruption in older animals (Yang et al., 2005).

Photoreceptor cilia are unique in that they have an extremely deep and expanded ciliary pocket, referred to as the periciliary ridge complex (PRC) that was first reported in amphibians (Peters et al., 1983). This is a highly specialized compartment of the apical inner segment plasma membrane that invaginates and closely surrounds the connecting ciliary axoneme (Liu et al., 2007). In amphibians the walls of the invagination have grooves separated by ridges. This region is central to the transport of opsin (Peters et al., 1983). It also acts as the docking site for trafficking vesicles arriving from the Golgi, at which point they are handed over to ciliary transport mechanisms (Ghossoub et al., 2011; Maerker et al., 2008; Papermaster et al., 1985; Papermaster, 2002).

An important difference between rodents and other species is the absence of calyceal processes. These are microvilli-like projections that extend from the apex of the inner segment to the proximal end of the outer segments. Calyceal processes are present in primates, amphibians and teleosts but are not present in rodents (Pagh-Roehl et al., 1992; Sahly et al., 2012). These actin filament-rich processes form a collar-like structure around the base of the outer segment. The polarized actin filaments extend into the inner segment forming thick actin roots. Both rods and cones possess calyceal processes, however, cones have a higher number of projections (between 14–16) that appear more regularly and are comprised of thicker actin bundles and thicker roots (Sahly et al., 2012). The calyceal processes are formed before the formation of the outer segments (Sahly et al., 2012). Although little is known about the function of these processes, they have been suggested to be required for shaping the growing rod outer segments (Marc, 1999). It has also been predicted that calcium within calyceal processes might regulate the plus-directed motor myosin VIIa (Inoue and Ikebe, 2003), thereby controlling mechanical tension (Sahly et al., 2012). Various Usher syndrome proteins have been localized to these structures, which suggests that they may be disrupted in human retinal degeneration disorders.

Rod and cone distribution

The distribution pattern of rods versus cones varies in the retina between species. Rodents and primates have a rod-dominant retina composed of 95–97% rods and just 3–5% of cones. Primates have a cone-rich macula and rod-free fovea, which forms the central region of the retina (Hendrickson, 2005). The human eye contains three different types of cones for perception of different colors; blue sensitive cones (short wave cones, i.e., S cones), green sensitive cones (middle wave cones, i.e., M cones) and red sensitive cones (long wave cones, i.e., L cones) which enable the photoreceptor to respond to different wavelengths of light (Solomon and Lennie, 2007). Many animals including fish, birds and reptiles have an additional type of cone sensitive for ultraviolet light with a sensitivity peak around 360 nm. In mouse the dorsal retina contains both S and M cones although it is rich in M cones, however the ventral retina exclusively contains S cones (Szél et al., 2000). Understanding the patterning and differences among the rods and cones is not only significant for color vision, but will help us to understand why rods are more prone to degeneration than cones.

Development of the Eye

Retina, retinal pigment epithelia and lens

The role of the cilium in the differentiation and development of photoreceptors is well established, however, whether the cilium or ciliary signaling is involved in other regions in the eye is less well understood. In cilia mutants that have been examined so far, development of the eye and ocular tissues other than the retina seems relatively well preserved. However, residual ciliary function is maintained in these models, as complete loss of cilia is incompatible with life. Generation of conditional mutants is also hampered by the fact that many ocular tissues and ciliary components are also required in many other developmental processes. For example, ciliary protein Arl13b plays an important role in the development of the cerebral cortex (Higginbotham et al., 2012).

The eye has a very complex multicellular organization. Different signaling events and transcription factors expressed in one cell type can either initiate or inhibit the formation of another cell type. After gastrulation, part of the neural tube evaginates to form the optic vesicle. This optic vesicle again evaginates until it comes into contact with the thickened ectoderm. At this location, the lens placode forms from the ectoderm. Next, coordinated invagination of the optic vesicle and lens placode takes place forming the optic cup and the lens pit (Chow and Lang, 2001; Cvekl and Piatigorsky, 1996). After the invagination of the optic cup, two layers of cells are established; the outer cell layer is destined to become the retinal pigment epithelium and the inner cell layer is destined to develop into the neural retina (Dollar and Sokol, 2007; Heavner and Pevny, 2012). Activin-like signals from the extraocular mesenchyme trigger the activation of transcription factors homeodomain-containing transcription factor (OTX2) and microphthalmia-associated transcription factor (MITF), which triggers the differentiation of the retinal pigment epithelium (Fuhrmann et al., 2000). Differentiation of the retinal pigment epithelium along with hedgehog signaling from the emerging retinal ganglion cell layer is responsible for neural retina proliferation, differentiation and retinogenesis (Jensen and Wallace, 1997; Strauss, 2005; Bharti et al., 2006). Patterning of the retina takes place throughout this process. A population of multipotent progenitor cells residing in the inner layer of the optic cup proliferate and differentiate into seven retinal cell types that become organized into the different cell layers, as described above. The differentiation of these cells occurs in a specific order, although there is considerable overlap (Belliveau and Cepko, 1999; Reese, 2010). Retinal ganglion cells and horizontal cells differentiate first, followed by cone photoreceptors, amacrine cells, rod photoreceptors, bipolar cells and Müller glia cells (Marquardt and Gruss, 2002; Reese, 2010).

Lens development takes place in parallel to retinogenesis. Emergence of the lens pit is followed by lens vesicle formation, which differentiates to form either lens fibers or lens epithelia. Lens epithelial cells produce new cells that exit the cell cycle and form secondary lens fibers that line up along the outer surface of the growing lens. Newly formed lens fibers are deposited over central cells and form a hexagonally packed fiber cell mass (Cvekl and Ashery-Padan, 2014). Alignment of the lens cells is important for lens packing and development. This process is regulated by planar cell polarity signaling (Sugiyama et al., 2011).

Although cilia have been observed on other cell types in the eye, the photoreceptor cilia have been most extensively studied. The process of photoreceptor ciliogenesis shares many similarities with regular ciliogenesis, but there are additional steps involved that are slightly different from regular cilia formation.

Development of photoreceptor cilia

Photoreceptor cilia grow outwards into the subretinal space between the developing retina and the retinal pigment epithelium. Similar to other types of cilia, photoreceptor ciliogenesis undergoes several different stages. The first three stages are identical to other cilia, whilst the final stages are specific to photoreceptors. The nomenclature of different stages is as per Pedersen et al. and Sedmak et al. (Pedersen et al., 2008; Sedmak and Wolfrum, 2011).

Cilia development in photoreceptor cells is initiated post terminal mitosis, when the paired centrioles migrate towards the plasma membrane. In stage one (S1) a solitary intracellular primary vesicle attaches to the distal end of the mother centriole (consisting of nine triplet microtubules) and thus becomes the basal body (Figure 2).

In the second stage (S2), Golgi-derived secondary vesicles fuse with the primary vesicle forming the ciliary vesicle. The ciliary vesicle forms the sheath of the protruding axonemal shaft. Microtubules assemble and the axoneme begins to emerge forming the transition zone, which in photoreceptor cells has been designated the connecting cilium (CC). As with regular transition zones, the connecting cilium has a 9 + 0 microtubule doublet arrangement. Similar to regular ciliary transition zones, doublet microtubules are arranged in a circular fashion and are connected to the ciliary membrane via Y-linkers, forming the ciliary necklace (Besharse et al., 1985; Röhlich, 1975). The connecting cilium continues to elongate along the growing primary vesicle (Greiner et al., 1981; Sedmak and Wolfrum, 2011).

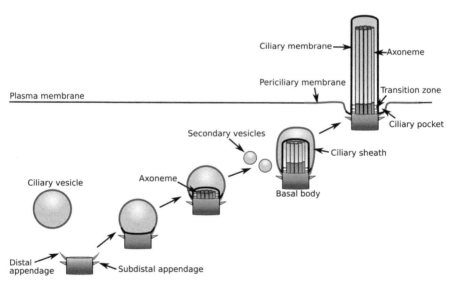

Figure 2. Schematic depicting the different stages of primary cilia formation. A primary vesicle attaches and fuses to the distal end of the mother centriole. This is followed by the addition of secondary vesicles that fuse with the primary vesicle to form the ciliary vesicle. The ciliary axoneme grows within the growing vesicle, giving rise to the ciliary shaft and ciliary sheath. The ciliary vesicle migrates and then fuses with the plasma membrane. This leads to the externalization of the cilium, which grows and protrudes from the surface of the cell.

In stage three (S3) the shaft and ciliary vesicle migrate to the apex of the cell. The ciliary vesicle fuses with the plasma membrane, protrudes from the surface of the cell and continues to lengthen (Sorokin, 1962; Pedersen et al., 2008; Sedmak and Wolfrum, 2011). The ciliary pocket forms at the base of the ciliary shaft. Accumulation of centriolar satellites at the base of the cilium is observed in the initial stages, whereas their numbers decrease between stage one and three and disappear completely in later stages, suggesting that centriolar satellites play a role in the transport of material required for ciliary biogenesis (Sedmak and Wolfrum, 2011).

In stage four (S4) outer segment disc morphogenesis is initiated at the base of the outer segment by expansion of the plasma membrane (Steinberg et al., 1981). The precise mechanism for this has not yet been fully elucidated. The plasma membrane at the distal end of the connecting cilium swells out and vesicles of different sizes are observed, which could be formed by pinocytosis. In mouse photoreceptors, a bulge at the tip of the connecting cilium is seen as early as post-natal day 5 (P5). By P6 this swelling has expanded and the rapid building of the discs takes place (Sedmak and Wolfrum, 2011).

In stage five (S5) some of these vesicles fuse to form larger vesicles and are compressed and flattened by an unknown mechanism, forming the outer segment discs, which align horizontal to the ciliary microtubules. In stage six (S6) the discs are partially stacked and randomly arranged (Obata and Usukura, 1992). The doublet microtubule axoneme emerges from the basal body and extends into the outer segment, at some point converting to singlet microtubules. In rods, the microtubule axoneme extends half way up the distal end of the outer segment; in cones, the microtubules extend to the tip of the outer segment (Roof et al., 1991).

In developing mouse photoreceptors, stage one to four occurs between postnatal day (P) 0-P3 whilst stages three to six are seen from P7 onwards (Sedmak and Wolfrum, 2011). Partially stacked discs can been seen between P6–P8 and regular stacking is observed by P14, where the partial stacks are rearranged to form the mature outer segments (Obata and Usukura, 1992). The formation of the cilium is critically dependent on a highly efficient ciliary transport machinery, discussed in the next section.

Intraflagellar Transport (IFT) in Photoreceptors

As with all cilia, growth, maintenance and function relies on highly efficient and organized cargo trafficking along the outer doublet microtubules, referred to as intraflagellar transport (IFT). In photoreceptors, intraflagellar transport not only plays an important role in photoreceptor synthesis, transport and maintenance, but also in phototransduction (Pazour et al., 2002; Wolfrum and Schmitt, 2000). Photoreceptors are highly metabolically active and the outer segments require constant membrane renewal, as the outermost discs are shed by phagocytosis of the retinal pigment epithelium, thereby creating a high need for protein replenishment (Young, 1967). Additionally, phototransduction occurs in the outer segment, yet protein synthesis occurs in the inner segment. The connecting cilium and outer segments do not have any transcriptional or translational machinery and therefore rely on protein synthesis in the inner segment and IFT trafficking to meet their protein demands. For this reason, there is a high demand for trafficking across the connecting cilium.

The growing end of the microtubules is referred to as the plus end and the nucleating end at the basal body is termed the minus end (Rosenbaum and Witman, 2002). Plus-end directed trafficking is referred to as anterograde transport, in which IFT particles carry cargo towards the tip of the cilium. Minus-end directed trafficking is termed retrograde transport, in which IFT particles transport cargo back to the base of the cilium. This bidirectional transport is mediated by two different types of microtubule-based motors: Kinesin-II motors responsible for anterograde trafficking and Dynein motors for retrograde trafficking (Rosenbaum and Witman, 2002; Scholey, 2003). There are two types of kinesin motors, heterotrimeric and homodimeric. Heterotrimeric Kinesin-II consists of two heterodimerized Kinesin-II motor subunits (KIF3A, KIF3B) and an accessory subunit, kinesin-associated protein (KAP) (Cole, 1999a; Cole, 1999b). Dyneins are multiprotein complexes that consist of four subunits, a heavy chain (DYNC2H1), a light intermediate chain (DYNC2LI1), an intermediate chain (IC) and a light chain (LC) (Rompolas et al., 2007; Hao et al., 2011). There are two sub-complexes of IFT particles which assemble to form 'rafts' to traffic cargo along the ciliary axoneme. Traditionally it was thought that the IFTA complex was associated with retrograde transport and IFTB anterograde transport, but this distinction is not clear cut. The IFTA complex consists of six core proteins IFT144, IFT140, IFT139, IFT122, IFT121 and IFT143 (Pedersen and Rosenbaum, 2008). The IFTB complex consists of 15 proteins IFT172, IFT88, IFT81, IFT80, IFT74, IFT72, IFT70, IFT57, IFT54, IFT52, IFT46, IFT27, IFT25, IFT22 and IFT20 (Taschner et al., 2012). All these IFT particles work together but have distinct roles in ciliary transport (Ishikawa and Marshall, 2011; Rosenbaum and Witman, 2002). IFT particles and their cargo accumulate and assemble at the basal body transition fibers before trafficking across the transition zone (connecting cilium) (Cole, 1999; Deane et al., 2001; Scholey, 2003, 2008). This process can be compared to a moving cargo train: the microtubules resemble the train tracks, Kinesin-II and Dynein are the wheels, IFT particles are the train compartments. Signaling molecules, membrane components, tubulin subunits and retrograde motors are all examples of cargo that need to be trafficked into and out of the cilium. IFT mechanisms have been elucidated to a great extent in recent years, but still the IFT complexes and their exact role are poorly understood. Although IFTA mutants share some phenotypes with dynein-2 mutants, both IFTA and IFTB complexes can travel in either direction. Recently it has been shown in *Caenorhabditis elegans* that IFT27, a component of the IFTB complex is involved in both retrograde and anterograde transport (Huet et al., 2014).

Protein localization studies in mice have shown that IFT is involved in all stages of photoreceptor ciliogenesis. IFT proteins are localized to the mother centriole or basal body during the initial stages of photoreceptor ciliogenesis. IFT52, IFT57, IFT88 and IFT140 have been seen not only localized at the base of the outer segment, but also in the evolving ciliary shaft and connecting cilium. In contrast, IFT20 only localizes to the basal part of the connecting cilium, where it is most likely required for the transport of vesicles for the formation of membrane discs (Sedmak and Wolfrum, 2011). Similar to other cilia mutant mouse models, Ift88/Tg737^{orpk} mutant mice have abnormal outer segments and show progressive photoreceptor degeneration (Pazour et al., 2002).

Diseases Relevant to Ciliary Defects in the Eye

Mutations in ciliary genes can cause isolated non-syndromic blindness or can be associated with multisystem defects with a wide range of clinical features. There are several types of isolated non-syndromic retinal dystrophies depending on the type of photoreceptors affected (Estrada-Cuzcano et al., 2012). These include cone dystrophy (Openshaw et al., 2008), cone-rod dystrophy (Hamel, 2007), macular dystrophy (Michaelides et al., 2003) or Leber's congenital amaurosis (LCA) (den Hollander et al., 2008). The most common syndromic ciliopathies include Bardet–Biedl syndrome (BBS), Ellis van Creveld syndrome (EVC), Meckel–Gruber syndrome (MKS), McKusick–Kaufman syndrome (MKKS), Joubert syndrome (JBTS), Senior-Løken syndrome (SLS), Nephronophthisis (NPHP), Usher syndrome (USH), Alström syndrome (ALMS), Oral-facial-digital syndrome type 1 (OFD1) and asphyxiating thoracic dystrophy (ATD). All of these syndromes are named after the clinician that first described them. The most common primary feature amongst syndromic ciliopathies is retinal degeneration, with other features including obesity, polydactyly, renal failure, hypogonadism, olfactory defects, learning difficulties and various secondary features (Table 1).

Retinal degeneration is most commonly attributed to retinitis pigmentosa (RP), a genetically inherited disease associated with progressive photoreceptor degeneration. Retinitis pigmentosa is one of the most common causes of blindness worldwide (Haim, 2002), with an incidence of 1:4,000 (Bunker and Berson, 1984; Hartong et al., 2006). Retinopathy is the most common clinical feature associated with syndromic ciliopathies. The concept of an association between ciliary defects and retinal degeneration first arose in the 1980s, when human nasal ciliary defects were observed in retinitis pigmentosa patients (Fox et al., 1980). It was thus suggested that there may be an association between ciliary defects and other diseases. However, "ciliary dysfunctions"

Table 1. Phenotypes associated with various ciliopathies.

	BBS	MKS	JBTS	JATD	OFD1	MKKS	SLS	NPHP	LCA	ALMS
Retinopathy	✓	✓	✓	✓			✓	✓	✓	✓
Polydactyly	✓	✓	✓		✓	✓				
Obesity	✓									✓
Kidney diseases	✓	✓	✓	✓	✓		✓	✓		
Situs Inversus	✓	✓	✓				✓			
Cognitive Impairment	✓	✓	✓		✓	.		✓	✓	✓

BBS: Bardet-Biedl syndrome; MKS: Meckel Gruber syndrome; JBTS: Joubert syndrome; JATD: Jeune syndrome; OFD1: Oro-facial-digital syndrome type 1; MKKS: McKusick-Kaufman syndrome; SLS: Senior-Loken syndrome; NPHP: Nephronophthisis; ALMS: Alstrom syndrome.

gained major attention at the turn of the century with the advent of human molecular genetics. Until now around 300 genes have been mapped that are associated with inherited retinal dystrophies (RetNet; http://www.sph.uth.tmc.edu/retnet/), the vast majority of which transcribe ciliary proteins.

As previously mentioned, the genes mutated in many ciliopathies exhibit great genetic overlap. For example, mutations in BBS2, BBS4 and BBS6 have been identified in patients with MKS, as well as Bardet-Biedl Syndrome (Gerdes et al., 2009). This results in complications in the identification of these diseases and makes the development of therapeutic approaches for these patients challenging.

Usher syndrome is the most common cause of deafness and blindness. It affects the neurosensory cells in the cochlea and the retina. USH is genetically and clinically heterogeneous, with 15 chromosomal loci assigned to 3 clinical types, USH1, USH2 and USH3. Although USH is not a classic ciliopathy, it is classified as a subset of ciliopathies since the phenotype of the patients overlaps with many of the ciliopathy phenotypes, and USH proteins have been shown to localize to cilia. Whilst most USH mouse models recapitulate human deafness, they do not recapitulate the retinal degeneration observed in patients. One hypothesis is that this is because of the lack of calycael processes in rodent photoreceptors (Friedman et al., 2011; Sahly et al., 2012).

Animal Models

In recent years various animal models such as mouse, rat, zebrafish, chicken and dog have contributed to dissecting the mechanisms and regulatory pathways governing the development and maintenance of the eye. These animal models have also been very useful for establishing preclinical trials for retinal diseases, particularly for gene-based therapies (Boye et al., 2013). Mouse and rat models are the most commonly used models and there are numerous naturally occurring mouse models of retinal degeneration, like the retinal degeneration 1 (*rd1*) mouse (Samardzija et al., 2010). A rapidly increasing number of genetically engineered mouse models currently exist, in which constructs have been used to silence or overexpress genes of interest. Transgenic mouse models of retinal degeneration will be discussed further in the section below and are summarized in Table 2.

Although mouse models are commonly used, large animal models such as dog, cat and pig are becoming increasing popular, not only to study disease mechanisms but also to assess the safety and efficacy of treatments in clinical trials. Mutant dog models for PDE6β, PDE6α, RPE65, RPGR, RPGRIP1, RHO, RD3 and NPHP4 show strong genotypic and phenotypic correlations with patients (Beltran, 2009). For several reasons, the transgenic pig is also emerging as an attractive alternative animal model. In many aspects the pig eye more closely resembles the human eye compared to the mouse. Additionally, pig eyes are well suited for retinal injections for gene therapy. So far, pig models for autosomal dominant retinitis pigmentosa, expressing human rhodopsin mutations have been created. The first such transgenic pig model expressed the Pro347Leu rhodopsin mutation and was created by Petters et al. (Petters et al., 1997). These pigs showed various characteristic features that are seen in patients with RP (Petters et al., 1997; Li et al., 1998; Ross et al., 2012).

Although there are several advantages to using large animal models, maintaining animal facilities is expensive and the phenotypes can be slow to develop. For this reason, the zebrafish (*Danio Rerio*) has become an attractive model of choice, given the ease with which genetic manipulations can be carried out. The zebrafish offers a relatively inexpensive and time efficient model, as the retina undergoes rapid development and they produce a large number of offspring. Early studies in zebrafish used forward genetics screens and various retinal degeneration associated gene mutations were thus identified (Malicki et al., 1996; Fadool and Dowling, 2008). Increasingly, transient morpholino knockdown models have been generated to study many different syndromic and non-syndromic retinal degeneration diseases. Although the morpholino knockdown approach is limited to six days post fertilization, emerging genomic editing tools such as transcription activator-like effector nuclease (TALEN) and clustered regularly interspaced short palindromic repeats associated protein 9 (CRISPR-Cas9) have made genetic manipulations more robust and long lasting. Therefore, transgenic zebrafish are increasingly used to study retinal degenerations (Raghupathy et al., 2013).

There are a few reports of avian models of retinal degeneration, including an LCA chicken model. Chickens have cone dominant retinas like humans and are sometimes used as an alternative model to study retinal degeneration (Ali et al., 2011; Ulshafer et al., 1984).

Anatomical differences among species lead to different susceptibility to retinal degenerations. One major issue is the absence of calyceal processes in rodent photoreceptors, which may be important in a subset of ciliopathy syndromes and in Usher syndrome. In this regard, other species such as primates, *Xenopus laevis*, teleosts or *Rana catesbeiana* (bullfrog) would be

Table 2. Best characterized ciliopathy genes and animal models with retinal degeneration.

Gene	Protein localization/ In the retina and/or primary	Mouse retinal phenotype/cell ciliary defect	Human disease	References
ALMS1	BB	Reduction in ONL at 24 weeks, mislocalization of rhodopsin, normal localization of cone opsin	ALMS	(Collin et al., 2005) (Jagger et al., 2011)
Ahi1/Jouberin	TZ, ONL, axoneme, horizontal, bipolar, and ganglion cell bodies	Fail to develop complete photoreceptors, apoptosis apparent by P22, significant loss at P24 and complete loss of photoreceptors by P200	JBST	(Westfall et al., 2010)
BBS1	BB	Intact CC, abnormal OS, aberrant ERG	BBS	(Mykytyn et al., 2002) (Davis et al., 2007)
BBS2	BB	Rhodopsin mislocalization and retinal degeneration due to apoptosis. 50% of ONL degeneration by 5 months and complete ONL degeneration at 10 months		(Nishimura et al., 2004)
BBS3/ARL6/ RP55	Photoreceptor and ganglion cell layer, nerve fiber layer	Disruption of the normal photoreceptor architecture	BBS	(Pretorius et al., 2010) (Abu Safieh et al., 2010)
BBS4	BB, IS, OPL	Disruption of intraflagellar transport of proteins; complete loss of photoreceptors by 7 months	BBS	(Davis et al., 2007) (Abd-el-barr et al., 2007) (Mykytyn et al., 2004)
BBS6/MKKS	BB	IS and OS reduced by 2 months of age, ONL completely lost by 8 months	BBS	(Fath et al., 2005)
BBS7	BB	Significant degeneration of IS, OS and ONL by 4–6 months	BBS	(Zhang et al., 2013)
BBS8/TTC8/ RP51	BB, CC, OS	Retinal degeneration and mislocalization of rhodopsin to photoreceptor inner segments	BBS	(Tadenev et al., 2011)
BBS10	BB	Loss of IS, OS and ONL thinning at 3 months. No mislocalization of rhodopsin, although reduction of rhodopsin content is observed	BBS	(Cognard et al., 2015)

BBS17/ LZTFL1	BB	Photoreceptor degeneration due to accumulation of non-OS proteins in OS	BBS	(Datta et al., 2015)
CEP290	CC/TZ	Mislocalization of rhodopsin and arrestin in rd16 mouse. Rapid photoreceptor degeneration, however the severity of degeneration and ciliogenesis differ, depending on the mouse strain	BBS, LCA, JBTS, NPHP, MKS, SLS	(Chang et al., 2006) (Cideciyan et al., 2011) (Cideciyan et al., 2007) (Rachel et al., 2015)
LCA5	CC	LCA5 mutant mice have mutated lebercilin. These mice display cone and rod opsins mislocalization	LCA	(Boldt et al., 2011)
NPHP1	CC	Mutants show defect in protein sorting. OS protein accumulation in IS, causing disorganized OS and also IFT defects	NPHP, SLS, JBTS	(Jiang et al., 2009)
NPHP4	CC	Mutants display severe photoreceptor degeneration. OS protein mislocalized to IS. OS do not develop. Synaptic ribbons develop normally but degenerate	NPHP, SLS	(Won et al., 2011)
RPGR	CC/TZ/BB	Slow retinal degeneration. Opsin mislocalization is observed by postnatal day 20, prominent retinal degeneration is not observed until 6 months of age	RP3	(Hong et al., 2000) (Huang et al., 2012) (Brunner et al., 2010)
RPGRIP1	CC/TZ	Mislocalization of OS proteins. Mutant mice develop oversized OS discs		(Won et al., 2009) (Pawlyk et al., 2010)

ALMS1: Alstrom syndrome protein 1; BBS: Bardet-Biedl syndrome; ARL6: ADP-ribosylation factor (ARF)-like-6; RP: Retinitis Pigmentosa; MKKS: McKusick-Kaufman syndrome; TTC8: Tetratricopeptide Repeat Domain 8; LZTFL1: Leucine Zipper Transcription Factor-Like 1; CEP290: Centrosomal protein 290; LCA: Leber's congenital amaurosis; NPHP: Nephronophthisis; RPGR: Retinitis pigmentosa GTPase regulator; RPGRIP1: Retinitis pigmentosa GTPase regulator interacting protein 1; RPGRIP1L: RPGRIP1-like protein; MKS: Meckel Gruber syndrome; JBTS: Joubert syndrome and SLS: Senior-Loken syndrome; BB: basal body; CC: connecting cilia; TZ: transition zone; OS: Outer segment; IN: inner segment; ONL: outer nuclear layer and IFT: Intraflagellar transport; P: postnatal day.

a better model as they all possess these structures (Sahly et al., 2012). A second issue is the absence of a fovea and macular in many animal models, particularly non-primates, such as rodents. Although the dog is an excellent model for retinitis pigmentosa, the lack of a fovea-macula region is one of the disadvantages in comparison to other animal models, particularly when studying diseases of the macula. However, it is also argued that dogs and some non-primates have a specialized central area referred to as the *area centralis*, which shares similar features to that of the primate fovea (Beltran et al., 2014; Mowat et al., 2008; Peichl, 2005).

Photoreceptor Protein Complexes and their Role in Ciliopathies

Ciliary proteins have diverse functional roles, depending on their localization and function in different compartments of the photoreceptor. The protein complexes involved in these compartments also have diverse functions. In an effort to identify the different protein complexes present in the separate compartments, Liu et al. performed comparative proteomics analysis on mouse photoreceptor inner and outer segments. They identified ~ 1200 proteins present in the OS and ~ 800 proteins present in the inner segment (Liu et al., 2007). In recent years many studies have begun mapping the complicated network of interactions among these proteins and their involvement in human disease. Below we group the different retinal ciliopathies caused by mutations in genes, based on the separate compartments they affect and their functions.

Outer segment proteins

Mutations in the gene for Retinitis Pigmentogase 1 (RP1) are the major cause of autosomal dominant retinitis pigmentosa (adRP) (Bowne et al., 1999). RP1 is a photoreceptor specific protein which localizes to the outer segment axoneme and is found at the sites of newly formed disc membranes (Guillonneau et al., 1999; Liu et al., 2002; Liu et al., 2004). It is also involved in the orientation and stacking of the outer segment discs (Liu et al., 2003). It is a microtubule associated protein (MAP) that stabilizes the photoreceptor axoneme and also plays an important role in controlling axoneme length (Liu et al., 2004). RP1 mutant mice show accumulation of rhodopsin in the inner segment. These mice experience rapid photoreceptor degeneration and have disorganized outer segment discs. Similarly, mutations in the genes encoding the β subunit of cGMP-phosphodiesterase (PDE6B) (Farber, 1995), RDS/peripherin (Arikawa et al., 2011) and rhodopsin (Sung et al., 1993) are also associated with retinitis pigmentosa and their protein products localize to the outer segments. In most of these cases, mutations disrupt the targeting of proteins from the inner to the outer segment. For example, rhodopsin mutations cause mislocalization of rhodopsin in photoreceptors, which leads to cell death, yet the biochemical properties rhodopsin are not altered (Concepcion and Chen, 2010).

Connecting cilia/transition zone proteins

The transition zone of photoreceptors is highly selective and acts as a diffusion barrier, regulating the movement of proteins in and out of the cilium. The connecting cilium measures around 200 to 500 nm in length and 170 nm in diameter. This space is tightly packed with the axoneme and Y-linkers. As all trafficking between the inner segment and outer segment takes place in this limited space, it is not surprising that the connecting cilium exerts regulatory properties on what is trafficked in and out of the cilium (Insinna and Besharse, 2008). For example, rhodopsin is effectively trafficked through the connecting cilium and enters the photoreceptor outer segment, however, the outward flow of rhodopsin back into the inner segment is restricted (Spencer et al., 1988).

There is a growing list of proteins that have been identified in the connecting cilium. Some of these will be discussed further below.

RPGR

Retinitis pigmentosa GTPase regulator (RPGR) is one of the key proteins that localizes to the connecting cilium. Mutations in the *RPGR* gene are a common cause of X-linked retinitis pigmentosa (XLRP) (Vervoort et al., 2000; Breuer et al., 2002). XLRP patients show an inconsistent pattern of rod and cone degeneration (Huang et al., 2012). There are various splice variants of the *RPGR* gene, however, all the variants encode an N-terminal RCC1-like domain (RLD) homologous to the RCC1 protein, a guanine nucleotide exchange factor for the small GTP-binding protein Ran (Renault et al., 1998). Although the exact role of RPGR has not yet been elucidated, the presence of the RLD domain suggests ciliary localization of RPGR (He et al., 2008) and raises the possibility of a trafficking role for RPGR. Various proteins such as RPGR-interacting protein 1 (RPGRIP1), RPGRIP1-like (RPGRIP1L), Ras-related proteins in brain (RAB8), Structural Maintenance of Chromosome 1/3 (SMC1/3) and Phosphodiesterase-δ PDEδ (Wätzlich et al., 2013) bind to the RLD domain (Patnaik et al., 2015). RPGR has been reported to primarily associate with the GDP-bound form of RAB8A and to stimulate GDP/GTP nucleotide exchange (Murga-Zamalloa et al., 2010). RAB8 enters the cilia and promotes extension of the ciliary membrane. RPGR also regulates the trafficking of rhodopsin to the outer segment of photoreceptors (Deretic et al., 1995; Moritz et al., 2001). There are two major forms of RPGR. Firstly RPGR[ex1-19] which has 19 exons and a C-terminal isoprenylation site, and secondly RPGR[ORF15], which contains exons 1 to 14 followed by a large ORF15 exon (Meindl et al., 1996; Roepman et al., 1996; Schwahn et al., 1998). The RPGR[ORF15] isoform is highly expressed in photoreceptors and most mutations occur in exon ORF15 (Kirschner et al., 1999; Vervoort et al., 2000). One of the reasons RPGR is so important is that it interacts with other ciliary proteins such as RPGRIP1, RPGRIP1L, CEP290, NPHP4, NPHP5, IFT88, KIF3A, PDEδ, Spermatogenesis associated 7 (SPATA7) and Whirlin, which are also implicated in ciliopathies (Boylan and Wright, 2000; Boylan and Wright, 2000; Patnaik et al., 2015). Most of these proteins play central roles in ciliary gating, as mutant mice show mislocalization of opsin molecules. There are two naturally occurring mutant dog models (Beltran et al., 2006; Zeiss et al., 1999; Zhang et al., 2002) and several murine models of RPGR (Brunner et al., 2010; Chang et al., 2002; Hong et al., 2000; Hong, 2004; Thompson et al., 2012). The RPGR knockout mouse shows relatively slow retinal degeneration and although mislocalization of opsin is seen by postnatal day 20, prominent retinal degeneration is not observed until six months (Hong et al., 2000). Surprisingly, recent studies of RPGR conditional mutants with a proximal promoter and exon1 deletion showed relatively early onset retinal degeneration (Huang et al., 2012). These discrepancies may result from different background strains or other unknown genetic modifiers.

RPGRIP1 and RPGRIP1L

RPGRIP1 is an interacting partner of RPGR (Boylan and Wright, 2000), which is often mutated in LCA patients (Meindl et al., 1996). It is a large multi-domain protein consisting of an N-terminal coiled coil domain, two protein kinase C (PKC; C2) domains (C2N and C2C) and a C-terminal RPGR-interacting domain (RID). Most of the LCA mutations in patients have been identified in the C2 domains (Dryja et al., 2001). Also, the mutant mouse model of RPGRIP1 shows retinal degeneration. In contrast to RPGR, RPGRIP1 knockout mice exhibit rapid retinal degeneration, similar to the phenotype observed in LCA patients (Zhao et al., 2003). Recently, it has been shown that SPATA7 is required for localization of RPGRIP1 to the connecting cilium and mutations in this gene cause LCA3 and juvenile retinitis pigmentosa (Eblimit et al., 2015).

Another protein homolog of RPGRIP1 is RPGRIP1L (RPGRIP1 like), which also interacts with RPGR (Khanna et al., 2009). Mutations in RPGRIP1L are associated with MKS and JBTS (Arts et al., 2007; Delous et al., 2007). RPGRIP1 mutations are mostly associated with non-syndromic retinal degeneration, however, RPGRIP1L mutations are involved in a wide range of ciliopathies. Knockout mouse models of RPGRIP1L show a characteristic resemblance to MKS fetuses, including retinitis pigmentosa, polydactyly, kidney cysts, defects in the neural tube and brain patterning, liver and heart defects, craniofacial malformation and planar cell polarity defects (Arts et al., 2007; Vierkotten et al., 2007; Devuyst and Arnould, 2008; Sang et al., 2011; Chen et al., 2015). Both RPGRIP1 and RPGRIP1L interact with NPHP4. Mutations in NPHP4 are associated with nephronophthisis (an autosomal recessive renal disease) and Senior Loken Syndrome (a combination of nephronophthisis and retinitis pigmentosa) (Roepman et al., 2005).

CEP290 (NPHP6)

Another key component of the connecting cilium is CEP290 (NPHP6). CEP290 is mutated in ~ 15–20% of patients with retinitis pigmentosa (Hollander et al., 2008). CEP290 interacts with several ciliary proteins and is thereby a component of many different protein complexes, e.g., it forms complexes with NPHP5, RPGR and BBS6 (Meindl et al., 1996; Rachel et al., 2012; Roepman et al., 1996). It is thought that CEP290 is a component of the Y linkers (Rachel et al., 2015). SSX2IP, a centriolar satellite protein has been shown to target CEP290 to the transition zone (Klinger et al., 2014). CEP290 mutations show a high degree of variable clinical manifestations in patients. Some mutations cause isolated LCA, however other mutations cause various types of ciliopathies, including MKS, NPHP, BBS, and JBST (den Hollander et al., 2008; Coppieters et al., 2010). CEP290 is mutated in the rd16 mouse, in which an in-frame deletion results in a truncated protein (Chang et al., 2006). CEP290$^{rd16/rd16}$ mice display normal photoreceptor formation but a rapid rod photoreceptor degeneration is observed as early as P14 (Chang et al., 2006). Similar to other retinal degeneration mutants, cone degeneration is slower than rod degeneration (Cideciyan et al., 2011; Cideciyan et al., 2007). Only few CEP290 null mice (CEP290$^{KO/KO}$) survive post weaning, and those that do lack connecting cilia and outer segments and display early vision loss (Rachel et al., 2015).

Ciliary transport proteins

Proteins involved in ciliary transport are indispensable for photoreceptor formation, function and survival. In addition to the IFT complex proteins, there are other proteins, which actively mediate ciliary protein trafficking. These include the BBS proteins. So far, 19 different genes have been identified which independently cause the disorder (Harville et al., 2010; Forsythe and Beales, 2013). Eight BBS proteins (BBS1, BBS2, BBS4, BBS5, BBS7, BBS8, BBS9) and BBS18 (BBIP10) form a complex referred to as the BBSome (Nachury et al., 2007; Loktev et al., 2008). The BBSome associates with the ciliary membrane and functions in vesicular trafficking of membrane proteins along the primary cilium (Jin et al., 2010). Mutations in the BBSome subunits cause mistargeting of organ-specific ciliary signaling receptors, leading to multi-organ pathology, which is characteristic of BBS (Nachury et al., 2007; Loktev et al., 2008). The second group of BBS proteins is a group of chaperonin-like proteins that regulate assembly of the BBSome. These are MKKS (BBS6) (Fath et al., 2005), BBS10 and BBS12 (Seo et al., 2010), which interact and are thought to function together. LZTFL1 (BBS17) negatively regulates

BBSome trafficking to the ciliary membrane (Seo et al., 2011). Photoreceptor degeneration preceded by rhodopsin mislocalization is a common feature of all the BBS mutant mouse models examined (BBS2 (Nishimura et al., 2004), BBS4 (Mykytyn et al., 2004), BBS6 (Fath et al., 2005), BBS7 (Zhang et al., 2013), BBS8 (Tadenev et al., 2011) and of the BBS9 morphant zebrafish (Veleri et al., 2012)).

Ciliary targeting

Usher syndrome proteins play a major role in protein trafficking and targeting to the cilium. They mostly localize to the photoreceptor periciliary region or to calyceal processes. These proteins are thought to be involved in molecular trafficking processes and are required for cargo delivery to the outer segment (Maerker et al., 2008; Cosgrove and Zallocchi, 2014). For example, in Myosin VIIa mutant mice, also known as Shaker-1 mice, opsins accumulate in the connecting cilium, suggesting a role in protein transport (Liu et al., 1999). Other Usher proteins are implicated in cell adhesion (Ahmed et al., 2003; Sahly et al., 2012; Cosgrove and Zallocchi, 2014). As discussed above, one of the major differences between primates and rodents is the presence (or absence) of calycael processes. Localization of several Usher proteins to this structure, coupled with the fact that Usher mutant mice do not exhibit photoreceptor degeneration, indicates that this region is particularly important in primate photoreceptor function. It also highlights the importance of other animal models such as *Xenopus laevis, Rana catesbeiana* (bullfrog) or the pig for studying Usher protein function and the disease mechanisms (Sahly et al., 2012).

Basal body

The mother centriole of the basal body has distal and subdistal appendages, which extend from the proximal end. Mutations in appendage proteins have also been found in ciliopathy patients. CEP164 is a distal appendage protein which is involved in the formation of transition fibers (also called alar sheets), once the basal body docks to the plasma membrane. Oral-facial-digital type I (OFD1), another basal body component, is required for the localization of CEP164 to the primary cilium (Singla et al., 2012) and mutations in this gene cause Oral-facial-digital syndrome and are associated with X-linked Joubert syndrome (Coene et al., 2009).

Cilia and Ciliary Signaling in Other Ocular Tissues

Although photoreceptor cilia have received the greatest attention in the last few decades, there are several reports of cilia in other ocular cell types. Primary cilia have been identified on other retinal cells, on lens fiber cells, in the trabecular meshwork and in the retinal pigment epithelium (Boycott and Hopkins, 1984; Chamling et al., 2014; Ennis and Kunz, 1986; Ferraro et al., 2014; Kim et al., 2013). One of the possible functions of primary cilia in these other cells types is to regulate cell signaling. The primary cilium is a coordinating hub for a growing number of signaling pathways such as canonical and non-canonical Wnt (Wallingford and Mitchell, 2011; May-Simera and Kelley, 2012), Sonic hedgehog (Shh) (Dafinger and Liebau, 2011; Goetz and Anderson, 2010; Rix et al., 2011), Fibroblast growth factor (FGF) (Hong and Dawid, 2009; Tanaka et al., 2005), Notch (Ezratty et al., 2011; Marcet et al., 2011), mechanistic target of rapamycin (mTOR) (Boehlke et al., 2010), Platelet derived growth factor receptor α (PDGFR α) (Schneider et al., 2005), and Hippo signaling pathways (Habbig et al., 2011). Sequestration of proteins at the cilium can influence signal transduction in two ways. Firstly, the transport of signaling proteins into the cilium is required for activation of downstream signaling molecules, conversely, proteins can be trapped in the cilium, thereby preventing their function in downstream signaling events. Although primary cilia signaling has been extensively studied in various other tissues, the signaling function of primary cilia in ocular tissues is largely unknown. However, recent studies have shown that cilia function is important in several parts of the eye as described below.

Role of Müller glia cilia in regeneration of the retina

The first report of cilia in Müller glia cells was published in the 1980s (Ennis and Kunz, 1986). In the teleost retina, a primary cilium can be observed in juvenile and adult Müller glia cells. Müller glia cells are the only cell type in the retina that possesses regeneration capacity in teleost. On retinal injury, Müller glia cells undergo reprogramming and are able to undifferentiate from a differentiated to a progenitor-like state, that can regenerate all major retinal cell types in zebrafish (Goldman, 2014). Similar to many cultured cells types, mature Müller glia primary cultures isolated from rat have primary cilia. Small interfering RNA (siRNA) mediated suppression of IFT20 in these cells prevents Müller cell proliferation that is controlled by Shh signaling. This highlights the importance of cilia in Müller cells and also their potential role in the process of retinal regeneration (Ferraro et al., 2014).

The mechanosensory role of cilia in the trabecular meshwork (TM)

The primary cilium has also been shown to regulate intraocular pressure in the eye. The trabecular meshwork (TM) maintains intraocular pressure and is responsible for drainage of aqueous fluid (Llobet et al., 2003). Any defects in sensing pressure changes cause an imbalance of the aqueous humor. Low ocular pressure causes structural changes, whereas high pressure results in optic nerve damage, a hallmark in patients with Glaucoma. Tumor Necrosis factor (TNF-α) and transforming growth factor (TGF-β) accumulate in the aqueous humor in these patients (Ozcan et al., 2004). Primary cilia in the trabecular meshwork have been shown to sense pressure changes in the eye (Luo et al., 2014). An important protein involved in this process is oculocerebrorenal syndrome of Lowe (OCRL), an inositol polyphosphate 5-phosphatase, which interacts with a transient receptor potential vanilloid 4 (TRPV4) channel. The two proteins function together in osmotic regulation and calcium influx. OCRL is a ciliary protein, which is required for localization and function of TRPV4 to the cilia. Mutations in OCRL cause a decrease in cilia length, and result in increased levels of TNF-α, TGF-β and GLI1, resulting in increased intraocular pressure (Luo et al., 2014).

Signaling functions of cilia in the lens

The lens consists of two major populations of cells, lens fiber cells and lens epithelial cells. Elongated fiber cells form the central mass, while epithelial cells surround the outside of the lens. The epithelial cells are proliferative in nature and form secondary lens fibers upon cell cycle exit (Cvekl and Ashery-Padan 2014). The characteristic feature of the lens is the coordinated alignment of cells and this process is regulated by planar cell polarity (PCP) signaling (Sugiyama et al., 2010; Sugiyama et al., 2011). Both lens fibers and epithelial cells are ciliated. The presence of an apically situated cilium which polarizes towards the anterior pole is a readout of planar cell polarity. Core components of the planar cell polarity pathway, such as Fz6, Vangl2, Dvl2, Dvl3, and inversin are localized in the lens fiber and regulate fiber cell alignment (Sugiyama et al., 2010; Sugiyama and McAvoy, 2012). One hypothesis is that cilia might be functioning to regulate the alignment of lens fibre cells via regulation of planar cell polarity signaling (Sugiyama et al., 2011). However, recent reports have shown that lens fibers are normally polarized in cilia mutants (Sugiyama et al., 2016).

Cilia in the retinal pigment epithelium

Cilia in retinal pigment epithelium derived cell lines such as hTERT-RPE1 and ARP19 cells have been well characterized. These cell lines have long been used in the cilia community to elucidate ciliogenesis and cilia function. However, these cell lines are far removed from native retinal pigment epithelium tissue. Early passage primary cultures of human retinal pigment epithelial cells show higher expression of proteins involved in cell migration, adhesion and the extracellular matrix, compared to the immortalized cells lines. Simultaneously, genes required for cell polarization are downregulated in immortalized cell lines (Alge et al., 2006). Furthermore, significant differences in gene expression profiles between hTERT-RPE1 and ARP19 cells highlight the importance of examining retinal pigment epithelium cilia *in vivo*. Few previous reports of cilia in the mammalian retinal pigment epithelium have been described, one in the cat (Fisher and Steinberg, 1982) and one in the rat (Nishiyama et al., 2002). A recent study in mice has shown that a single primary cilium emerges at embryonic day (E)14 and is important for regulating Wnt signaling in the developing retinal pigment epithelium (May-Simera et al., under review). This study also reported retinal pigment epithelium defects in cilia mutant mice, which might contribute to the rate of retinal degeneration.

Therapeutic Approaches

Better understanding of the genetic basis and mechanisms underlying ciliopathies has massively improved diagnosis and therapeutic approaches over the last decade. Furthermore, the accessibility of the retina coupled with its ocular immune privilege, by which the eye attempts to limit local immune and inflammatory responses, makes this tissue an ideal target for gene therapy. The concept of retinal gene therapy was reported in the early 1990s. This principle involves introducing a therapeutic gene into the retinal pigment epithelium or photoreceptors with the help of viral or non-viral vectors (Lem et al., 1992; Bok, 1993). Various modified viruses like retroviruses, lentiviruses, adenoviruses and adeno-associated viruses (AAVs) are used to deliver genes (Kay, 2011). Currently adeno-associated viruses are the most commonly used vehicles for retinal gene therapy, as they can target both the retinal pigment epithelium and the photoreceptors depending on which serotype is used (Becquet, 2000). Successful gene therapy in mice and canine models has hugely advanced this field, leading to clinical trials in patients (Acland et al., 2001; Pawlyk et al., 2010; Lhériteau et al., 2014). Gene replacements by sub-retinal delivery of RPE65 cDNA in LCA patients with RPE65 mutations, improved visual acuity (Bainbridge et al., 2008). Currently, four clinical trials for the treatment of LCA are ongoing, highlighting the efficacy of using adeno-associated virus gene therapy as a treatment method (Dalkara and Sahel, 2014). Other clinical trials that use gene therapy as a tool for retinal degeneration include Stargardt's (ABC4 gene) and USH (MYO7A) (Boye et al., 2013; McClements and MacLaren, 2013). Despite the advancements in the field of gene therapy, it remains challenging to prevent disease progression, as photoreceptor degeneration continues even after gene therapy (Cideciyan et al., 2013; Dalkara and Sahel, 2014). In addition to this, many retinal ciliopathies exhibit a high degree of genetic heterogeneity and phenotypic variability, making it difficult and extremely expensive to treat individuals with specific gene mutations.

Broader therapeutic strategies include treatment with neuroprotective, anti-apoptotic or anti-oxidant drugs in order to delay or prevent retinal degeneration. The use of translational read-through drugs (TRIDS) looks increasingly promising. TRIDS target in-frame nonsense mutations and are shaping up to be an attractive alternative for gene therapy (Nagel-Wolfrum et al., 2014). Drug re-profiling has also been suggested as an effective and safe treatment in some patients. Clinical trials have shown the therapeutic potential of drugs, such as topical unoprostone isopropyl for treating patients with retinitis pigmentosa (Yamamoto et al., 2012; Akiyama et al., 2014). Alternative approaches for improving gene therapy administration include the use of lipid nanoparticle-based therapy (Wang et al., 2015).

In recent years, progress has been made in the field of stem cell transplantation therapy in retinal degeneration. Embryonic stem cell derived retinal pigment epithelia cell replacement therapy is already in clinical trials (Schwartz et al., 2015), and photoreceptor progenitor cells replacement therapy has been successful in the rodent retina (Gonzalez-Cordero et al., 2013). Although many researchers have attempted to derive mature photoreceptors from induced pluripotent stem cells (iPS cells), none have managed to successfully grow photoreceptors in culture. Although there have been successful stem cell studies in animal models, there are many issues that need to be considered when transferring these studies to human patients (Megaw et al., 2015). The risk of rejection by the immune system poses one of the greatest problems (He et al., 2014). Other limitations include teratoma formation, risk of oncogene expression and low survival rate of transplanted cells (He et al., 2014; Roeder and Radtke, 2009).

Although, great success has been achieved in the field of gene therapy using animal models, many challenges still need to be addressed before gene therapy becomes routine treatment for human patients. Combinational therapies with the above mentioned strategies may be the best approach to restore vision and delay retinal degeneration.

Conclusion

The importance of primary cilia in the development and homeostasis of many different tissues and organs is becoming increasingly apparent. While the role of cilia in the development and function of photoreceptors is well established, their role in other ocular tissues is not well known and has not been given much attention yet. Considering that cilia have been shown to regulate signaling pathways, one can imagine that such a role could be extended to various parts of the eye. Further studies are required to see if there are signaling defects in patients with retinal degeneration. Current approaches to address this issue involve investigating the cilia related signaling pathways in patient-derived iPS cells and using animal models.

Ultrastructural localization of proteins using super-resolution microscopy techniques along with protein interaction profiles will help us understand the functional role of ciliary proteins and unravel their complicated signaling networks. A growing list of proteins have been identified in the transition zone and basal body, however, the mechanism by which these proteins regulate cilia structure or function remains unsolved. It is also important to define the precise localization and function of these proteins. Also, further studies are necessary to evaluate the role of these proteins in rods versus cones, as the pattern of rod degeneration and cone degeneration differs in various retinal ciliopathies. This will help to identify the different molecular mechanisms in rods versus cones and allows for more precise therapeutic targeting.

There are several factors to consider when choosing an animal model for molecular analysis and therapeutic development. Some mouse models do not recapitulate the human ciliopathy phenotypes and differences are often observed between stains due to varying genetic backgrounds. For example, rod dominant degeneration is observed in C57BL/6 mice harboring an RPGR mutation, whereas the same mutation in BALB/c mice shows a cone-dominant phenotype (Brunner et al., 2010). Furthermore, dissimilarities in the structure of mouse and human retina (classic examples being the absence of calyceal processes and fovea in mouse) favour the use of other animal models. For this reason, many people are beginning to explore the option of generating transgenic pigs, whose eyes more closely resemble the human eye.

Photoreceptor degeneration is a major cause of retinal degeneration in patients, yet little is known about the molecular mechanism of photoreceptor degeneration, i.e., if cell death is caused by apoptosis, necrosis or necroptosis. Considering that signaling pathways differ in each instance, elucidation of these pathways is crucial for developing therapeutic options. Combining the various therapeutic options may also be considered. Elucidation of these pathways in animal models will shed light on the mechanism of photoreceptor death that is the underlying cause of retinal degeneration caused by ciliary defects in patients.

Acknowledgments

The authors would like to thank Christopher Brinson for help generating the figures and Deva Kusuluri and Sandra Schneider for critical reading of this manuscript. The authors are supported by the Alexander von Humboldt foundation (VK, HMS).

References

Abd-el-barr, M. M., Sykoudis, K., Andrabi, S., Eichers, E. R., Mark, E., Tan, P. L., Wilson, J. H., Katsanis, N., Lupski, J. R. and Wu, S. M. 2007. Impaired photoreceptor protein transport and synaptic transmission in a mouse model of Bardet-Biedl syndrome. Vision Res. 47: 3394–3407.

Abu Safieh, L., Aldahmesh, M. A., Shamseldin, H., Hashem, M., Shaheen, R., Alkuraya, H., Al Hazzaa, S. A., Al-Rajhi, A. and Alkuraya, F. S. 2010. Clinical and molecular characterisation of Bardet-Biedl syndrome in consanguineous populations: the power of homozygosity mapping. J. Med. Genet. 47: 236–41.

Acland, G. M., Aguirre, G. D., Ray, J., Zhang, Q., Aleman, T. S., Cideciyan, A. V., Pearce-Kelling, S. E., Anand, V., Zeng, Y., Maguire, A. M. et al. 2001. Gene therapy restores vision in a canine model of childhood blindness. Nat. Genet. 28: 92–95.

Ahmed, Z. M., Riazuddin, S. and Wilcox, E. R. 2003. The molecular genetics of Usher syndrome. Clin. Genet. 63: 431–44.

Akiyama, M., Ikeda, Y., Yoshida, N., Notomi, S., Murakami, Y., Hisatomi, T., Enaida, H. and Ishibashi, T. 2014. Therapeutic efficacy of topical unoprostone isopropyl in retinitis pigmentosa. Acta Ophthalmol. 92: e229–e234.

Alge, C. S., Hauck, S. M., Priglinger, S. G., Kampik, A. and Ueffing, M. 2006. Differential protein profiling of primary versus immortalized human RPE cells identifies expression patterns associated with cytoskeletal remodeling and cell survival. J. Proteome Res. 5: 862–78.

Ali, M., Hocking, P. M., McKibbin, M., Finnegan, S., Shires, M., Poulter, J. A., Prescott, K., Booth, A., Raashid, Y., Jafri, H. et al. 2011. Mpdz null allele in an avian model of retinal degeneration and mutations in human leber congenital amaurosis and retinitis pigmentosa. Invest. Ophthalmol. Vis. Sci. 52: 7432–7440.

Arikawa, K., Molday, L. L., Molday, R. S. and Williams, D. S. 2011. Localization of of peripherin/rds in disk membranes of cone and rod photoreceptors: relationship to disk membrane morphogenesis and retinal degeneration. J. Cell Biol. 116: 659–667.

Arts, H. H., Doherty, D., van Beersum, S. E. C., Parisi, M. A., Letteboer, S. J. F., Gorden, N. T., Peters, T. A., Märker, T., Voesenek, K., Kartono, A. et al. 2007. Mutations in the gene encoding the basal body protein RPGRIP1L, a nephrocystin-4 interactor, cause Joubert syndrome. Nat. Genet. 39: 882–888.

Bainbridge, J. W. B., Smith, A. J., Barker, S. S., Robbie, S., Henderson, R., Balaggan, K., Viswanathan, A., Holder, G. E., Stockman, A., Tyler, N. et al. 2008. Effect of gene therapy on visual function in Leber's congenital amaurosis. N. Engl. J. Med. 358: 2231–2239.

Belliveau, M. J. and Cepko, C. L. 1999. Extrinsic and intrinsic factors control the genesis of amacrine and cone cells in the rat retina. Development 126: 555–566.

Beltran, W. A., Hammond, P., Acland, G. M. and Aguirre, G. D. 2006. A frameshift mutation in RPGR exon ORF15 causes photoreceptor degeneration and inner retina remodeling in a model of X-linked retinitis pigmentosa. Investig. Ophthalmol. Vis. Sci. 47: 1669–1681.

Beltran, W. A. 2009. The use of canine models of inherited retinal degeneration to test novel therapeutic approaches. Vet. Ophthalmol. 12: 192–204.

Beltran, W. A., Cideciyan, A. V., Guziewicz, K. E., Iwabe, S., Swider, M., Scott, E. M., Savina, S. V., Ruthel, G., Stefano, F., Zhang, L. et al. 2014. Canine retina has a primate fovea-like bouquet of cone photoreceptors which is affected by inherited macular degenerations. PLoS One 9: 11–17.

Besharse, J. C., Forestner, D. M. and Defoe, D. M. 1985. Membrane assembly in retinal photoreceptors. III. Distinct membrane domains of the connecting cilium of developing rods. J. Neurosci. 5: 1035–1048.

Besharse, J. and Horst, C. 1990. The photoreceptor connecting cilium a model for the transition zone. Ciliary and Flagellar Membranes 978: 389–417.

Bharti, K., Nguyen, M. T. T., Skuntz, S., Bertuzzi, S. and Arnheiter, H. 2006. The other pigment cell: Specification and development of the pigmented epithelium of the vertebrate eye. Pigment Cell Res. 19: 380–394.

Boehlke, C., Kotsis, F., Patel, V., Braeg, S., Voelker, H., Bredt, S., Beyer, T., Janusch, H., Hamann, C., Gödel, M. et al. 2010. Primary cilia regulate mTORC1 activity and cell size through Lkb1. Nat. Cell Biol. 12: 1115–22.

Bok, D. 1993. Retinal transplantation and gene therapy. Investig. Ophthalmol. Vis. Sci. 34: 473–476.

Boldt, K., Mans, D. A., Won, J., van Reeuwijk, J., Vogt, A., Kinkl, N., Letteboer, S. J. F., Hicks, W. L., Hurd, R. E., Naggert, J. K. et al. 2011. Disruption of intraflagellar protein transport in photoreceptor cilia causes Leber congenital amaurosis in humans and mice. J. Clin. Invest. 121: 2169–80.

Bowne, S. J., Daiger, S. P., Hims, M. M., Sohocki, M. M., Malone, K. A., McKie, A. B., Heckenlively, J. R., Birch, D. G., Inglehearn, C. F., Bhattacharya, S. S. et al. 1999. Mutations in the RP1 gene causing autosomal dominant retinitis pigmentosa. Hum. Mol. Genet. 8: 2121–8.

Boycott, B. B. and Hopkins, J. M. 1984. A neurofibrillar method stains solitary (primary) cilia in the mammalian retina: their distribution and age-related changes. J. Cell Sci. 66: 95–118.

Boye, S. E., Boye, S. L., Lewin, A. S. and Hauswirth, W. W. 2013. A comprehensive review of retinal gene therapy. Mol. Ther. 21: 509–519.

Boylan, J. and Wright, A. 2000. Identification of a novel protein interacting with RPGR. Hum. Mol. Genet. 9: 2085–93.

Breuer, D. K., Yashar, B. M., Filippova, E., Hiriyanna, S., Lyons, R. H., Mears, A. J., Asaye, B., Acar, C., Vervoort, R., Wright, A. F. et al. 2002. A comprehensive mutation analysis of RP2 and RPGR in a North American cohort of families with X-linked retinitis pigmentosa. Am. J. Hum. Genet. 70: 1545–54.

Brunner, S., Skosyrski, S., Kirschner-Schwabe, R., Knobeloch, K. P., Neidhardt, J., Feil, S., Glaus, E., Luhmann, U. F. O., Rüther, K. and Berger, W. 2010. Cone versus rod disease in a mutant RPGR mouse caused by different genetic backgrounds. Investig. Ophthalmol. Vis. Sci. 51: 1106–1115.

Bunker, C., Berson, E., Bromley, W. C., Hayes, R. P. and Roderick, T. H. 1984. Prevalence of retinitis pigmentosa in Maine. Am. J. Ophthalmol. 97: 357–365.

Chamling, X., Seo, S., Searby, C. C., Kim, G., Slusarski, D. C. and Sheffield, V. C. 2014. The centriolar satellite protein AZI1 interacts with BBS4 and regulates ciliary trafficking of the BBSome. PLoS Genet. 10: e1004083.

Chang, B., Hawes, N. L., Hurd, R. E., Davisson, M. T., Nusinowitz, S. and Heckenlively, J. R. 2002. Retinal degeneration mutants in the mouse. Vision Res. 42: 517–525.

Chang, B., Khanna, H., Hawes, N., Jimeno, D., He, S., Lillo, C., Parapuram, S. K., Cheng, H., Scott, A., Hurd, R. E. et al. 2006. In-frame deletion in a novel centrosomal/ciliary protein CEP290/NPHP6 perturbs its interaction with RPGR and results in early-onset retinal degeneration in the rd16 mouse. Hum. Mol. Genet. 15: 1847–1857.

Chen, J., Laclef, C., Moncayo, A., Snedecor, E. R., Yang, N., Li, L., Takemaru, K. I., Paus, R., Schneider-Maunoury, S. and Clark, R. A. 2015. The ciliopathy gene Rpgrip1l is essential for hair follicle development. J. Invest. Dermatol. 135: 701–709.

Chow, R. L. and Lang, R. A. 2001. Early eye development in vertebrates. Annu. Rev. Cell Dev. Biol. 17: 255–296.

Cideciyan, A. V., Aleman, T. S., Jacobson, S. G., Khanna, H., Sumaroka, A., Aguirre, G. K., Schwartz, S. B., Windsor, E. A. M., He, S., Chang, B. et al. 2007. Centrosomal-ciliary geneCEP290/NPHP6 mutations result in blindness with unexpected sparing of photoreceptors and visual brain: implications for therapy of Leber congenital amaurosis. Hum. Mutat. 28: 1074–1083.

Cideciyan, A. V., Rachel, R. A., Aleman, T. S., Swider, M., Schwartz, S. B., Sumaroka, A., Roman, A. J., Stone, E. M., Jacobson, S. G. and Swaroop, A. 2011. Cone photoreceptors are the main targets for gene therapy of NPHP5 (IQCB1) or NPHP6 (CEP290) blindness: generation of an all-cone Nphp6 hypomorph mouse that mimics the human retinal ciliopathy. Hum. Mol. Genet. 20: 1411–23.

Cideciyan, A. V., Jacobson, S. G., Beltran, W. A., Sumaroka, A., Swider, M., Iwabe, S., Roman, A. J., Olivares, M. B., Schwartz, S. B., Komáromy, A. M. et al. 2013. Human retinal gene therapy for Leber congenital amaurosis shows advancing retinal degeneration despite enduring visual improvement. Proc. Natl. Acad. Sci. 110: E517–25.

Coene, K. L., Roepman, R., Doherty, D., Afroze, B., Kroes, H. Y., Letteboer, S. J., Ngu, L. H., Budny, B., van Wijk, E., Gorden, N. T. et al. 2009. OFD1 is mutated in X-linked Joubert syndrome and interacts with LCA5-encoded leberilin. Am. J. Hum. Genet. 85: 465–481.

Cognard, N., Scerbo, M. J., Obringer, C., Yu, X., Costa, F., Haser, E., Le, D., Stoetzel, C., Roux, M. J., Moulin, B. et al. 2015. Comparing the Bbs10 complete knockout phenotype with a specific renal epithelial knockout one highlights the link between renal defects and systemic inactivation in mice. Cilia 4: 10.

Cole, D. G. 1999a. Kinesin-II, coming and going. J. Cell Biol. 147: 463–476.

Cole, D. G. 1999b. Kinesin-II, the heteromeric kinesin. Cell. Mol. Life Sci. 56: 217–26.

Collin, G. B., Cyr, E., Bronson, R., Marshall, J. D., Gifford, E. J., Hicks, W., Murray, S. A., Zheng, Q. Y., Smith, R. S., Nishina, P. M et al. 2005. Alms1-disrupted mice recapitulate human Alstrom syndrome. Hum. Mol. Genet. 14: 2323–2333.

Concepcion, F. and Chen, J. 2010. Q344ter mutation causes mislocalization of rhodopsin molecules that are catalytically active: A mouse model of Q344ter-induced retinal degeneration. PLoS One 5: e10904.

Coppieters, F., Casteels, I., Meire, F., De Jaegere, S., Hooghe, S., van Regemorter, N., Van Esch, H., Matulevičienė, A., Nunes, L., Meersschaut, V. et al. 2010. Genetic screening of LCA in Belgium: predominance of CEP290 and identification of potential modifier alleles in AHI1 of CEP290-related phenotypes. Hum. Mutat. 31: E1709–E1766.

Cosgrove, D. and Zallocchi, M. 2014. Usher protein functions in hair cells and photoreceptors. Int. J. Biochem. Cell Biol. 46: 80–9.

Cvekl, A. and Piatigorsky, J. 1996. Lens development and crystallin gene expression: many roles for Pax-6. BioEssays 18: 621–630.

Cvekl, A. and Ashery-Padan, R. 2014. The cellular and molecular mechanisms of vertebrate lens development. Development 141: 4432–4447.

Dafinger, C., Liebau, M. C., Elsayed, S. M., Hellenbroich, Y., Boltshauser, E., Korenke, G. C., Fabretti, F., Janecke, A. R., Ebermann, I., Nürnberg, G. et al. 2011. Mutations in KIF7 link Joubert syndrome with Sonic Hedgehog signaling and microtubule dynamics. J. Clin. Invest. 121: 2662–7.

Dalkara, D. and Sahel, J. A. 2014. Gene therapy for inherited retinal degenerations. C. R. Biol. 337: 185–192.

Datta, P., Allamargot, C., Hudson, J. S., Andersen, E. K., Bhattarai, S., Drack, A. V., Sheffield, V. C. and Seo, S. 2015. Accumulation of non-outer segment proteins in the outer segment underlies photoreceptor degeneration in Bardet-Biedl syndrome. Proc. Natl. Acad. Sci. 112: E4400–E4409.

Davis, R. E., Swiderski, R. E., Rahmouni, K., Nishimura, D. Y., Mullins, R. F., Agassandian, K., Philp, A. R., Searby, C. C., Andrews, M. P., Thompson, S. et al. 2007. A knockin mouse model of the Bardet-Biedl syndrome 1 M390R mutation has cilia defects, ventriculomegaly, retinopathy, and obesity. Proc. Natl. Acad. Sci. 104: 19422–19427.

Deane, J. A., Cole, D. G., Seeley, E. S., Diener, D. R. and Rosenbaum, J. L. 2001. Localization of intraflagellar transport protein IFT52 identifies basal body transitional fibers as the docking site for IFT particles. Curr. Biol. 11: 1586–1590.

Delamere, N. A. 2005. Ciliary body and ciliary epithelium. Adv. Organ. Biol. 10: 127–148.

Delous, M., Baala, L., Salomon, R., Laclef, C., Vierkotten, J., Tory, K., Golzio, C., Lacoste, T., Besse, L., Ozilou, C. et al. 2007. The ciliary gene RPGRIP1L is mutated in cerebello-oculo-renal syndrome (Joubert syndrome type B) and Meckel syndrome. Nat. Genet. 39: 875–81.

Deretic, D., Huber, L. A., Ransom, N., Mancini, M., Simons, K. and Papermaster, D. S. 1995. Rab8 in retinal photoreceptors may participate in rhodopsin transport and in rod outer segment disk morphogenesis. J. Cell Sci. 108: 215–224.

Devuyst, O. and Arnould, V. J. 2008. Mutations in RPGRIP1L: extending the clinical spectrum of ciliopathies. Nephrol. Dial. Transplant. 23: 1500–1503.

Dollar, G. L. and Sokol, S. Y. 2007. Wnt signaling and the establishment of cell polarity. Adv. Dev. Biol. 17: 61–94.

Dryja, T. P., Adams, S. M., Grimsby, J. L., McGee, T. L., Hong, D. H., Li, T., Andréasson, S. and Berson, E. L. 2001. Null RPGRIP1 alleles in patients with Leber congenital amaurosis. Am. J. Hum. Genet. 68: 1295–8.

Eblimit, A., Nguyen, T. M. T., Chen, Y., Esteve-Rudd, J., Zhong, H., Letteboer, S., Van Reeuwijk, J., Simons, D. L., Ding, Q., Wu, K. M. et al. 2015. Spata7 is a retinal ciliopathy gene critical for correct RPGRIP1 localization and protein trafficking in the retina. Hum. Mol. Genet. 24: 1584–1601.

Ebrey, T. and Koutalos, Y. 2001. Vertebrate photoreceptors. Prog. Retin. Eye Res. 20: 49–94.

Ennis, S. and Kunz, Y. W. 1986. Differentiated retinal Müller glia are ciliated—Ultrastructural evidence in the teleost Poecilia reticulata P. Cell Biol. Int. Rep. 10: 611–622.

Estrada-Cuzcano, A., Roepman, R., Cremers, F. P. M., den Hollander, A. I. and Mans, D. A. 2012. Non-syndromic retinal ciliopathies: translating gene discovery into therapy. Hum. Mol. Genet. 21: R111–24.

Ezratty, E. J., Stokes, N., Chai, S., Shah, A. S., Williams, S. E. and Fuchs, E. 2011. A role for the primary cilium in Notch signaling and epidermal differentiation during skin development. Cell 145: 1129–41.

Fadool, J. M. and Dowling, J. E. 2008. Zebrafish: a model system for the study of eye genetics. Prog. Retin. Eye Res. 27: 89–110.

Farber, D. B. 1995. From mice to men: The cyclic GMP phosphodiesterase gene in vision and disease: The proctor lecture. Investig. Ophthalmol. Vis. Sci. 36: 263–275.

Fath, M. A., Mullins, R. F., Searby, C., Nishimura, D. Y., Wei, J., Rahmouni, K., Davis, R. E., Tayeh, M. K., Andrews, M., Yang, B. et al. 2005. Mkks-null mice have a phenotype resembling Bardet-Biedl syndrome. Hum. Mol. Genet. 14: 1109–18.

Federman, M. and Nichols, G. 1974. Bone cell cilia: vestigial or functional organelles? Calcif. Tissue Res. 17: 81–85.

Ferraro, S., Gomez-Montalvo, A. I., Olmos, R., Ramirez, M. and Lamas, M. 2014. Primary cilia in rat mature müller glia: Downregulation of IFT20 expression reduces sonic hedgehog-mediated proliferation and dedifferentiation potential of müller glia primary cultures. Cell Mol. Neurobiol. 35: 533–542.

Fisher, S. K. and Steinberg, R. H. 1982. Origin and organization of pigment epithelial apical projections to cones in cat retina. J. Comp. Neurol. 206: 131–145.

Fliegauf, M., Benzing, T. and Omran, H. 2007. When cilia go bad: cilia defects and ciliopathies. Nat. Rev. Mol. Cell Biol. 8: 880–93.

Forsythe, E. and Beales, P. L. 2013. Bardet-Biedl syndrome. Eur. J. Hum. Genet. 21: 8–13.

Fox, B., Bull, T. B. and Arden, G. B. 1980. Variations in the ultrastructure of human nasal cilia including abnormalities found in retinitis pigmentosa. J. Clin. Pathol. 33: 327–35.

Friedman, T. B., Schultz, J. M., Ahmed, Z. M., Tsilou, E. T. and Brewer, C. C. 2011. Usher syndrome: Hearingloss with vision loss. Adv. Otorhinolaryngol. 70: 56–65.

Fuhrmann, S., Levine, E. M. and Reh, T. A. 2000. Extraocular mesenchyme patterns the optic vesicle during early eye development in the embryonic chick. Development 127: 4599–4609.

Gerdes, J. M., Davis, E. E. and Katsanis, N. 2009. The vertebrate primary cilium in development, homeostasis, and disease. Cell 137: 32–45.

Ghossoub, R., Molla-Herman, A., Bastin, P. and Benmerah, A. 2011. The ciliary pocket: a once-forgotten membrane domain at the base of cilia. Biology of the Cell 103: 131–144.

Goetz, S. C. and Anderson, K. V. 2010. The primary cilium: a signalling centre during vertebrate development. Nat. Rev. Genet. 11: 331–44.

Goldman, D. 2014. Müller glial cell reprogramming and retina regeneration. Nat. Rev. Neurosci. 15: 431–442.

Gonzalez-Cordero, A., West, E. L., Pearson, R. A., Duran, Y., Carvalho, L. S., Chu, C. J., Naeem, A., Blackford, S. J. I., Georgiadis, A., Lakowski, J. et al. 2013. Photoreceptor precursors derived from three-dimensional embryonic stem cell cultures integrate and mature within adult degenerate retina. Nat. Biotech. 31: 741–7.

Greiner, J. V., Weidman, T. A., Bodley, H. D. and Greiner, C. A. 1981. Ciliogenesis in photoreceptor cells of the retina. Exp. Eye Res. 33: 433–446.

Guillonneau, X., Piriev, N. I., Danciger, M., Kozak, C. A., Cideciyan, A. V., Jacobson, S. G. and Farber, D. B. 1999. A nonsense mutation in a novel gene is associated with retinitis pigmentosa in a family linked to the RP1 locus. Hum. Mol. Genet. 8: 1541–1546.

Habbig, S., Bartram, M. P., Müller, R. U., Schwarz, R., Andriopoulos, N., Chen, S., Sägmüller, J. G., Hoehne, M., Burst, V., Liebau, M. C. et al. 2011. NPHP4, a cilia-associated protein, negatively regulates the Hippo pathway. J. Cell Biol. 193: 633–42.

Haim, M. 2002. Epidemiology of retinitis pigmentosa in Denmark. Acta Ophthalmol. 233: 1689–1699.

Hamel, C. P. 2007. Cone rod dystrophies. Orphanet J. Rare Dis. 2: 7.

Hao, L., Efimenko, E., Swoboda, P. and Scholey, J. M. 2011. The retrograde IFT machinery of C. elegans cilia: two IFT dynein complexes? PloS One 6: e20995.

Hartong, D. T., Berson, E. L. and Dryja, T. P. 2006. Retinitis pigmentosa. Lancet 368: 1795–809.

Harville, H., Held, S. and Diaz-Font, A. 2010. Identification of 11 novel mutations in 8 BBS genes by high-resolution homozygosity mapping. J. Med. Genet. 47: 262–267.

He, S., Parapuram, S. K., Hurd, T. W., Behnam, B., Margolis, B., Swaroop, A. and Khanna, H. 2008. Retinitis Pigmentosa GTPase Regulator (RPGR) protein isoforms in mammalian retina: insights into X-linked Retinitis Pigmentosa and associated ciliopathies. Vision Res. 48: 366–76.

He, Y., Zhang, Y., Liu, X., Ghazaryan, E., Li, Y., Xie, J. and Su, G. 2014. Recent advances of stem cell therapy for retinitis pigmentosa. Int. J. Mol. Sci. 15: 14456–14474.

Heavner, W. and Pevny, L. 2012. Eye development and retinogenesis. Cold Spring Harb. Perspect. Biol. 4: a008391.

Hendrickson, A. 2005. Organization of the adult primate fovea. Macular Degeneration pp. 1–20.

Higginbotham, H., Eom, T. Y., Mariani, L. E., Bachleda, A., Hirt, J., Gukassyan, V., Cusack, C. L., Lai, C., Caspary, T. and Anton, E. S. 2012. Arl13b in primary cilia regulates the migration and placement of interneurons in the developing cerebral cortex. Dev. Cell 23: 925–938.

Hildebrand, G. D. and Fielder, A. R. 2011. Pediatric Retina. Edited by J. Reynolds and S. Olitsky. Berlin, Heidelberg: Springer Berlin Heidelberg.

Hollander, A. I. D., Roepman, R., Koenekoop, R. K. and Cremers, F. P. M. 2008. Leber congenital amaurosis: Genes, proteins and disease mechanisms. Prog. Retin. Eye Res. 27: 391–419.

Hong, D. H., Pawlyk, B. S., Shang, J., Sandberg, M. A., Berson, E. L. and Li, T. 2000. A retinitis pigmentosa GTPase regulator (RPGR)-deficient mouse model for X-linked retinitis pigmentosa (RP3). Proc. Natl. Acad. Sci. 97: 3649–54.

Hong, D. H., Pawlyk, B. S., Adamian, M. and Li, T. 2004. Dominant, gain-of-function mutant produced by truncation of RPGR. Invest. Ophthalmol. Vis. Sci. 45: 36–41.

Hong, S.-K. and Dawid, I. B. 2009. FGF-dependent left-right asymmetry patterning in zebrafish is mediated by Ier2 and Fibp1. Proc. Natl. Acad. Sci. 106: 2230–5.

Huang, W. C., Wright, A. F., Roman, A. J., Cideciyan, A. V., Manson, F. D., Gewaily, D. Y., Schwartz, S. B., Sadigh, S., Limberis, M. P., Bell, P. et al. 2012. RPGR-associated retinal degeneration in human X-linked RP and a murine model. Investig. Ophthalmol. Vis. Sci. 53: 5594–5608.

Huet, D., Blisnick, T., Perrot, S. and Bastin, P. 2014. The GTPase IFT27 is involved in both anterograde and retrograde intraflagellar transport. eLife 2014: 1–25.

Hughes, M. 2004. Anatomy of the anterior eye for ocularists. J. Ophthalmic Prosthetics pp. 25–35.

Inoue, A. and Ikebe, M. 2003. Characterization of the motor activity of mammalian myosin VIIA. J. Biol. Chem. 278: 5478–5487.

Insinna, C. and Besharse, J. C. 2008. Intraflagellar transport and the sensory outer segment of vertebrate photoreceptors. Dev. Dyn. 237: 1982–1992.

Ishikawa, H. and Marshall, W. F. 2011. Ciliogenesis: building the cell's antenna. Nature reviews. Mol. Cell Biol. 12: 222–34.

Jagger, D., Collin, G., Kelly, J., Towers, E., Nevill, G., Longo-Guess, C., Benson, J., Halsey, K., Dolan, D., Marshall, J. et al. 2011. Alström syndrome protein ALMS1 localizes to basal bodies of cochlear hair cells and regulates cilium-dependent planar cell polarity. Hum. Mol. Genet. 20: 466–81.

Jensen, A. M. and Wallace, V. A. 1997. Expression of sonic hedgehog and its putative role as a precursor cell mitogen in the developing mouse retina. Development 124: 363–371.

Jiang, S. -T., Chiou, Y. -Y., Wang, E., Chien, Y. -L., Ho, H. -H., Tsai, F. -J., Lin, C. -Y., Tsai, S. -P. and Li, H. 2009. Essential role of nephrocystin in photoreceptor intraflagellar transport in mouse. Hum. Mol. Genet. 18: 1566–77.

Jin, H., White, S. R., Shida, T., Schulz, S., Aguiar, M., Gygi, S. P., Bazan, J. F. and Nachury, M. V. 2010. The conserved bardet-biedl syndrome proteins assemble a coat that traffics membrane proteins to Cilia. Cell 141: 1208–1219.

Katz, M. L. and Gao, C. L. 1995. Vitamin A incorporation into lipofuscin-like inclusions in the retinal pigment epithelium. Mech. Ageing Dev. 84: 29–38.

Kay, M. A. 2011. State-of-the-art gene-based therapies: the road ahead. Nat. Rev. Genet. 12: 316–328.

Kevany, B. M. and Palczewski, K. 2010. Phagocytosis of retinal rod and cone photoreceptors. Physiology 25: 8–15.

Khanna, H., Davis, E. E., Murga-Zamalloa, C. A., Estrada-Cuzcano, A., Lopez, I., Hollander, A. I., Zonneveld, M. N., Othman, M. I., Waseem, N., Chakarova, C. F. et al. 2009. A common allele in RPGRIP1L is a modifier of retinal degeneration in ciliopathies. Nat. Genet. 41: 739–45.

Kim, Y.-K., Kim, J. H., Yu, Y. S., Ko, H. W. and Kim, J. H. 2013. Localization of primary cilia in mouse retina. Acta Histochem. 115: 789–794.

Kirschner, R., Rosenberg, T., Schultz-Heienbrok, R., Lenzner, S., Feil, S., Roepman, R., Cremers, F. P., Ropers, H. H. and Berger, W. 1999. RPGR transcription studies in mouse and human tissues reveal a retina-specific isoform that is disrupted in a patient with X-linked retinitis pigmentosa. Hum. Mol. Genet. 8: 1571–8.

Klinger, M., Wang, W., Kuhns, S., Bärenz, F., Dräger-Meurer, S., Pereira, G. and Gruss, O. J. 2014. The novel centriolar satellite protein SSX2IP targets Cep290 to the ciliary transition zone. Mol. Biol. Cell 25: 495–507.

Lem, J., Flannery, J. G., Li, T., Applebury, M. L., Farber, D. B. and Simon, M. I. 1992. Retinal degeneration is rescued in transgenic rd mice by expression of the cGMP phosphodiesterase beta subunit. Proc. Natl. Acad. Sci. 89: 4422–4426.

Lhériteau, E., Petit, L., Weber, M., Le Meur, G., Deschamps, J. -Y., Libeau, L., Mendes-Madeira, A., Guihal, C., François, A., Guyon, R. et al. 2014. Successful gene therapy in the RPGRIP1-deficient dog: a large model of cone-rod dystrophy. Mol. Ther. 22: 265–77.

Li, Z. Y., Wong, F., Chang, J. H., Possin, D. E., Hao, Y., Petters, R. M. and Milam, A. H. 1998. Rhodopsin transgenic pigs as a model for human retinitis pigmentosa. Investig. Ophthalmol. Vis. Sci. 39: 808–819.

Liu, Q., Zhou, J., Daiger, S. P., Farber, D. B., Heckenlively, J. R., Smith, J. E., Sullivan, L. S., Zuo, J., Milam, A. H. and Pierce, E. A. 2002. Identification and subcellular localization of the RP1 protein in human and mouse photoreceptors. Investig. Ophthalmol. Vis. Sci. 43: 22–32.

Liu, Q., Lyubarsky, A., Skalet, J., Pugh Jr., E. N. and Pierce, E. A. 2003. RP1 Is required for the correct stacking of outer segment discs. Investig. Ophthalmol. Vis. Sci. 44: 4171–4183.

Liu, Q., Zuo, J. and Pierce, E. A. 2004. The retinitis pigmentosa 1 protein is a photoreceptor microtubule-associated protein. J. Neurosci. 24: 6427–6436.

Liu, Q., Tan, G., Levenkova, N., Li, T., Pugh Jr., E. N., Rux, J. J., Speicher, D. W. and Pierce, E. A. 2007. The proteome of the mouse photoreceptor sensory cilium complex. Mol. Cell. Proteomics 6: 1299–1317.

Liu, X., Udovichenko, I. P., Brown, S. D., Steel, K. P. and Williams, D. S. 1999. Myosin VIIa participates in opsin transport through the photoreceptor cilium. J. Neurosci. 19: 6267–74.

Liu, X., Bulgakov, O. V., Darrow, K. N., Pawlyk, B., Adamian, M., Liberman, M. C. and Li, T. 2007. Usherin is required for maintenance of retinal photoreceptors and normal development of cochlear hair cells. Proc. Natl. Acad. Sci. 104: 4413–4418.

Llobet, A., Gasull, X. and Gual, A. 2003. Understanding trabecular meshwork physiology: a key to the control of intraocular pressure? News Physiol. Sci. 18: 205–209.

Luo, N., Conwell, M. D., Chen, X., Kettenhofen, C. I., Westlake, C. J., Cantor, L. B., Wells, C. D., Weinreb, R. N., Corson, T. W., Spandau, D. F. et al. 2014. Primary cilia signaling mediates intraocular pressure sensation. Proc. Natl. Acad. Sci. 111: 12871–12876.

Maerker, T., van Wijk, E., Overlack, N., Kersten, F. F. J., Mcgee, J., Goldmann, T., Sehn, E., Roepman, R., Walsh, E. J., Kremer, H. et al. 2008. A novel Usher protein network at the periciliary reloading point between molecular transport machineries in vertebrate photoreceptor cells. Hum. Mol. Genet. 17: 71–86.

Malicki, J., Neuhauss, S. C., Schier, A. F., Solnica-Krezel, L., Stemple, D. L., Stainier, D. Y., Abdelilah, S., Zwartkruis, F., Rangini, Z. and Driever, W. 1996. Mutations affecting development of the zebrafish retina. Development 123: 263–273.

Marc, R. E. 1999. The structure of vertebrate retinas. Retin. Basis Vis. 1: 3–19.

Marcet, B., Chevalier, B., Luxardi, G., Coraux, C., Zaragosi, L. -E., Cibois, M., Robbe-Sermesant, K., Jolly, T., Cardinaud, B., Moreilhon, C. et al. 2011. Control of vertebrate multiciliogenesis by miR-449 through direct repression of the Delta/Notch pathway. Nat. Cell Biol. 13: 693–9.

Marquardt, T. and Gruss, P. 2002. Generating neuronal diversity in the retina: One for nearly all. Trends Neurosci. 25: 32–38.

Masland, R. H. 2001. The fundamental plan of the retina. Nature Neuroscience 4: 877–886.

May-Simera, H. L. and Kelley, M. W. 2012. Cilia, Wnt signaling, and the cytoskeleton. Cilia 1: 7.

McCaa, C. S. 1982. The eye and visual nervous system: anatomy, physiology and toxicology. Environ. Health Perspect. 44: 1–8.

McClements, M. E. and MacLaren, R. E. 2013. Gene therapy for retinal disease. Transl. Res. 161: 241–254.

Megaw, R. D., Soares, D. C. and Wright, A. F. 2015. RPGR: Its role in photoreceptor physiology, human disease, and future therapies. Exp. Eye Res. 138: 1–10.

Meindl, A., Dry, K., Herrmann, K., Manson, E., Ciccodicola, A., Edgar, A., Carvalho, M. R. S., Achatz, H., Hellebrand, H., Lennon, A. et al. 1996. A gene (RPGR) with homology to the RCC1 guanine nucleotide exchange factor is mutated in X-linked retinitis pigmentosa (RP3). Nat. Genet. 13: 35–42.

Michaelides, M., Hunt, D. M. and Moore, A. T. 2003. The genetics of inherited macular dystrophies. J. Med. Genet. 40: 641–50.

Moritz, O. L., Tam, B. M., Hurd, L. L., Peränen, J., Deretic, D. and Papermaster, D. S. 2001. Mutant rab8 Impairs docking and fusion of rhodopsin-bearing post-Golgi membranes and causes cell death of transgenic Xenopus rods. Mol. Biol. Cell 12:. 2341–2351.

Mowat, F. M., Petersen-Jones, S. M., Williamson, H., Williams, D. L., Luthert, P. J., Ali, R. R. and Bainbridge, J. W. 2008. Topographical characterization of cone photoreceptors and the area centralis of the canine retina. Mol. Vis. 14: 2518–2527.

Muresan, V., Joshi, H. C. and Besharse, J. C. 1993. Gamma-tubulin in differentiated cell types: localization in the vicinity of basal bodies in retinal photoreceptors and ciliated epithelia. J. Cell Sci. 104 : 1229–1237.

Murga-Zamalloa, C. A., Atkins, S. J., Peranen, J., Swaroop, A. and Khanna, H. 2010. Interaction of retinitis pigmentosa GTPase regulator (RPGR) with RAB8A GTPase: implications for cilia dysfunction and photoreceptor degeneration. Hum. Mol. Genet. 19: 3591–8.

Mustafi, D., Engel, A. H. and Palczewski, K. 2009. Structure of cone photoreceptors. Prog. Retin. Eye Res. 28: 289–302.

Mykytyn, K., Nishimura, D. Y., Searby, C. C., Shastri, M., Yen, H., Beck, J. S., Braun, T., Streb, L. M., Cornier, A. S., Cox, G. F. et al. 2002. Identification of the gene (BBS1) most commonly involved in Bardet-Biedl syndrome, a complex human obesity syndrome. Nat. Genet. 31: 435–8.

Mykytyn, K., Mullins, R. F., Andrews, M., Chiang, A. P., Swiderski, R. E., Yang, B., Braun, T., Casavant, T., Stone, E. M. and Sheffield, V. C. 2004. Bardet-Biedl syndrome type 4 (BBS4)-null mice implicate Bbs4 in flagella formation but not global cilia assembly. Proc. Natl. Acad. Sci. 101: 8664–8669.

Nagel-Wolfrum, K., Möller, F., Penner, I. and Wolfrum, U. 2014. Translational read-through as an alternative approach for ocular gene therapy of retinal dystrophies caused by in-frame nonsense mutations. Vis. Neurosci. 31: 1–8.

Nickla, D. L. and Wallman, J. 2010. The multifunctional choroid. Prog. Retin. Eye Res. 29: 144–168.

Nischler, C., Michael, R., Wintersteller, C., Marvan, P., Van Rijn, L. J., Coppens, J. E., Van Den Berg, T. J. T. P., Emesz, M. and Grabner, G. 2013. Iris color and visual functions. Graefe's Arch. Clin. Exp. Ophthalmol. 251: 195–202.

Nishimura, D. Y., Fath, M., Mullins, R. F., Searby, C., Andrews, M., Davis, R., Andorf, J. L., Mykytyn, K., Swiderski, R. E., Yang, B. et al. 2004. Bbs2-null mice have neurosensory deficits, a defect in social dominance, and retinopathy associated with mislocalization of rhodopsin. Proc. Natl. Acad. Sci. 101: 16588–93.

Nishiyama, K., Sakaguchi, H., Hu, J. G., Bok, D. and Hollyfield, J. G. 2002. Claudin localization in cilia of the retinal pigment epithelium. Anat. Rec. 267: 196–203.

Obata, S. and Usukura, J. 1992. Morphogenesis of the photoreeeptor outer segment during postnatal development in the mouse (BALB/c) retina. Cell Tissue Res. 269: 39–48.

Omran, H. 2010. NPHP proteins: gatekeepers of the ciliary compartment. J. Cell Biol. 190: 715–717.

Openshaw, A., Branham, K. and Heckenlively, J. 2008. Understanding cone dystrophy. pp. 1–33.

Ozcan, A. A., Ozdemir, N. and Canataroglu, A. 2004. The aqueous levels of TGF-β 2 in patients with glaucoma. Int. Ophthalmol. 25: 19–22.

Pagh-Roehl, K., Wang, E. and Burnside, B. 1992. Shortening of the calycal process actin cytoskeleton is correlated with myoid elongation in teleost rods. Exp. Eye Res. 55: 735–746.

Palczewski, K. 2006. G Protein-coupled receptor rhodopsin. Annu. Rev. Biochem. 75: 743–767.

Panda-Jonas, S., Jonas, J. B. and Jakobczyk-Zmija, M. 1996. Retinal pigment epithelial cell count, distribution, and correlations in normal human eyes. Am. J. Ophthalmol. 121: 181–189.

Pang, J. J., Lei, L., Dai, X., Shi, W., Liu, X., Dinculescu, A. and McDowell, J. H. 2000. AAV-mediated gene therapy in mouse models of recessive retinal degeneration. Curr. Mol. Med. 23: 101–2.

Papermaster, D. S., Schneider, B. G. and Besharse, J. C. 1985. Vesicular transport of newly synthesized opsin from the Golgi apparatus toward the rod outer segment. Ultrastructural immunocytochemical and autoradiographic evidence in Xenopus retinas. Investig. Ophthalmol. Vis. Sci. 26: 1386–1404.

Papermaster, D. S. 2002. The birth and death of photoreceptors: The Friedenwald Lecture. Investig. Ophthalmol. Vis. Sci. 43: 1300–1309.

Patnaik, S. R., Raghupathy, R. K., Zhang, X., Mansfield, D. and Shu, X. 2015. The role of RPGR and its interacting proteins in ciliopathies. J. Ophtalmol. 1–10.

Pawlyk, B. S., Bulgakov, O. V., Liu, X., Xu, X., Adamian, M., Sun, X., Khani, S. C., Berson, E. L., Sandberg, M. A. and Li, T. 2010. Replacement gene therapy with a human RPGRIP1 sequence slows photoreceptor degeneration in a murine model of Leber congenital amaurosis. Hum. Gene Ther. 21: 993–1004.

Pazour, G. J., Baker, S. A., Deane, J. A., Cole, D. G., Dickert, B. L., Rosenbaum, J. L., Witman, G. B. and Besharse, J. C. 2002. The intraflagellar transport protein, IFT88, is essential for vertebrate photoreceptor assembly and maintenance. J. Cell Biol. 157: 103–113.

Pazour, G. J. and Rosenbaum, J. L. 2002. Intraflagellar transport and cilia-dependent diseases. Trends Cell Biol. 12: 551–555.

Pazour, G. J., Agrin, N., Leszyk, J. and Witman, G. B. 2005. Proteomic analysis of a eukaryotic cilium. J. Cell Biol. 170: 103–113.

Pedersen, L. B. and Rosenbaum, J. L. 2008. Intraflagellar Transport (IFT). Role in ciliary assembly, resorption and signalling. Curr. Top. Dev. Biol. 85: 23–61.

Pedersen, L. B., Veland, I. R., Schrøder, J. M. and Christensen, S. T. 2008. Assembly of primary cilia. Dev. Dyn. 237: 1993–2006.

Peichl, L. 2005. Diversity of mammalian photoreceptor properties: Adaptations to habitat and lifestyle? Anatomical Record—Part A. Discov. Mol. Cell. Evol. Biol. 287: 1001–1012.

Peters, K. R., Palade, G. E., Schneider, B. G. and Papermaster, D. S. 1983. Fine structure of a periciliary ridge complex of frog retinal rod cells revealed by ultrahigh resolution scanning electron microscopy. J. Cell Biol. 96: 265–276.

Petters, R. M., Alexander, C. A., Wells, K. D., Collins, E. B., Sommer, J. R., Blanton, M. R., Rojas, G., Hao, Y., Flowers, W. L., Banin, E. et al. 1997. Genetically engineered large animal model for studying cone photoreceptor survival and degeneration in retinitis pigmentosa. Nat. Biotech. 15: 965–70.

Pignatelli, V. and Strettoi, E. 2004. Bipolar cells of the mouse retina: a gene gun, morphological study. J. Comp. Neurol. 476: 254–266.

Pretorius, P. R., Baye, L. M., Nishimura, D. Y., Searby, C. C., Bugge, K., Yang, B., Mullins, R. F., Stone, E. M., Sheffield, V. C. and Slusarski, D. C. 2010. Identification and functional analysis of the vision-specific BBS3 (ARL6) long isoform. PLoS Genet. 6: e1000884.

Rachel, R. A., May-Simera, H. L., Veleri, S., Gotoh, N., Choi, B. Y., Murga-Zamalloa, C., McIntyre, J. C., Marek, J., Lopez, I., Hackett, A. N. et al. 2012. Combining Cep290 and Mkks ciliopathy alleles in mice rescues sensory defects and restores ciliogenesis. J. Clin. Invest. 122: 1233–45.

Rachel, R. A., Yamamoto, E. A., Dewanjee, M. K., May-Simera, H. L., Sergeev, Y. V., Hackett, A. N., Pohida, K., Munasinghe, J., Gotoh, N., Wickstead, B. et al. 2015. CEP290 alleles in mice disrupt tissue-specific cilia biogenesis and recapitulate features of syndromic ciliopathies. Hum. Mol. Genet. 24: 3775–3791.

Rachel, R., Yamamote, E., Mrinal, D., May-Simera, H., Sergeev, Y., Hackett, A., Pohida, K., Munasinghe, J., Gotoh, N., Wickstead, B. et al. 2015. CEP290 alleles in mice disrupt tissue-specific cilia biogenesis and recapitulate features of syndromic ciliopathies. Hum. Mol. Genet. 24: 3775–3791.

Raghupathy, R. K., Patnaik, S. R. and Shu, X. 2013. Transgenic zebrafish models for understanding retinitis pigmentosa. Cloning & Transgenes. 2: 2.

Reese, B. E. 2010. Development of the retina and optic pathway. Vision Res. 51: 613–632.

Reiter, J. F., Blacque, O. E. and Leroux, M. R. 2012. The base of the cilium: roles for transition fibres and the transition zone in ciliary formation, maintenance and compartmentalization. EMBO Rep. 13: 608–618.

Renault, L., Nassar, N., Vetter, I., Becker, J., Klebe, C., Roth, M. and Wittinghofer, A. 1998. The 1.7 A crystal structure of the regulator of chromosome condensation (RCC1) reveals a seven-bladed propeller. Nature 392: 97–101.

Rix, S., Calmont, A., Scambler, P. J. and Beales, P. L. 2011. An Ift80 mouse model of short rib polydactyly syndromes shows defects in hedgehog signalling without loss or malformation of cilia. Hum. Mol. Genet. 20: 1306–14.

Rizzo III, J. F. 2005. Embryology, anatomy and physiology of the afferent visual pathway. Clinical Neuro-opthalmol. 1: 3–82.

Robertis, E. D. E. 1956. Electron microscope observations on the submicroscopic organization of the retinal rods. J. Biophys. Biochem. 2: 319–330.

Roeder, I. and Radtke, F. 2009. Stem cell biology meets systems biology. Development 136: 3525–3530.

Roepman, R., van Duijnhoven, G., Rosenberg, T., Pinckers, A. J., Bleeker-Wagemakers, L. M., Bergen, A. A., Post, J., Beck, A., Reinhardt, R., Ropers, H. H. et al. 1996. Positional cloning of the gene for X-linked retinitis pigmentosa 3: homology with the guanine-nucleotide-exchange factor RCC1. Hum. Mol. Genet. 5: 1035–41.

Roepman, R., Letteboer, S. J. F., Arts, H. H., van Beersum, S. E. C., Lu, X., Krieger, E., Ferreira, P. A. and Cremers, F. P. M. 2005. Interaction of nephrocystin-4 and RPGRIP1 is disrupted by nephronophthisis or leber congenital amaurosis-associated mutations. Proc. Natl. Acad. Sci. 102: 18520–5.

Röhlich, P. 1975. The sensory cilium of retinal rods is analogous to the transitional zone of motile cilia. Cell Tissue Res. 161: 421–30.

Rompolas, P., Pedersen, L. B., Patel-King, R. S. and King, S. M. 2007. Chlamydomonas FAP133 is a dynein intermediate chain associated with the retrograde intraflagellar transport motor. J. Cell Sci. 120: 3653–3665.

Roof, D., Adamian, M., Jacobs, D. and Hayes, A. 1991. Cytoskeletal specializations at the rod photoreceptor distal tip. J. Comp. Neurol. 305: 289–303.

Rosenbaum, J. L. and Witman, G. B. 2002. Intraflagellar transport. Nat. Rev. Mol. Cell Biol. 3: 813–825.

Ross, J. W., Fernandez de Castro, J. P., Zhao, J., Samuel, M., Walters, E., Rios, C., Bray-Ward, P., Jones, B. W., Marc, R. E., Wang, W. et al. 2012. Generation of an inbred miniature pig model of retinitis pigmentosa. Investig. Ophthalmol. Vis. Sci. 53: 501–507.

Sahly, I., Dufour, E., Schietroma, C., Michel, V., Bahloul, A., Perfettini, I., Pepermans, E., Estivalet, A., Carette, D., Aghaie, A. et al. 2012. Localization of usher 1 proteins to the photoreceptor calyceal processes, which are absent from mice. J. Cell Biol. 199: 381–399.

Samardzija, M., Neuhauss, S. C. F., Joly, S., Kurz-levin, M. and Grimm, C. 2010. Animal models for retinal degeneration. Neuromethods 46: 51–79.

Sang, L., Miller, J. J., Corbit, K. C., Giles, R. H., Brauer, M. J., Otto, E. A., Baye, L. M., Wen, X., Scales, S. J., Kwong, M. et al. 2011. Mapping the NPHP-JBTS-MKS protein network reveals ciliopathy disease genes and pathways. Cell 145: 513–528.

Schneider, L., Clement, C. A., Teilmann, S. C., Pazour, G. J., Hoffmann, E. K., Satir, P. and Christensen, S. T. 2005. PDGFRalphaalpha signaling is regulated through the primary cilium in fibroblasts. Curr. Biol. 15: 1861–6.

Scholey, J. M. 2003. Intraflagellar transport. Annu. Rev. Cell Dev. Biol. 19: 423–443.

Scholey, J. M. 2008. Intraflagellar transport motors in cilia: Moving along the cell's antenna. J. Cell Biol. 180: 23–29.

Schwahn, U., Lenzner, S., Dong, J., Feil, S., Hinzmann, B., van Duijnhoven, G., Kirschner, R., Hemberger, M., Bergen, A. A., Rosenberg, T. et al. 1998. Positional cloning of the gene for X-linked retinitis pigmentosa 2. Nat. Genet. 19: 327–332.

Schwartz, R. and Hildebrandt, F. 2011. Ciliopathies. N. Engl. J. Med. 364: 1533–1543.

Schwartz, S. D., Regillo, C. D., Lam, B. L., Eliott, D., Rosenfeld, P. J., Gregori, N. Z., Hubschman, J. -P., Davis, J. L., Heilwell, G., Spirn, M. et al. 2015. Human embryonic stem cell-derived retinal pigment epithelium in patients with age-related macular degeneration and Stargardt's macular dystrophy: follow-up of two open-label phase 1/2 studies. Lancet 385: 509–516.

Sedmak, T. and Wolfrum, U. 2011. Intraflagellar transport proteins in ciliogenesis of photoreceptor cells. Biol. Cell 103: 449–466.

Seo, S., Baye, L. M., Schulz, N. P., Beck, J. S., Zhang, Q., Slusarski, D. C. and Sheffield, V. C. 2010. BBS6, BBS10, and BBS12 form a complex with CCT/TRiC family chaperonins and mediate BBSome assembly. Proc. Natl. Acad. Sci. 107: 1488–93.

Seo, S., Zhang, Q., Bugge, K., Breslow, D. K., Searby, C. C., Nachury, M. V. and Sheffield, V. C. 2011. A novel protein LZTFL1 regulates ciliary trafficking of the BBSome and smoothened. PLoS Genet. 7: e1002358.

Sorokin, S. 1962. Centrioles and the formation of rudimentary cilia by fibroblasts and smooth muscle cells. J. Cell Biol. 15: 363–377.

Spencer, M., Detwiler, P. B. and Bunt-Milam, A. H. 1988. Distribution of membrane proteins in mechanically dissociated retinal rods. Investig. Ophthalmol. Vis. Sci. 29: 1012–20.

Steinberg, R. H., Fisher, S. K. and Anderson, D. H. 1981. Disc morphogenesis in vertebrate photoreceptors. Vision Res. 21: 1725–1981.

Strauss, O. 2005. The retinal pigment epithelium in visual function. Physiol. Rev. 85: 845–881.

Sugiyama, Y., Stump, R. J. W., Nguyen, A., Wen, L., Chen, Y., Wang, Y., Murdoch, J. N., Lovicu, F. J. and McAvoy, J. W. 2010. Secreted frizzled-related protein disrupts PCP in eye lens fiber cells that have polarised primary cilia. Dev. Biol. 338: 193–201.

Sugiyama, Y., Lovicu, F. J. and McAvoy, J. W. 2011. Planar cell polarity in the mammalian eye lens. *Organogenesis*, 7: 191–201.

Sugiyama, Y. and McAvoy, J. W. 2012. Analysis of PCP defects in mammalian eye lens. *Methods Mol. Biol.*, 839: 147–156.

Sugiyama, Y., Shelley, E. J., Yoder, B. K., Kozmik, Z., May-Simera, H. L., Beales, P. L., Lovicu, F. J. and McAvoy, J. W. 2016. Non-essential role of cilia in coordinating precise alignment of lens fibres. Mech. Dev. 139: 10–17.

Summers Rada, J. A., Shelton, S. and Norton, T. T. 2006. The sclera and myopia. Experimental Eye Research 82: 185–200.

Sung, C. H., Davenport, C. M. and Nathans, J. 1993. Rhodopsin mutations responsible for autosomal dominant retinitis pigmentosa: Clustering of functional classes along the polypeptide chain. J. Biol. Chem. 268: 26645–26649.

Szél, A., Lukáts, A., Fekete, T., Szepessy, Z. and Röhlich, P. 2000. Photoreceptor distribution in the retinas of subprimate mammals. J. Opt. Soc. Am. A. Opt. Image Sci. Vis. 17: 568–579.

Szymanska, K. and Johnson, C. A. 2012. The transition zone: an essential functional compartment of cilia. Cilia 1: 1–8.

Tadenev, A. L. D., Kulaga, H. M., May-Simera, H. L., Kelley, M. W., Katsanis, N. and Reed, R. R. 2011. Loss of Bardet-Biedl syndrome protein-8 (BBS8) perturbs olfactory function, protein localization, and axon targeting. Proc. Natl. Acad. Sci. 108: 10320–5.

Tamm, E. R. and Lütjen-Drecoll, E. 1996. Ciliary body. Microsc. Res. Tech. 33: 390–439.

Tanaka, Y., Okada, Y. and Hirokawa, N. 2005. FGF-induced vesicular release of Sonic hedgehog and retinoic acid in leftward nodal flow is critical for left-right determination. Nature 435: 172–177.

Taschner, M., Bhogaraju, S. and Lorentzen, E. 2012. Architecture and function of IFT complex proteins in ciliogenesis. Differentiation 83: S12–S22.

Thompson, D. A., Khan, N. W., Othman, M. I., Chang, B., Jia, L., Grahek, G., Wu, Z., Hiriyanna, S., Nellissery, J., Li, T., Khanna, H., Colosi, P., Swaroop, A. and Heckenlively, J. R. 2012. Rd9 is a naturally occurring mouse model of a common form of retinitis pigmentosa caused by mutations in RPGR-ORF15. PloS One 7: p. e35865.

Troutt, L. L., Wang, E., Pagh-Roehl, K. and Burnside, B. 1990. Microtubule nucleation and organization in teleost photoreceptors: Microtubule recovery after elimination by cold. J. Neurocytol. 19: 213–223.

Ulshafer, R. J., Allen, C., Dawson, W. W. and Wolf, E. D. 1984. Hereditary retinal degeneration in the Rhode Island Red chicken. I. Histology and ERG. Exp. Eye Res. 39: 125–135.

Veleri, S., Bishop, K., Dalle Nogare, D. E., English, M. A., Foskett, T. J., Chitnis, A., Sood, R., Liu, P. and Swaroop, A. 2012. Knockdown of Bardet-Biedl syndrome gene BBS9/PTHB1 leads to cilia defects. PloS One 7: p. e34389.

Vervoort, R., Lennon, A., Bird, A. and Tulloch, B. 2000. Mutational hot spot within a new RPGR exon in X-linked retinitis pigmentosa. Nat. Genet. 25: 462–476.

Vierkotten, J., Dildrop, R., Peters, T., Wang, B. and Rüther, U. 2007. Ftm is a novel basal body protein of cilia involved in Shh signalling. Development 134: 2569–77.

Wallingford, J. B. and Mitchell, B. 2011. Strange as it may seem: the many links between Wnt signaling, planar cell polarity, and cilia. Genes Dev. 25: 201–13.

Wang, Y., Rajala, A. and Rajala, R. 2015. Lipid nanoparticles for ocular gene delivery. J. Funct. Biomater. 6: 379–394.

Wätzlich, D., Vetter, I., Gotthardt, K., Miertzschke, M., Chen, Y. -X., Wittinghofer, A. and Ismail, S. 2013. The interplay between RPGR, PDEδ and Arl2/3 regulate the ciliary targeting of farnesylated cargo. EMBO Rep. 14: 465–72.

Webber, W. A. and Lee, J. 1975. Fine structure of mammalian renal cilia. Anat. Rec. 182: 339–43.

Westfall, J. E., Hoyt, C., Liu, Q., Hsiao, Y. -C., Pierce, E. A., Page-McCaw, P. S. and Ferland, R. J. 2010. Retinal degeneration and failure of photoreceptor outer segment formation in mice with targeted deletion of the Joubert syndrome gene, Ahi1. J. Neurosci. 30: 8759–8768.

Willoughby, C. E., Ponzin, D., Ferrari, S., Lobo, A., Landau, K. and Omidi, Y. 2010. Anatomy and physiology of the human eye: effects of mucopolysaccharidoses disease on structure and function—a review. Clin. Experiment. Ophthalmol. 38: 2–11.

Wolfrum, U. and Schmitt, A. 2000. Rhodopsin transport in the membrane of the connecting cilium of mammalian photoreceptor cells. Cell Motil. Cytoskeleton 46: 95–107.

Won, J., Gifford, E., Smith, R. S., Yi, H., Ferreira, P. A., Hicks, W. L., Li, T., Naggert, J. K. and Nishina, P. M. 2009. RPGRIP1 is essential for normal rod photoreceptor outer segment elaboration and morphogenesis. Hum. Mol. Genet. 18: 4329–4339.

Won, J., de Evsikova, C. M., Smith, R. S., Hicks, W. L., Edwards, M. M., Longo-Guess, C., Li, T., Naggert, J. K. and Nishina, P. M. 2011. NPHP4 is necessary for normal photoreceptor ribbon synapse maintenance and outer segment formation, and for sperm development. Hum. Mol. Genet. 20: 482–496.

Yamamoto, S., Sugawara, T., Murakami, A., Nakazawa, M., Nao-i, N., Machida, S., Wada, Y., Mashima, Y. and Myake, Y. 2012. Topical isopropyl unoprostone for retinitis pigmentosa: microperimetric results of the phase 2 clinical study. Ophthalmol. Ther. 1: 1–16.

Yang, J., Liu, X., Yue, G., Adamian, M., Bulgakov, O. and Li, T. 2002. Rootletin, a novel coiled-coil protein, is a structural component of the ciliary rootlet. J. Cell Biol. 159: 431–440.

Yang, J., Gao, J., Adamian, M., Wen, X. -H., Pawlyk, B., Zhang, L., Sanderson, M. J., Zuo, J., Makino, C. L. and Li, T. 2005. The ciliary rootlet maintains long-term stability of sensory cilia. Mol. Cell Biol. 25: 4129–4137.

Yang, J. and Li, T. 2006. Focus on molecules: Rootletin. Exp. Eye Res. 83: 1–2.

Young, R. W. 1967. The renewal of photoreceptor cell outer segments. J. Cell Biol. 33: 61–72.

Zeiss, C. J., Acland, G. M. and Aguirre, G. D. 1999. Retinal atrophy, the locus homologue of RP3. Investig. Ophthalmol. Vis. Sci. 40: 3292–3304.

Zhang, Q., Acland, G. M., Wu, W. X., Johnson, J. L., Pearce-Kelling, S., Tulloch, B., Vervoort, R., Wright, A. F. and Aguirre, G. D. 2002. Different RPGR exon ORF15 mutations in Canids provide insights into photoreceptor cell degeneration. Hum. Mol. Genet. 11: 993–1003.

Zhang, Q., Nishimura, D., Vogel, T., Shao, J., Swiderski, R., Yin, T., Searby, C., Carter, C. S., Kim, G., Bugge, K. et al. 2013. BBS7 is required for BBSome formation and its absence in mice results in Bardet-Biedl syndrome phenotypes and selective abnormalities in membrane protein trafficking. J. Cell Sci. 126: 2372–80.

Zhao, Y., Hong, D. -H., Pawlyk, B., Yue, G., Adamian, M., Grynberg, M., Godzik, A. and Li, T. 2003. The retinitis pigmentosa GTPase regulator (RPGR)-interacting protein: subserving RPGR function and participating in disk morphogenesis. Proc. Natl. Acad. Sci. 100: 3965–70.

CHAPTER 8

The Role of Cilia in the Auditory System

Helen May-Simera

INTRODUCTION

The auditory system is one of the most remarkable organs in vertebrate organisms. Not only is the auditory system required for our perception of sound, but also for our perception of spatial orientation. Although much is known about how this system develops and functions, little is known about the role of cilia during these processes, which is unfortunate considering that auditory dysfunction is one of the phenotypes of several human ciliopathy syndromes. In this chapter, we summarise what is currently known about cilia function in the auditory system with a particular focus on the mammalian cochlea.

Anatomy and Physiology of the Ear

The auditory system is required for the acquisition and transduction of sound, as well as for maintaining balance and perceiving spatial orientation. These complex processes involve numerous tissues and cell types, which are derived from the three germ layers: endoderm, mesoderm and ectoderm (Bailey and Streit, 2006). The vertebrate ear is comprised of three distinct regions, the outer ear, middle ear and inner ear (Figure 1). The outer ear includes the pinna and auditory meatus (auditory canal), the middle ear contains the three bony ossicles and the inner ear is made up of the vestibular and cochlea compartments. Sound collected by the pinna of the outer ear is funnelled towards the tympanic membrane, the so-called eardrum, which

Institute of Molecular Physiology, Johannes-Gutenberg University, Mainz, 55128, Germany.
  E-mail: hmaysime@uni-mainz.de

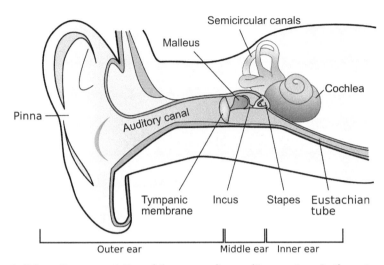

Figure 1. Schematic representation of the mammalian auditory system. In the outer ear, the pinna funnels sound waves along the auditory canal. The tympanic membrane marks the beginning of the middle ear. The tympanic membrane transmits sound to the ossicles, a series of delicate bones, the Malleus, Incus and Stapes. Vibration of the ossicles transmits sound through the middle ear towards the fluid filled inner ear, which is responsible for sound and balance detection. Sensory cells in the cochlea are responsible for detecting sound, whilst sensory cells in the vestibular system detect balance.

separates the outer and middle ear. Vibration of the tympanic membrane leads to oscillation of the three bony ossicles, the so-called hammer, anvil and stapes, in the fluid filled middle ear. These three bony ossicles are the smallest bones found in the vertebrate body and are vital for the transmission of sound impulses from the outer ear to the inner ear. One end of the hammer is directly connected to the tympanic membrane; the other end forms a lever-like hinge with the anvil. The opposite end of the anvil is fused with the stapes, which connects with the oval window at the base of the cochlea spiral. The footplate of the stapes is only loosely attached and can move in and out of the oval window of the cochlea in a piston like motion. This motion generates vibrations in the fluid-filled inner ear, which relays the signal to the brain in response to sound (Neely, 1998).

The vestibular system is responsible for maintaining balance and perceiving spatial orientation and acceleration (Figure 2) (Lopez, 2016). Since movement consists of both rotations and accelerations, the vestibular system has two apparatuses. Firstly, the semi-circular canals detect rotational movement. Each of the three semi-circular canals contain a crista ampullaris, a cone shaped structure covered in sensory epithelia, which transduces the mechanical movement of fluid through the semi-circular canals into electrical signals. Secondly, the patches of sensory epithelia in the utricle and saccule indicate linear accelerations and gravity. The utricle and saccule are referred to as the otolith organs, because of the calcium carbonate crystals, so called

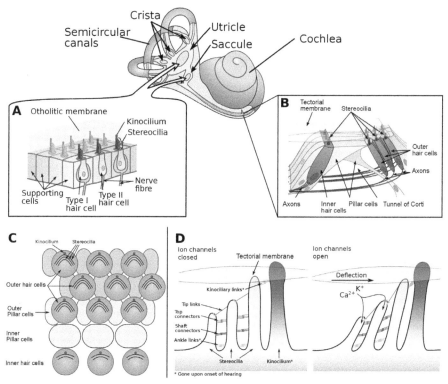

Figure 2. Sensory hair cells in the inner ear. Schematic representation of the semi-circular canals of the vestibular and the cochlea in the inner ear. (**A**) Patches of sensory hair cells in the vestibular contain two types of hair cells, both surrounded by supporting cells. Stereociliary bundles projecting from the apical surface of hair cells increase in length as they approach the kinocilium. In the vestibular, the kinocilium does not retract with age. Stereocilia and kinocilia are embeded in the otholitic membrane. (**B**) A cross section through the sensory epithelia of the organ of Corti. One row of inner hair cells and three rows of outer hair cells are separated via the inner and outer pillar cells. The pillar cells form the tunnel of Corti. Stereociliary bundles project from the apical surface of each hair cell. In developing cochlea, a kinocilium is connected to the tallest row of stereocilia, which retracts upon onset of hearing. The tectorial membrane lies above stereociliary bundles and causes them to deflect upon incoming sound waves. Neurotransmitter release at the base of the hair cell signals to the spiral ganglion neurons. (**C**) Wholemount view of the organ of Corti. Chevron shaped stereociliary bundles are composed of three rows of stereocilia increasing in height. The kinocilium attaches to the tallest row. (**D**) Deflection of the stereocilia towards the tallest row opens ion channels allowing Ca^{2+} and K^+ to entre. Various links connect individual stereocilia. Ankle links and kinociliary links are only seen in developing cochlea.

otoconoia, that are embedded in a gelatinous membrane lying above the sensory epithelia (Lindeman, 1973). Movement causes the otoconia to shift this membrane relative to the underlying sensory epithelia, which in turn stimulates the sensory neurons (Figure 2A). The vestibular system primarily sends neuronal signals responsible for controlling eye movements and the musculature in order to keep an individual upright.

Although each system is somewhat specialized, the patches of sensory epithelia in the cochlea and vestibule all contain mechanosensory hair cells and non-sensory supporting cells (Figure 2B,C). Mechanosensitivity of the hair cells is achieved via the deflection of stereociliary bundles that project from the lumenal surface (Figure 2D). Although called stereociliary bundles, these bundles of modified stiff microvilli are actin-based and are not to be confused with true cilia. However, each stereociliary bundle is associated with a true microtubule-based cilium, the specialized so-called kinocilium, which is attached at the vertex of each bundle. The structure of mechanosensory hair cells will be described in more detail below, using the cochlea hair cells as an example. The mechanosensory hair cells transduce sound waves or physical motion into graded responses and provoke neurotransmitter release. These signals are then conveyed to the central nervous system via neuronal innervation of the VIIIth nerve (Whitfield, 2015).

Most auditory defects arise from disruptions to the inner ear; as such, auditory research has focused on understanding the precise function and molecular mechanisms within the inner ear. Since the inner ear has several cell types that display primary cilia and various ciliopathies are associated with auditory dysfunction and balance problems, this chapter will place its focus on the inner ear.

The Inner Ear

The inner ear develops predominantly from the otic placode, a thickening of the dorsal ectoderm located adjacent to the developing hindbrain (Morsli et al., 1998; Whitfield, 2015; Driver and Kelley, 2009; Puligilla and Kelley, 2009). As the placode proliferates, it sinks inwards towards the hindbrain and constricts at the ectodermal surface. Eventually this structure pinches off to form a fluid filled otocyst. Formation of the otocyst is followed by a series of elaborate morphogenetic changes resulting in the development of the various structures within the inner ear. The epithelial cells that line the otocyst and all of its derivatives give rise to both the sensory patches containing mechanosensory hair cells and supporting cells and the non-sensory cells that comprise the remainder of the inner ear.

The cochlea

The cochlea is the auditory portion of the inner ear in mammals. It begins as a ventral out-pocketing from the otocyst that extends and coils over time. It is a spiralled, hollow chamber, encased in the bony otic capsule, which is embedded in the temporal bone. Within it lies the organ of Corti, the sensorineural organ for hearing. In mouse, the extending cochlear duct more than doubles in length between embryonic day (E) 12.5 and post natal day (P) 0 (Morsli et al., 1998). The organ of Corti is comprised of polarised epithelial cells, sensory and non-sensory cells of placodal origin, which are spatially organised in an orderly pattern, spiralling from the base of the cochlea to the apex. In the mouse organ of Corti, the progenitor pool of cells that will give rise to the sensory hair cells and supporting cells becomes post mitotic between E13 and E14 (Chen et al., 2002; Chen and Segil, 1999; Ruben, 1967; Wang et al., 2005), however, non-sensory cells continue to proliferate. As the derivatives from this precursor population extend along the entire length of the cochlear duct, this pro-sensory domain undergoes significant extension and rearrangement.

There are two distinct types of sensory cells in the cochlea, inner hair cells and outer hair cells, so called because of their morphology and location. In addition, there is a range of non-sensory cell types, referred to collectively as supporting cells, which play auxiliary and supportive roles (Wan et al., 2013). A single row of inner hair cells is separated from the three rows of outer hairs cells by the inner and outer pillar cells (Figure 2B,C). Support cells have been suggested to function by monitoring a homeostatic environment in which the hair cells operate (Raphael and Altschuler, 2003). Hair cells are responsible for converting sound waves, that have travelled from the outer ear via the middle ear and through the cochlea spiral, into electrical signals that are relayed to the auditory brainstem (Raphael and Altschuler, 2003). Since hair cells have an excess of negatively charged ions inside the cell and an excess of positively charged ions outside, slight deflection of the stereocilia opens transduction ion channels allowing K^+- and Ca^{2+}-ions into the cell, causing depolarization (Figure 2B,D) (Roberts et al., 1988). This causes the hair cell to release neurotransmitters at the base, initiating signalling towards the brain.

The inner hair cell is the true sensory cell type and sends impulses via the auditory nerve upon stimulation. The outer hair cells enhance the performance of the cochlea by increasing the selectivity and sensitivity of its response. Often referred to as the cochlear amplifier, the outer hair cells display a dual response upon stimulation (Reichenbach and Hudspeth, 2014). The release of neurotransmitter at the base of the cell is accompanied by a change in length and stiffness of the outer hair cell. This motile response allows for a regional specific amplification of movement in the organ of Corti, which in turn enhances the transduction of the inner hair cell at that specific region of the cochlear spiral. Expression of the transmembrane motor protein Prestin, at the lateral plasma membrane of outer hair cells, enables outer hair cells to rapidly alter their stiffness and length (Zheng et al., 2000). In most mammals there are three rows of outer hair cells, however some species that have adapted for low frequency hearing can have a fourth or fifth row (e.g., the mole rat (Raphael et al., 1991)).

The non-sensory cells in the organ of Corti are referred to as supporting cells. These are highly differentiated epithelial cells with distinctive morphological features that surround and support the hair cells (Gale and Jagger, 2010). The inner hair cells are surrounded by inner and outer phalangeal cells, whilst the outer hair cells are in contact with the inner and outer pillar cells and lie above and between the supporting Deiter's cells.

Deiter's cells also make contact basolaterally. More laterally along the organ of Corti are the Hensen's cells. It is thought that the support cells' functions go beyond that of structural support, as there is evidence that they participate in regulating the ionic environment within and around the organ of Corti (Raphael and Altschuler, 2003; Jagger and Forge, 2014).

Stereocilia

The stereociliary bundles protruding from the apical surface of mechanosensory hair cells enable these particular cell types to carry out their diverse and specialised roles. Their deflection opens ion channels depolarising the cell, which produces an electrical signal or a receptor potential. Hair cell transduction and auditory perception are dependent on the uniform orientation of the stereociliary bundles and disruption to these can cause hearing and/or balance defects, such as in Usher syndrome (Self et al., 1998; Alagramam et al., 2001; Di Palma et al., 2001; Johnson et al., 2003).

As described above, during extension of the cochlear duct, individual cells begin to differentiate into mechanosensory hair cells and non-sensory support cells. Early on, the lumenal surface of these cells is covered with a lawn of short, actin-based microvilli with a single microtubule-based primary cilium protruding from the apex, as is the case for virtually all epithelial cells post terminal mitosis. Specifically in the differentiating mechanosensory hair cells, this single kinocilium begins to migrate towards the lateral edge of each cell (Frolenkov et al., 2004). As it migrates, the microvilli located adjacent to this cilium also begin to elongate and give rise to individual stereocilia. The hair bundle is comprised of many individual stereocilia. These are membrane bound cellular projections comprised of actin and arranged in a staircase-like array at the lateral edge of hair cells (Frolenkov et al., 2004). At the base of each stereocilium, there is an electron-dense rootlet embedded in the actin fibre network of the cuticular plate. Inner hair cells usually have two main rows of stereocilia, whilst outer hair cells have three. They are arranged in rows of increasing height with the tallest positioned laterally. All stereociliary bundles are arranged in a staircase-like array, yet the shape of the stereociliary bundles varies based on location. In the cochlea, the stereocilia form a chevron shaped bundle with the kinocilium at the vertex. In inner hair cells, the bundles appear flatter. The functional significance of these modifications in morphology are unclear but have been suggested to be linked to the overall structure of the organ of Corti. In the vestibular epithelia, bundles are generally round with the kinocilium located at one edge. The staircase arrangement of stereocilia is maintained by a complex array of protein links (Hackney and Furness, 2013; Nayak et al. 2007). These links are extracellular filaments that span between adjacent stereocilia. Tip links, horizontal top connectors and shaft connectors (also termed lateral links) connect adjacent stereocilia (Reviewed in Hackney and Furness, 2013). Kinociliary links connect the kinocilium to the tallest row of stereocilia. In developing hair bundles, ankle links connect adjacent stereocilia at their base, although they are lost upon onset of hearing (Goodyear et al., 2005; Michalski et al., 2007). Although the precise role of ankle links is not entirely known, it is likely that they are required for the stabilisation of stereocilia during development of the stereociliary bundle (Boeda et al., 2002). During this time, stereocilia do not express actin crosslinking filaments and therefore require additional support for stabilisation.

The most widely accepted model is that deflection of stereocilia in the direction of the tallest row puts tension on the tip links, which mechanically opens ion channels allowing K^+- and Ca^{2+}-ions to enter the cell, thus changing the ionic potential and depolarising the cell (Roberts et al., 1988). It is thought that the transduction channels, possibly located in the proximity of the tip link filaments, are mechanically gated and that a spring-like mechanism transmits forces for opening and closing these channels (Fettiplace, 2006). The end result is the release of neurotransmitter at the base of the hair cell. Opposing deflection relaxes the tension placed on the tip links and closes the channels leading to hyper-polarization of the cell (Furness et al., 2010). Only deflection of the bundle in the direction of the tallest stereocilia leads to increased channel opening and depolarization. All of the mechanosensory epithelia within the inner ear contain morphological specializations that take advantage of the directional nature of the stereociliary bundles. In the coiled cochlear sensory epithelium of mammals, all stereociliary bundles are oriented towards the lateral edge of the coil. In response to sound waves, the overlying tectorial membrane is deflected closer and laterally relative to the lumenal surface. This motion applies a lateral force to the stereociliary bundles, leading to channel opening and cellular depolarization. In contrast, some vestibular epithelia contain arrays of hair cells which are orientated exactly opposite to one another. Thus deflection of an overlying membrane in one direction, leads to depolarization of some cells and hyper polarization of others, which is thought to enhance sensitivity.

The exact composition of these links, and the properties and function of the transduction channels is still being debated. Recently, cadherin 23 and protocadherin 15, defects of which cause the sensorineural disease Usher syndrome, have been shown to be components of the tip links (Kazmierczak et al., 2007; Lagziel et al., 2005; Ahmed et al., 2006; Kazmierczak et al., 2007). Two other Usher syndrome proteins, Usherin (USH2a) and VLGR1 (USH2C) are thought to be components of the ankle links (McGee et al., 2006; Michalski et al., 2007). A number of candidates have been considered as key proteins of the transduction ion channel, including the epithelial sodium channel (ENaC) and acid-sensing ion channel (ASIC), members of the DEG/ENaC superfamily of amiloride-sensitive sodium channels. However, the transduction ion channel is most likely thought to consist of transient receptor potential (TRP) channels, such as TRPN1, TRPV4, TRPML3 and TRPA1 (Corey, 2006) although this is still under debate.

Although stereocilia are not "true" cilia, their arrangement is under the control of a true microtubule based cilium, the kinocilium, which moves across the hair cell during its development and lies behind the row of the tallest stereocilia. In mammals the kinocilium disappears soon after birth, whilst in reptiles it is maintained throughout life (Denman-Johnson and Forge, 1999; Frolenkov et al., 2004). The kinocilium and other primary cilia will be discussed below.

Vestibular hair cells

Although there are fewer studies on the development of vestibular hair cells in general, their developmental progression is similar to that of cochlea hair cells, if somewhat less co-ordinated (Denman-Johnson and Forge, 1999). As the kinocilium begins to elongate it moves in a non-random fashion towards one edge of the lumenal surface. In contrast to the chevron-shaped stereocilia bundles in the cochlea, in the vestibular epithelia, bundles are generally round, with the kinocilia located at one edge.

The overall process of polarization in the vestibular is more difficult to define, because hair cells are not generated in a distinct pattern, as they are in the cochlea (from apex to base). Hair cells seem to arise at random positions within the epithelium over the course of several days. In the utricle and saccule, hair cells are oriented either towards or away from the striola, a reversal line that runs across the centre of the epithelia. This results in a "reversal zone" in which bundle orientations switch by 180°. This arrangement is thought to result in heightened sensitivity, because deflection of an overlying membrane in one direction leads to depolarization of some cells and hyper polarization of others.

Primary Cilia in the Inner Ear

Most epithelial cell types display primary cilia at some point during development. Therefore, it is not surprising that virtually all cell types lining the developing cochlea duct are ciliated. In most of these cell types the primary cilium retracts upon onset of hearing, and as yet little is known about their function. The exception to this is the primary cilium on mechanosensory hair cells, the so-called kinocilium, which has been characterised to some extent. In the vestibular organs, other than the presence of kinocilia, there have been no reports of primary cilia to date.

Kinocilia

All mechanosensory hair cells, both cochlea and vestibular hair cells, have a kinocilium. The term 'kino' (Greek for movement) is thought to be in reference to the fact that the kinocilium moves across the surface of the hair cells. Most studies describe the kinocilium as a primary, non-motile cilium with a 9 + 0 microtubule composition (Kikuchi et al., 1989, 1988; Sobkowicz et al., 1995), although some studies reported the presence of 9 + 2 or even a more unusual 8 + 1 organisation (Arima et al., 1986; Kelley et al., 1992; Sobkowicz et al., 1995). These differences are not necessarily species dependent, as all three types of microtubule organisation have been seen in mouse. Although the kinocilium is retained throughout life in the vestibular, it retracts upon onset of hearing in cochlear hair cells.

Numerous studies have shown that the kinocilium is important for stereociliary bundle development during cochlear maturation (Denman-Johnson and Forge, 1999; Jones et al., 2008; Ross et al., 2005). The unique organisation of stereocilia that is required for mechanotransduction and their precise arrangement appears to be evolutionarily conserved (Wiederhold, 1976). This arrangement is dependent on the kinocilium. Midway through gestation the kinocilium emerges on the apex of all hair cells. It migrates to the apical side of the cell, from where the stereocilia subsequently elongate to form their staircase-like bundle (Frolenkov et al., 2004). This migration of the kinocilium is thought to determine the eventual orientation of the bundle. Thus the kinocilium is crucial for establishing the morphological polarity, which defines functional polarity, since deflections of the stereocilia along the line of polarity optimally stimulate the hair cell (as described above).

In human cochlea hair cells, ciliogenesis is observed in the 11th and 12th week of gestation. By the 22nd week the stereociliary bundles take on the adult formation, although complete maturation is achieved in the last trimester (Igarashi, 1980; Lavigne-Rebillard and Pujol, 1986). In mouse, the regression of the kinocilium begins at P8 and disappears by P12. Thus, the kinocilium has long been considered as a transient embryonic feature without any functional significance in the adult inner ear, however, the residual basal body may still be required to maintain bundle orientation, structure and function. Remarkably, in adult, post-traumatic hair cells that have survived mechanical injury are able to regrow their kinocilia and do so regardless of age. Similar to what happens during ear development, as the re-kinocilium grows, the stereociliary bundles reform (Sobkowicz et al., 1995).

The kinocilium is attached to stereocilia via kinocilial links, which have similar morphological and biochemical properties as stereocilial links (Goodyear and Richardson, 2003). Apart from the customary components of all primary cilia, the composition of the kinocilium has not been elucidated. A protein containing putative transmembrane domains, Kinocilin, was found to localise mainly to the kinocilium (Leibovici et al., 2005). Little is known about this protein but it has been postulated that it may stabilise dense microtubular networks or play a role in vesicular trafficking. Somewhat surprisingly, Kinocilin knockout mice show no structural phenotype or auditory abnormalities (M. Leibovici, personal communication). Intriguingly, myosin VIIa was also shown to localise to kinocilia in developing hair cells in mouse (Wolfrum et al., 1998), and myosin Ic has been observed to localise to kinocilia in frog vestibular hair cells (Cyr et al., 2002; Hasson et al., 1997). The function of these myosins within the kinocilium is still unknown.

Several conditional mouse mutants in which the kinocilium does not function correctly display rotated, disrupted and mis-shapen stereociliary bundles. These mutants and their phenotypes will be described in detail below.

Other Cilia in the Cochlea

Mechanosensory hair cells are not the only cells in the cochlea that are ciliated. As the cochlea develops, the cells lining the cochlear duct become regionalised and form specialised domains. These include the organ of Corti, a neuro-epithelium which contains the mechanosensory hair cells; the inner sulcus and outer sulcus which flank the organ of Corti; stria vascularis, a vascularised ion-transporting tissue at the lateral wall of the cochlear duct; and Reissner's membrane which separates scala media from scala vestibuli. Scala media is an endolymph-filled cavity surrounded by all of these specialized domains. Considering that the cells that line scala media are of epithelial origin, it is perhaps not surprising that most bear cilia whilst the cochlea is developing. These include the cells of the developing organ of Corti, the epithelial layer of Reissner's membrane, and the marginal cells of stria vascularis. In the description of the cells lining scala media below, the timings are mostly from studies of cochlear development in altricial mammals such as mice, gerbils and rats.

In addition to the mechanosensory hair cells, the organ of Corti also comprises non-sensory supporting cells (Gale and Jagger, 2010; Jagger and Forge, 2014). These include the inner and outer pillar cells that separate the inner and outer hair cells and the Deiters cells that provide mechanical and homeostatic support to the outer hair cells. As the organ of Corti develops, all of these cells have primary cilia that similar to the hair cell kinocilium, retract once the tissue has matured. The timing of this retraction has not been specifically documented, but it seems to occur around the onset of hearing. The developing organ of Corti is flanked by patches of ciliated epithelial cells. On the medial side in the greater epithelial ridge is a body of columnar cells termed Kölliker's organ. Kölliker's organ extends between the bony spiral limbus and the organ of Corti during development, yet it begins to retract prior to hearing onset and is no longer evident by P21. As Kölliker's organ retracts, it is replaced by squamous inner sulcus cells, which do not bear cilia. On the lateral side of the developing organ of Corti in the lesser epithelial ridge, lie the Hensen's, Claudius and outer sulcus cells. All of these cells are ciliated early on, but lose their cilia upon maturation of the tissue.

Reissner's membrane is comprised of an epithelial and a mesothelial cell layer. The epithelial cell layer is in contact with the endolymph within the scala media. Intriguingly, these cells may retain primary cilia into adulthood (Jagger et al., 2011). This might reflect a role for cilia in regulating the ionic balance of the endolymph. In mature tissue, the composition of endolymph must be highly regulated. This extracellular fluid has a unique composition, containing high levels of K^+-ions, yet low levels of Na^+-ions and Ca^+-ions. This creates a driving force for the influx of K^+-ions during mechano-transduction, and contributes to generation of the endocochlear potential (EP) (Hibino and Kurachi, 2006). Along the lateral wall of scala media lies stria vascularis. It is thought that the stria secretes K^+ ions into the endolymph thereby contributing to the EP. The high K^+ concentration of endolymph is almost adult-like by P7 (Bosher and Warren, 1971). EP begins to develop after P7 though, and it reaches adult-like levels of +80–100 mV around P21 (Souter and Forge, 1998). The stria is comprised of three cell layers, marginal cells, intermediate cells and basal cells. The cells of the outermost layer, namely the marginal cells, display primary cilia in the neonatal mouse. By P30 these have disappeared. A role for these cilia is not known, yet it is possible that they play a part in gating K^+-ion channels during establishment of the EP. The membrane of the ciliary axoneme contains numerous ion channels, which may help the cilium function as a sensory organelle regulating and responding to ionic changes in concentration. The intermediate cells are pigment-containing melanocytes. Considering that both the intermediate and the basal cells are embedded in other tissue it is difficult to discern if these contain cilia. As will be discussed below, the intermediate cells of the stria vascularis appear to be affected in a basal body mutant, suggesting that cilia function might be of relevance to them (Jagger et al., 2011).

Little is known about the specific roles primary cilia may play during development of the cochlea or in the maturation of auditory function. Similarly, with the exception of the kinocilia of vestibular hair cells, primary cilia in other vestibular cells have not been described, yet they may also be involved in the development and function of vestibular organs.

Signalling Pathways Relevant to the Inner Ear

In recent years, numerous signalling pathways have been shown to be regulated via the primary cilium. Although many cells in the developing cochlear duct are ciliated, surprisingly few ciliary related signalling pathways have been studied in the relation to auditory development or function. The notable exception to this is the non-canonical branch of Wnt signalling.

Planar cell polarity signalling in cilia mutants

The non-canonical branch of Wnt signalling also referred to as planar cell polarity (PCP) signalling, co-ordinates orientation of cellular structures within the plane of an epithelium such that the structures are all orientated the same way (Fanto and McNeill, 2004). It is a fundamental aspect of development, conserved through evolution and involves a core set of structurally related proteins. This process was first identified in Drosophila, in which the PCP pathway directs position and direction of wing hairs on the surface of fly wings. In vertebrates, this signalling pathway appears to regulate specific cell movements known as convergent extension during gastrulation and neurulation (Copp et al., 2003). The mammalian cochlea has long been used as a model system to study vertebrate non-canonical Wnt signalling since as the cochlea develops, extension of the cochlea duct involves convergent extension movements, during which the mechanosensory hair cells must orient themselves correctly (Dabdoub and Kelley, 2005). Several of these processes have been shown to be disrupted upon loss of ciliary function.

Between E12.5 and P0, the cochlea duct undergoes significant growth, more than doubling in length (Morsli et al., 1998). Although non-sensory cells within the duct continue to proliferate during this time, the progenitor pool that gives rise to hair cells and support cells within the organ of Corti, becomes post mitotic between E13 and E14 (Chen et al., 2002; Chen and Segil, 1999; Ruben, 1967; Wang et al., 2005). Derivatives from this precursor population (termed the prosensory domain) extend along the entire length of the cochlear duct by P0. Therefore, as these cells are differentiating, the prosensory domain undergoes a period of extension and cellular rearrangement. Morphological studies in which the distribution of these cells was analysed at different developmental time points indicated cellular rearrangements that are consistent with the process of convergence and extension (Chen and Segil, 1999; McKenzie et al., 2004). In addition, a by-product of these cell rearrangements is the alignment of both hair cells and support cells into the highly ordered rows that extend along the entire length of the cochlea. As the hair cells arise from surrounding support cells, they differentiate and re-orient to their final arrangement. The polarity of hair-bundles on different hair cells with respect to each other is also highly organized. They show a distinct 'orientation' related to that of their immediate neighbours and their position in the sensory patch. The correct coordination of these re-orientation events is essential for hearing.

It has been well documented that PCP signalling controls these developmental processes and mouse mutants for PCP genes display convergent extension defects and mis-orientated stereociliary bundles. The cascade is first triggered by Wnt molecules, which bind to membrane bound receptors, such as Fz3/6, Vangl1/2 and Celsr1 (Curtin et al., 2003; Montcouquiol et al., 2003; Torban et al., 2008; Wang et al., 2006). These then trigger multiple downstream intracellular pathways via association with Dvl1-3 (Etheridge et al., 2008), which ultimately result in cytoskeletal rearrangements. Protein localization studies have shown asymmetric membrane targeting of Vangl2, Fz3, and Fz6 to the medial membranes of cochlear hair cells, whereas Dvl1-3 are localized to the lateral sides of the same hair cells. It was initially thought that this asymmetric protein localisation was a determinant of establishing polarity, however, discrepancies in asymmetric localisation of another core cytoplasmic PCP protein, Prickle2, have challenged this belief. Although asymmetrically localised in cochlea hair cells, in vestibular hair cells Prickle2 is restricted to the same side across the entire sensory epithelia (Deans et al., 2007), regardless of which side of the reversal zone the hair cells are situated, even though the orientation of the stereociliary bundles is rotated by 180°. This suggests that stereociliary bundle localisation is not always directly correlated with asymmetric localisation of PCP proteins. It also highlights that establishment of polarity appears to be a multifaceted affair, which may also require additional cell autonomous pathways (see below).

The first indication that primary cilia may in some way be connected to PCP signalling came after identification of PCP phenotypes in ciliary mutant cochlea. Cilia mutant mice displayed shortened cochlea ducts and mis-oriented stereociliary bundles (Jones et al., 2008; May-Simera et al., 2015; Ross et al., 2005; Sipe and Lu, 2011). Further support to this hypothesis came from the identification of PCP phenotypes in other tissues coupled with a genetic interaction with other PCP proteins in both cilia mutant mice and zebrafish (Gerdes et al., 2007; Ross et al., 2005). More recent support for cilia regulating PCP signalling comes from studies involving ciliary protein TMEM67/MKS3 (Abdelhamed et al., 2015). Not only do Tmem67 mutant mice exhibit PCP-like phenotypes, but TMEM67 was shown to co-localise with non-canonical Wnt receptor Ror2 to the ciliary transition zone, where it is also thought to function as a receptor mediated by Wnt5a. Experimental evidence for the role of cilia in PCP signalling is controversial since some studies in zebrafish and mouse mutants have suggested no connection between cilia and Wnt signalling (Borovina and Ciruna, 2013; Huang and Schier, 2009; Ocbina et al., 2009). Discrepancies in the data may reflect species, tissue, or temporal-dependent differences in ciliary contributions towards Wnt signalling and might also reflect variations between individual ciliary components and their precise function. More recently, mis-localisation of core PCP components in some ciliary mutant cochlea (Bbs8 and Ift20) suggest alternative 'non ciliary' roles, which may influence tissue polarity in a cilia-independent manner (May-Simera et al., 2015).

Surprisingly, while cochlea tissue polarity is disrupted in cilia mutants, utricular hair cells do not seem to be affected (Jones et al., 2008; Sipe and Lu, 2011). This may represent differences in the propagation of planar cell polarity signals between the hearing and balance organs, which could be based on the distinct patterning of sensory versus support cells (Deans et al., 2007; Hackett et al., 2002).

Cell intrinsic polarity pathways

The establishment of polarity is a complex affair. In addition to signalling via the PCP pathway, several cell intrinsic polarity pathways have been identified in cochlea hair cells. Although not much is known about these as yet, they have been shown to associate with ciliary proteins. A recent paper showed that G protein-dependent signalling controls the migration of the kinocilium and affects hair bundle orientation and shape in a cell-autonomous manner

(Ezan et al., 2013). Asymmetric localisation of G-protein alpha-i subunit 3 (Gαi3) and its interacting partner mPins that does not overlap with the aPKC/Par-3/Par-6b and Vangl2 expression domains, is disrupted in Bbs8 and Bbs6 knockout cochlea (Ezan et al., 2013; May-Simera et al., 2015). This strengthens the role of ciliary proteins in defining the localization of polarity molecules.

The activity of Rho GTPases has been shown to mediate cell polarity via the regulation of cytoskeletal reorganization, formation of junctional complexes, and activation and localization of polarity proteins (Schlessinger et al., 2009). Specifically in the mouse cochlea, the small GTPase Rac1 together with PAK (p21-activated kinase) appear to be required for basal body positioning and stereociliary bundle morphology in cochlear hair cells (Grimsley-Myers et al., 2009; Sipe and Lu, 2011). Although PCP mutants do not show any defects in Rac-PAK signalling, loss of the ciliary motor protein Kif3a does appear to influence basal body positioning via the regulation of cortical pPAK activity (Sipe and Lu, 2011). Kif3a is a component of Kinesin-II, the motor for anterograde IFT required for ciliogenesis (Goetz and Anderson, 2010), and is also required for intracellular vesicle trafficking (Marszalek and Goldstein, 2000; Nishimura et al., 2004). Not only does loss of Kif3a result in the elimination of the kinocilium, shortening of the cochlea duct and flattened stereociliary bundles, but the basal body positioning at the apex of the developing hair cell and the coupling of the basal body to stereociliary bundles is also disrupted. The latter is thought to be controlled in a non-ciliary manner, by regulation of the activation of Rac GTPases at cortical locations.

More recently, another Rho GTPase, Cdc42, has also been shown to regulate polarity and cellular patterning in the developing cochlea (Kirjavainen et al., 2015). It is unknown if this is in any way connected to ciliary function, however, Cdc42 has been shown to facilitate ciliogenesis in cultured kidney cells (Zuo et al., 2011).

Disruption of Signalling Pathways and Other Ciliary Defects in Mouse Models

Signalling defects in cilia mutants

Since the early 2000s, the inner ear, specifically the cochlear epithelium, has been accepted as one of the best models in which to study multiple aspects of PCP signalling in mammals. Around the same time the first mouse models harbouring mutations in primary cilia genes emerged. Because of the PCP-like phenotype observed in the cochlea of many of these mutants, the suggestion that primary cilia may influence polarity and polarity signalling arose. Table 1 gives an overview of the various cilia mouse models examined so far, with a breakdown of cochlea PCP phenotypes observed in each model.

Table 1. Cochlea PCP phenotype in cilia mutants.

	Bbs4 KO	Bbs6 KO	Bbs8 KO	Ift20 cKO	Ift25 KO	Ift27 KO	Gmap210 KO	Ift88 cKO	Kif3a cKO	Alms1 KO	Tmem67 (Mks3) KO	Mks1 del64-323	Jbts17 Hug	Wdpcp Cys40
Duct lenth	Unchanged	Unchanged	Unchanged	Shortened	Unchanged	Slightly shortened	Unchanged	Shortened	Shortened	Unchanged	Unchanged	N/A	Unchanged	N/A
Cellular Patterning	Normal	Normal	Normal	Abnormal	Normal	Normal	Normal	Disrupted	Disrupted	Normal	Normal	N/A	Normal	N/A
Flattened bundels	Yes	Yes	Yes	Yes	No	Yes	Yes	Yes	Yes	No	Mild	Yes	No	Yes
Rotated bundles	Yes	Yes	Yes	Yes	No	No	No	Yes	Yes	Mild	Yes	Mild	No	Yes
Splayed bundles	Mildly	No	Yes	No	No	No	No	No	Yes	No	No	Yes	No	Yes
Circular bundles	No	No	No	Observed	No	No	No	Observed	No	No	No	Yes	No	N/A
Central basal bodies/no migration	N/A	N/A	N/A	N/A	N/A	N/A	N/A	Yes	No	No	No	N/A	N/A	N/A
Uncoupled bb/kino from bundle	N/A	Yes	Yes	Yes	N/A	N/A	N/A	Yes	Yes	No	Yes	Yes	N/A	Yes
Apical basal defects of bb	N/A	N/A	N/A	Yes	N/A	N/A	N/A	N/A	Yes	No	N/A	N/A	N/A	N/A
FM143 uptake	N/A	Normal (unpub)	N/A	N/A	N/A	N/A	N/A	N/A	Normal	N/A	N/A	N/A	N/A	N/A
ABR	N/A	N/A	Normal	N/A	N/A	N/A	N/A	N/A	N/A	Elevated in older animals	N/A	N/A	N/A	N/A
OAE	N/A	No response	Normal	N/A	N/A	N/A	N/A	N/A	N/A	Lost in older animals	N/A	N/A	N/A	N/A
Preyer Reflex	Absent	Absent	N/A	N/A	N/A	N/A	N/A	N/A	N/A	N/A	N/A	N/A	N/A	N/A
Reference	May-Simera, 2009	Ross, 2005	May-Simera, 2015	May-Simera, 2015	May-Simera, 2015	May-Simera, 2015	May-Simera, 2015	Jones, 2008	Sipe, 2011	Jagger, 2011	Abdelhamed, 2015	Cui, 2011	Damerla, 2015	Cui, 2013
	Ross, 2005										Leightner, 2013			

Considering that aspects of the phenotype can be difficult to quantify, characterization of the cilium/PCP phenotype has often been incorrectly credited in the literature. This can be attributed to several factors. Firstly, stereociliary hair cell bundle morphology can be challenging to quantify, particularly when encountered with various facets of bundle dysmorphology (i.e., rotated, flatted, splayed bundles). Secondly, especially in early reports, the number of affected hair cells might have been overestimated. When comparing similarities versus differences in mutant phenotypes, and trying to extrapolate a role in signalling, it is important to consider that age, genetic background and the type of cilia mutation all factor heavily in shaping the cochlea phenotype.

Some of the seminal findings include the first description of disrupted cochlea bundles in Mkks knockout mice (Ross et al., 2005). This was nicely followed up by the first description of a conditional knockout in which the cilium was completely abolished in developing hair cells, disrupting basal body positioning and bundle morphology, including the appearance of circular stereociliary bundles positioned in the centre of hair cells (Jones et al., 2008). A detailed study of a conditional cilia motor mutant, Kif3a, in which ciliogenesis is also abolished, began teasing apart ciliary vs non-ciliary hair cell functions, uncoupling hair bundle orientation from basal body positioning (Sipe and Lu, 2011). In comparison with Ift88 mutants, Kif3a mutants did not display circular hair bundles or centrally positioned basal bodies. Hence, hair bundle orientation was no longer coupled with basal body position and basal bodies were mis-positioned along both apico-basal and planar polarity axes, a phenotype not observed in core PCP mutants. The authors suggested that these results reflect a coupling of hair bundle orientation and basal body positioning that, at least in part, may be mediated in a non-ciliary function. Subsequent studies have found similar phenotypes in other cilia mutants (May-Simera et al., 2015), which may support the hypothesis that cilia-related proteins have additional cellular functions, or may broaden the spectrum of ciliary function. In support of this, mis-localisation of core PCP protein Vangl2, in two cilia mutants (Bbs8 and Ift20), suggests that these proteins may regulate membrane targeting.

Although numerous ciliopathy mouse models are available, few have thus far been examined for the presence of cochlear PCP defects. Embryonic lethality, a common phenotype of cilia mutants, does not have to be the limiting factor considering that many aspects of the cochlea PCP phenotype can be observed embryonically, and conditional mutants can easily be generated (May-Simera, 2016).

Another aspect that needs additional investigation is the extent of cross talk between signalling pathways in developing cochlea hair cells. It has already been shown that loss of Mkks (Bbs6) and Bbs8 affects G protein-dependent signalling (Ezan et al., 2013) and Kif3a has been shown to regulate the activation of Rac GTPases (see above). There are undoubtedly other signalling pathways involved, which are likely to function in concert and their coordination may be regulated by the kinocilium.

Other defects in cilia mutants

Not much is known about a ciliary phenotype in other tissues of the auditory system. In Bbs8 knockouts, it was shown that the orientation of utricle hair cells and overall shape of the sensory epithelia was perturbed (May-Simera et al., 2015). However, none of the cilia mutants described thus far have balance disorders, which could be attributed to vestibular dysfunction.

Surprisingly, loss of Tmem67 (Mks3) a ciliary protein that localizes to the transition zone, also resulted in primary cilia loss in the lateral part of the organ of Corti, namely the Deiters' cells and outer pillar cells (Abdelhamed et al., 2015). The consequence of this is not known, as Tmem67 knockouts do not survive postnatally.

Alms1 is a basal body protein, mutations in which cause Alstrom syndrome (described in detail in the next section; Collin et al., 2012; Hearn et al., 2002), the only classic ciliopathy in which patients have a significant auditory phenotype (described below). Loss of Alms1 in mice introduced large 'spaces' in the stria vascularis, particularly in the intermediate cell layer (Jagger et al., 2011). This suggests that in addition to loss of outer hair cells, degeneration of the stria vascularis may also be contributing to decreased hearing sensitivity as these animals age. Nevertheless, the endocochlear potential is normal and protein expression of endocochlear potential proteins is unchanged. This is particularly intriguing considering that the stria vascularis did not appear disrupted upon loss of another basal body protein, Mkks (Bbs6) (May-Simera et al., 2009). The Alms1 knockout mouse is also the only ciliopathy mouse model in which loss of outer hair cells has been described in older animals, although it must be noted that most cilia mutants have not been examined for this in later life.

Similar to other cilia mutant mice, Bbs4 knockout cochlea display mis-oriented and mis-shapened stereocilia hair bundles (May-Simera et al., 2009). However, Bbs4 closely associates with developing microtubules in the developing organ of Corti and Bbs4 knockouts display defects in microtubule dynamics, possible via Bbs4's association with the scaffold protein pericentriolar material 1 (PCM1) (May-Simera et al., 2009).

Auditory Assessments of Cilia Mutants

Although there is only limited data published, despite the disruption of stereociliary bundles, ciliopathy mutant mice do not always exhibit auditory dysfunction (see Table 1). Consistent with this, hearing defects are not a common human ciliopathy phenotype, with the exception of Alström syndrome (discussed below).

Loss of Bbs8 caused one of the more extreme disruptions to stereociliary bundle morphology, yet auditory brainstem responses (ABR) or otoacoustic emission (OAE; a measurement of outer hair cell function) are not significantly different between mutant and control mice (May-Simera et al., 2015). This suggests that stereociliary bundle morphology is not necessarily linked to auditory dysfunction in cilia mutants, potentially because of corrective reorientation of bundles later in development, as it has been reported for Vangl2 conditional knockout (CKO) mutants (Copley et al., 2013).

It is thus apparent that a more thorough examination of auditory phenotypes in cilia mutants is required, especially when taking into consideration that existing studies show that onset of hearing dysfunction may come with age (Jagger et al., 2011).

Other Animal Models of Auditory Defects

Most of the work in this field has been done in mouse models, partly because of the similarities between the mouse and human auditory systems and also the availability of gene knockouts. A less common model is *Danio rerio*, the Zebrafish. Fish are also able to perceive auditory and vestibular stimuli. The otic vesicles contain otolith organs (utricle and saccule) (Stooke-Vaughan et al., 2012). The otolith organs are so called because of the otoliths (calcium carbonate crystals), which cause the deflection of hair cell bundles. Similar to mouse, deflection of sensory hair cells found in the otoliths translates into electrical signals. It has been shown that the size of the otolith is crucial for the differential perception of sound and balance and that fish regulate otolith size to improve their sensitivity to sound (Inoue et al., 2013). Otolith abnormalities have also been described in ciliary zebrafish mutants (May-Simera et al., 2010), suggesting that cilia are important for normal otolith development. This might be due to a defect in otolith seeding. Otolith seeding is initiated via tethering of precursor particles to the tips of hair cell kinocilia (Stooke-Vaughan et al., 2012). This process is perturbed in zebrafish mutants, in which the otogelin (otog) gene is disrupted (Stooke-Vaughan et al., 2015). Human mutations in OTOG have been shown to be causative for deafness and vestibular dysfunction (Schraders et al., 2012). These examples highlight the usefulness of zebrafish as a model in which to study auditory dysfunction caused by defective cilia.

Diseases Relevant to Ciliary Defects in the Auditory System

Defects in cilia formation and function result in a spectrum of clinical syndromes collectively termed ciliopathies (Fliegauf et al., 2007). Most ciliopathies are syndromic disorders characterized by a wide spectrum of clinical features. With the exception of Alström syndrome, an auditory phenotype is not common, and only few reports have been published.

However there is another notable exception, namely Usher syndrome, the most prevalent cause of congenital deaf-blindness.

Worldwide approximately 250 million people are thought to suffer from hearing impairment, a number that is expected to rise as the global population ages. Genetic hearing loss is thought to contribute to about half of all cases (Bitner-Glindzicz, 2002). Genetic hearing impairment can broadly be divided into two categories, non-syndromic and syndromic hearing impairment, depending on the absence (or presence respectively) of other symptoms. So far, most cilia genes that have been identified to influence human auditory function are also responsible for other clinical phenotypes and are commonly observed in various ciliopathies.

Bardet-Biedl syndrome

Bardet-Biedl syndrome (BBS) is considered the archetypical ciliopathy, as patients exhibit most of the phenotypes associated with primary cilia dysfunction. Although deafness is not a primary feature of the human phenotype, it has been reported that ~ 3% of patients suffer from sensorineural deafness (Beales et al., 1999; Burn, 1950) Chronic otitis media ("glue ear"), causing conductive hearing loss, has been found to be common in children (~ 20%), but it seems to be largely resolved by puberty (Beales et al., 1999). In more recent clinical findings, audiometric assessments in 19 BBS patients at 1 kHz and 4 kHz showed that 53% and 84% of subjects respectively, had a threshold that was > 2 S.D. above the expected value for their age, indicative of a subclinical hearing loss (Ross et al., 2005). In an unpublished patient survey recently conducted, 45% of patients who responded said they had 'problems with their ears'. This survey was sent to affected individuals who volunteered to participate, therefore selection was not unbiased. A high prevalence of middle ear issues was reported, with 23% continuing to have problems with their 'ears and hearing'. Although a previous study had shown increased incidence of 'glue ear' in younger children, in the patient survey there did not seem to be an association between reports of auditory concerns and age of affected individuals. Anecdotally, some parents have reported that their affected children are hypersensitive to sound (P. Beales, personal communication).

Alström syndrome

Alström syndrome is the only ciliopathy in which hearing impairment is a defined phenotype. It was first described by a swedish doctor Carl-Henry Alström in 1959 as a specific syndrome distinct from Bardet-Biedl syndrome (Alstrom et al., 1959). Alström syndrome is caused by mutations in *ALMS1*, the protein of which localizes to the basal body. Most patients develop mild-to-moderate bilateral sensorineural hearing loss that usually develops between one and ten years of age. The hearing impairment is initially in the high frequency range (Marshall et al., 2007). About 10% of patients become profoundly deaf. Hearing loss develops gradually and the onset is post-lingual, therefore most children do not experience the speech and language problems often associated with deafness. Similar to Bardet-Biedl Syndrome, there is high incidence of otitis media and fluid retention along with high susceptibility to glue ear, which compounds the existing sensorineural impairment (Marshall et al., 2007). Vestibular function can also been affected. As described above, loss of Alms1 in a mouse model introduced large 'spaces' in the stria vascularis, particularly in the intermediate cell layer (Jagger et al., 2011). This led to the suggestion that in addition to outer hair cell loss, degeneration of the stria vascularis may also be contributing to the auditory phenotype.

Joubert syndrome

A recent study examined whether hearing loss was associated with another ciliopathy, Joubert syndrome. They found no evidence for significant hearing loss, although it must be noted that they were only able to perform audiological examinations on 14 affected individuals, ranging in age between 3–40 years (Kroes et al., 2010). Similar to other Bardet-Biedl syndrome patients, middle ear infections were common in young children and a few parents reported that their children were hypersensitive to sound.

Usher syndrome

If the ciliopathy spectrum is widened to include Usher syndrome (USH), the most common congenital cause of deaf-blindness, then auditory dysfunction becomes highly relevant. Usher syndrome, an autosomal recessive condition involving sensorineural hearing impairment and progressive retinal dystrophy, is the most common cause of deaf-blindness in humans and is estimated to affect 3–6.2 per 100,000 live births (Cohen et al., 2007).

There are three main clinical subtypes (USH1, USH2, USH3), which differ in terms of phenotypic manifestation. USH1 patients have severe to profound congenital hearing impairment, vestibular areflexia and onset of retinitis pigmentosa (retinal degeneration) within the first decade of life. USH2 patients have moderate to severe hearing loss, normal vestibular function and onset of retinitis pigmentosa within the second decade of life. USH3 is the mildest subtype, as hearing loss is progressive, vestibular dysfunction sporadic and onset of retinitis pigmentosa variable. Ten causative genes and three additional loci (modifiers) have been identified so far (Aparisi et al., 2014). Although identified USH genes encode proteins from very different protein families, most localize to ciliary related structures in retina photoreceptors (Sorusch et al., 2014), suggesting that they are required for primary cilia function. However, in the auditory system, most of the USH proteins do not localize to the cilium, but rather to the actin-based stereociliary bundles of hair cells. The one exception to this is Sans, which was found to be abundantly expressed at the base of the kinocilium, and also along the ciliary axoneme (Adato et al., 2005). Loss of Sans was shown to disrupt ciliogenesis via regulation of endocytosis at the base of the cilium (Bauß et al., 2014). How USH proteins may be functioning in a ciliary context in the cochlea is yet to be elucidated. It is known that the true microtubule-based kinocilium is required for establishing stereociliary bundle morphology and mutations in USH genes cause disorganized stereociliary bundles. Biochemical studies have shown that many of the USH proteins form protein complexes with other bona fide ciliary proteins (Sorusch et al., 2014). Usher patients do not exhibit many of the other features common of more traditional ciliopathies, therefore, it is often not included as a classical ciliopathy. This may change as recent findings have suggested that USH proteins are involved in olfaction (Jansen et al., 2016), a processes that is known to be affected in ciliopathy patients (Kulaga et al., 2004).

Concluding Remarks

When it comes to the role of primary cilia in the auditory system, there are still many outstanding questions that have to be addressed. In addition to the kinocilium on mechanosensory hair cells, there are several other cell types that bear primary cilia, though not much is known about their function. The role of the kinocilium itself needs further clarification. Most evidence suggests that it is required for the formation of the stereociliary bundle, however, loss of the kinocilium does not give a uniform phenotype and a small percentage of hair cells adopt a 'circular' appearance (Jones et al., 2008; May-Simera et al., 2015; Sipe and Lu, 2011). A major concern has been the role of the primary cilium in the regulation of non-canonical Wnt (PCP) signalling. Although numerous studies identified a connection between cilia and PCP signalling (Wallingford and Mitchell, 2011), some studies in both zebrafish and mouse mutants have suggested no connection between the two (Borovina and Ciruna, 2013; Huang and Schier, 2009; Ocbina et al., 2009). Discrepancies in the data may reflect species, tissue, or temporal-dependent differences. In addition to this, redundancy of ciliary and PCP genes and the sensitivity of cell polarity to generalized cellular abnormalities can make it difficult to directly link a mutation to PCP-specific defects. One of the readouts of PCP signalling is the positioning of the basal body and the primary cilium, therefore segregating primary from secondary defects is challenging. Furthermore, in many of the cilia mutants, the ciliary axoneme is missing, therefore some degree of signalling may still be retained if basal bodies or remnants of the ciliary transition zone remain functional. In support of this, it has been shown that transition zone protein TMEM67/MKS3, mutations in which cause the most severe ciliopathy, Meckel-Gruber syndrome, functions as a receptor for Wnt5a in non-canonical Wnt signalling

(Abdelhamed et al., 2015). Another question that remains to be answered is to what extent, and if so how, the ciliated support cells assist the hair cells in terms of differentiation and function. Intriguingly some types of support cells do retain cilia throughout life, which might suggest some function in homeostasis.

Although cilia research is expanding, little is known about cilia function in the auditory system. Many ciliopathies have an auditory component and most ciliopathy animal models have an auditory defect, therefore examination of cilia function in a tissue specific content is critical.

Acknowledgments

We would like to thank Viola Kretschmer for help generating the figures and Daniel Jagger and Zoe Mann for critical reading of this manuscript. The author is supported by the Alexander von Humboldt foundation.

References

Abdelhamed, Z. A., Natarajan, S., Wheway, G., Inglehearn, C. F., Toomes, C., Johnson, C. A. and Jagger, D. J. 2015. The Meckel-Gruber syndrome protein TMEM67 controls basal body positioning and epithelial branching morphogenesis in mice via the non-canonical Wnt pathway. Dis. Model. Mech. 8: 527–41.

Adato, A., Michel, V., Kikkawa, Y., Reiners, J., Alagramam, K. N., Weil, D., Yonekawa, H., Wolfrum, U., El-Amraoui, A. and Petit, C. 2005. Interactions in the network of Usher syndrome type 1 proteins. Hum. Mol. Genet. 14: 347–356.

Alstrom, C. H., Hallgren, B., Nilsson, L.B. and Asander, H. 1959. Retinal degeneration combined with obesity, diabetes mellitus, and neurogenous deafness. A specific syndrome distinct from Laurence-Moon-Biedl syndrome, a clinical endocrinological and genetic examination based on a large pedigree. Acta Psychiatr. Neurol. Scand. 34: 1–35.

Aparisi, M. J., Aller, E., Fuster-García, C., García-García, G., Rodrigo, R., Vázquez-Manrique, R. P., Blanco-Kelly, F., Ayuso, C., Roux, A.-F., Jaijo, T. et al. 2014. Targeted next generation sequencing for molecular diagnosis of Usher syndrome. Orphanet. J. Rare. Dis. 9: 168.

Arima, T., Masuda, H. and Uemura, T. 1986. Structural similarities between kinocilium of vestibular hair cell and tracheal motile cilium in the guinea pig. Auris. Nasus. Larynx 13 Suppl 2: S15–9.

Bailey, A. P. and Streit, A. 2006. Sensory organs: making and breaking the pre-placodal region. Curr. Top. Dev. Biol. 72: 167–204.

Bauß, K., Knapp, B., Jores, P., Roepman, R., Kremer, H., Wijk, E. V., Märker, T., Wolfrum, U., Knapp, B., Jores, P. et al. 2014. Phosphorylation of the Usher syndrome 1G protein SANS controls Magi2-mediated endocytosis. Hum. Mol. Genet. 23: 3923–42.

Beales, P. L., Elcioglu, N., Woolf, A. S., Parker, D. and Flinter, F. A. 1999. New criteria for improved diagnosis of Bardet-Biedl syndrome: results of a population survey. J. Med. Genet. 36: 437–446.

Bitner-Glindzicz, M. 2002. Hereditary deafness and phenotyping in humans. Br. Med. Bull. 63: 73–94.

Boeda, B., El-Amraoui, A., Bahloul, A., Goodyear, R., Daviet, L., Blanchard, S., Perfettini, I., Fath, K. R., Shorte, S., Reiners, J. et al. 2002. Myosin VIIa, harmonin and cadherin 23, three Usher I gene products that cooperate to shape the sensory hair cell bundle. EMBO J. 21: 6689–6699.

Borovina, A. and Ciruna, B. 2013. IFT88 plays a cilia- and PCP-independent role in controlling oriented cell divisions during vertebrate embryonic development. Cell Rep. 5: 37–43.

Bosher, S. K. and Warren, R. L. 1971. A study of the electrochemistry and osmotic relationships of the cochlear fluids in the neonatal rat at the time of the development of the endocochlear potential. J. Physiol. 212: 739–61.

Burn, R. A. 1950. Deafness and the Laurence-Moon-Biedl syndrome. Brit. J. Ophthal. 34: 65–88.

Cohen, M., Bitner-Glindzicz, M. and Luxon, L. 2007. The changing face of Usher syndrome: clinical implications. Int. J. Audiol. 46: 82–93.

Collin, G. B., Marshall, J. D., King, B. L., Milan, G., Maffei, P., Jagger, D. J. and Naggert, J. K. 2012. The alstrom syndrome protein, ALMS1, interacts with alpha-actinin and components of the endosome recycling pathway. PLoS One 7: e37925.

Copley, C. O., Duncan, J. S., Liu, C., Cheng, H. and Deans, M. R. 2013. Postnatal refinement of auditory hair cell planar polarity deficits occurs in the absence of Vangl2. J. Neurosci. 33: 14001–14016.

Copp, A. J., Greene, N. D. and Murdoch, J. N. 2003. The genetic basis of mammalian neurulation. Nat. Rev. Genet. 4: 784–793.

Corey, D. P. 2006. What is the hair cell transduction channel? J. Physiol. 576: 23–28.

Curtin, J. A., Quint, E., Tsipouri, V., Arkell, R. M., Cattanach, B., Copp, A. J., Henderson, D. J., Spurr, N., Stanier, P., Fisher, E. M. et al. 2003. Mutation of Celsr1 disrupts planar polarity of inner ear hair cells and causes severe neural tube defects in the mouse. Curr. Biol. 13: 1129–1133.

Cyr, J. L., Dumont, R. A. and Gillespie, P. G. 2002. Myosin-1c interacts with hair-cell receptors through its calmodulin-binding IQ domains. J. Neurosci. 22: 2487–95.

Dabdoub, A. and Kelley, M. W. 2005. Planar cell polarity and a potential role for a Wnt morphogen gradient in stereociliary bundle orientation in the mammalian inner ear. J. Neurobiol. 64: 446–457.

Deans, M. R., Antic, D., Suyama, K., Scott, M. P., Axelrod, J. D. and Goodrich, L. V. 2007. Asymmetric distribution of prickle-like 2 reveals an early underlying polarization of vestibular sensory epithelia in the inner ear. J. Neurosci. 27: 3139–3147.

Denman-Johnson, K. and Forge, A. 1999. Establishment of hair bundle polarity and orientation in the developing vestibular system of the mouse. J. Neurocytol. 28: 821–835.

Ezan, J., Lasvaux, L., Gezer, A., Novakovic, A., May-Simera, H., Belotti, E., Lhoumeau, A.-C. C., Birnbaumer, L., Beer-Hammer, S., Borg, J.-P. P. et al. 2013. Primary cilium migration depends on G-protein signalling control of subapical cytoskeleton. Nat. Cell Biol. 15: 1107–15.

Fanto, M. and McNeill, H. 2004. Planar polarity from flies to vertebrates. J. Cell Sci. 117: 527–33. doi:10.1242/jcs.00973.

Fettiplace, R. 2006. Active hair bundle movements in auditory hair cells. J. Physiol. 576: 29–36.

Fliegauf, M., Benzing, T. and Omran, H. 2007. When cilia go bad: cilia defects and ciliopathies. Nat. Rev. Mol. Cell Biol. 8: 880–93.

Frolenkov, G. I., Belyantseva, I. A., Friedman, T. B. and Griffith, A. J. 2004. Genetic insights into the morphogenesis of inner ear hair cells. Nat. Rev. Genet. 489–498.

Furness, D. N., Hackney, C. M. and Evans, M. G. 2010. Localisation of the mechanotransducer channels in mammalian cochlear hair cells provides clues to their gating. J. Physiol. 588: 765–72.

Gale, J. and Jagger, D. 2010. Cochlear Supporting Cells.

Gerdes, J. M., Liu, Y., Zaghloul, N. A., Leitch, C. C., Lawson, S. S., Kato, M., Beachy, P. A., Beales, P. L., Demartino, G. N., Fisher, S. et al. 2007. Disruption of the basal body compromises proteasomal function and perturbs intracellular Wnt response. Nat. Genet. 39: 1350–1360.

Goetz, S. C. and Anderson, K. V. 2010. The primary cilium: a signalling centre during vertebrate development. Nat. Rev. Genet. 11: 331–344.

Goodyear, R. J. and Richardson, G. P. 2003. A novel antigen sensitive to calcium chelation that is associated with the tip links and kinocilial links of sensory hair bundles. J. Neurosci. 23: 4878–87.

Goodyear, R. J., Marcotti, W., Kros, C. J. and Richardson, G. P. 2005. Development and properties of stereociliary link types in hair cells of the mouse cochlea. J. Comp. Neurol. 485: 75–85.

Grimsley-Myers, C. M., Sipe, C. W., Geleoc, G. S. and Lu, X. 2009. The small GTPase Rac1 regulates auditory hair cell morphogenesis. J. Neurosci. 29: 15859–15869.

Hackett, L., Davies, D., Helyer, R., Kennedy, H., Kros, C., Lawlor, P., Rivolta, M. N. and Holley, M. 2002. E-cadherin and the differentiation of mammalian vestibular hair cells. Exp. Cell Res. 278: 19–30.

Hackney, C. M. and Furness, D. N. 2013. The composition and role of cross links in mechanoelectrical transduction in vertebrate sensory hair cells. J. Cell Sci. 126: 1721–31.

Hasson, T., Gillespie, P. G., Garcia, J. A., MacDonald, R. B., Zhao, Y., Yee, A. G., Mooseker, M. S. and Corey, D. P. 1997. Unconventional myosins in inner-ear sensory epithelia. J. Cell Biol. 137: 1287–307.

Hearn, T., Renforth, G. L., Spalluto, C., Hanley, N. A., Piper, K., Brickwood, S., White, C., Comnnolly, V., Taylor, J. F. N., Russell-Eggitt, I. et al. 2002. Mutation of ALMS1, a large gene with a tandem repeat encoding 47 amino acids, causes Alstrom syndrome. Nat. Genet. 31: 79–83.

Hibino, H. and Kurachi, Y. 2006. Molecular and physiological bases of the K+ circulation in the mammalian inner ear. Physiology (Bethesda) 21: 336–45.

Huang, P. and Schier, A. F. 2009. Dampened hedgehog signaling but normal Wnt signaling in zebrafish without cilia. Development 136: 3089–3098.

Igarashi, Y. 1980. Cochlea of the human fetus: a scanning electron microscope study. Arch. Histol. Jpn. 43: 195–209.

Inoue, M., Tanimoto, M. and Oda, Y. 2013. The role of ear stone size in hair cell acoustic sensory transduction. Sci. Rep. 3: 2114.

Jagger, D., Collin, G., Kelly, J., Towers, E., Nevill, G., Longo-Guess, C., Benson, J., Halsey, K., Dolan, D., Marshall, J. et al. 2011. Alstrom syndrome protein ALMS1 localizes to basal bodies of cochlear hair cells and regulates cilium-dependent planar cell polarity. Hum. Mol. Genet. 20: 466–481.

Jagger, D. J. and Forge, A. 2014. Connexins and gap junctions in the inner ear. Its not just about K+ recycling. Cell Tissue Res.

Jansen, F., Kalbe, B., Scholz, P., Mikosz, M., Wunderlich, K. A., Kurtenbach, S., Nagel-Wolfrum, K., Wolfrum, U., Hatt, H. and Osterloh, S. 2016. Impact of the Usher syndrome on olfaction. Hum. Mol. Genet. 25: 524–33.

Jones, C., Roper, V. C., Foucher, I., Qian, D., Banizs, B., Petit, C., Yoder, B. K. and Chen, P. 2008. Ciliary proteins link basal body polarization to planar cell polarity regulation. Nat. Genet. 40: 69–77.

Kazmierczak, P., Sakaguchi, H., Tokita, J., Wilson-Kubalek, E. M., Milligan, R. A., Muller, U. and Kachar, B. 2007. Cadherin 23 and protocadherin 15 interact to form tip-link filaments in sensory hair cells. Nature 449: 87–91.

Kelley, M. W., Ochiai, C. K. and Corwin, J. T. 1992. Maturation of kinocilia in amphibian hair cells: growth and shortening related to kinociliary bulb formation. Hear. Res. 59: 108–15.

Kikuchi, T., Tonosaki, A. and Takasaka, T. 1988. Development of apical-surface structures of mouse otic placode. Acta Otolaryngol. 106: 200–207.

Kikuchi, T., Takasaka, T., Tonosaki, A. and Watanabe, H. 1989. Fine structure of guinea pig vestibular kinocilium. Acta Otolaryngol. 108: 26–30.

Kirjavainen, A., Laos, M., Anttonen, T. and Pirvola, U. 2015. The Rho GTPase Cdc42 regulates hair cell planar polarity and cellular patterning in the developing cochlea. Biol. Open 4: 516–26.

Kroes, H. Y., Van Zanten, B. G. A., De Ru, S. A., Boon, M., Mancini, G. M. S., Van der Knaap, M. S., Poll-The, B. T. and Lindhout, D. 2010. Is hearing loss a feature of Joubert syndrome, a ciliopathy? Int. J. Pediatr. Otorhinolaryngol. 74: 1034–8.

Kulaga, H. M., Leitch, C. C., Eichers, E. R., Badano, J. L., Lesemann, A., Hoskins, B. E., Lupski, J. R., Beales, P. L., Reed, R. R. and Katsanis, N. 2004. Loss of BBS proteins causes anosmia in humans and defects in olfactory cilia structure and function in the mouse. Nat. Genet. 36: 994–998.

Lavigne-Rebillard, M. and Pujol, R. 1986. Development of the auditory hair cell surface in human fetuses. A scanning electron microscopy study. Anat. Embryol. (Berl). 174: 369–77.

Leibovici, M., Verpy, E., Goodyear, R. J., Zwaenepoel, I., Blanchard, S., Laine, S., Richardson, G. P. and Petit, C. 2005. Initial characterization of kinocilin, a protein of the hair cell kinocilium. Hear Res. 203: 144–153.

Lindeman, H. H. 1973. Anatomy of the otolith organs. Adv. Otorhinolaryngol. 20: 405–33.

Lopez, C. 2016. The vestibular system: balancing more than just the body. Curr. Opin. Neurol. 29: 74–83.

Marshall, J. D., Beck, S., Maffei, P. and Naggert, J. K. 2007. Alstrom syndrome. Eur. J. Hum. Genet. 15: 1193–1202.

Marszalek, J. R. and Goldstein, L. S. 2000. Understanding the functions of kinesin-II. Biochim. Biophys. Acta 1496: 142–50.

May-Simera, H. 2016. Evaluation of planar-cell-polarity phenotypes in ciliopathy mouse mutant cochlea. J. Vis. Exp. 53559.

May-Simera, H. L., Ross, A., Rix, S., Forge, A., Beales, P. L. and Jagger, D. J. 2009. Patterns of expression of Bardet-Biedl syndrome proteins in the mammalian cochlea suggest noncentrosomal functions. J. Comp. Neurol. 514: 174–188.

May-Simera, H. L., Kai, M., Hernandez, V., Osborn, D. P. S., Tada, M. and Beales, P. L. 2010. Bbs8, together with the planar cell polarity protein Vangl2, is required to establish left-right asymmetry in zebrafish. Dev. Biol. 345: 215–25.

May-Simera, H. L., Petralia, R. S., Montcouquiol, M., Wang, Y.-X. X., Szarama, K. B., Liu, Y., Lin, W., Deans, M. R., Pazour, G. J. and Kelley, M. W. 2015. Ciliary proteins Bbs8 and Ift20 promote planar cell polarity in the cochlea. Development 142: 555–566.

McGee, J., Goodyear, R. J., McMillan, D. R., Stauffer, E. A., Holt, J. R., Locke, K. G., Birch, D. G., Legan, P. K., White, P. C., Walsh, E. J. et al. 2006. The very large G-protein-coupled receptor VLGR1: a component of the ankle link complex required for the normal development of auditory hair bundles. J. Neurosci. 26: 6543–6553.

Michalski, N., Michel, V., Bahloul, A., Lefèvre, G., Barral, J., Yagi, H., Chardenoux, S., Weil, D., Martin, P., Hardelin, J.-P. P. et al. 2007. Molecular characterization of the ankle-link complex in cochlear hair cells and its role in the hair bundle functioning. J. Neurosci. 27: 6478–6488.

Montcouquiol, M., Rachel, R. A., Lanford, P. J., Copeland, N. G., Jenkins, N. A. and Kelley, M. W. 2003. Identification of Vangl2 and Scrb1 as planar polarity genes in mammals. Nature 423: 173–177.

Morsli, H., Choo, D., Ryan, A., Johnson, R. and Wu, D. K. 1998. Development of the mouse inner ear and origin of its sensory organs. J. Neurosci. 18: 3327–3335.

Nayak, G. D., Ratnayaka, H. S., Goodyear, R. J. and Richardson, G. P. 2007. Development of the hair bundle and mechanotransduction. Int. J. Dev. Biol. 51: 597–608.

Neely, S. T. 1998. From Sound to Synapse: Physiology of the Mammalian Ear. Ear Hear.

Nishimura, T., Kato, K., Yamaguchi, T., Fukata, Y., Ohno, S. and Kaibuchi, K. 2004. Role of the PAR-3-KIF3 complex in the establishment of neuronal polarity. Nat. Cell Biol. 6: 328–34.

Ocbina, P. J., Tuson, M. and Anderson, K. V. 2009. Primary cilia are not required for normal canonical Wnt signaling in the mouse embryo. PLoS One 4: e6839.

Raphael, Y., Lenoir, M., Wroblewski, R. and Pujol, R. 1991. The sensory epithelium and its innervation in the mole rat cochlea. J. Comp. Neurol. 314: 367–382.

Raphael, Y. and Altschuler, R. A. 2003. Structure and innervation of the cochlea. Brain Res. Bull. 60: 397–422.

Reichenbach, T. and Hudspeth, A. J. 2014. The physics of hearing: fluid mechanics and the active process of the inner ear. Rep. Prog. Phys. 77: 76601.

Roberts, W. M., Howard, J. and Hudspeth, A. J. 1988. Hair cells: transduction, tuning, and transmission in the inner ear. Annu. Rev. Cell Biol. 4: 63–92.

Ross, A. J., May-Simera, H., Eichers, E. R., Kai, M., Hill, J., Jagger, D. J., Leitch, C. C., Chapple, J. P., Munro, P. M., Fisher, S. et al. 2005. Disruption of Bardet-Biedl syndrome ciliary proteins perturbs planar cell polarity in vertebrates. Nat. Genet. 37: 1135–1140.

Schlessinger, K., Hall, A. and Tolwinski, N. 2009. Wnt signaling pathways meet Rho GTPases. Genes Dev. 23: 265–277.

Schraders, M., Ruiz-Palmero, L., Kalay, E., Oostrik, J., del Castillo, F. J., Sezgin, O., Beynon, A. J., Strom, T. M., Pennings, R. J. E., Zazo Seco, C. et al. 2012. Mutations of the gene encoding otogelin are a cause of autosomal-recessive nonsyndromic moderate hearing impairment. Am. J. Hum. Genet. 91: 883–9.

Sipe, C. W. and Lu, X. 2011. Kif3a regulates planar polarization of auditory hair cells through both ciliary and non-ciliary mechanisms. Development 138: 3441–3449.

Sobkowicz, H. M., Slapnick, S. M. and August, B. K. 1995. The kinocilium of auditory hair cells and evidence for its morphogenetic role during the regeneration of stereocilia and cuticular plates. J. Neurocytol. 24: 633–653.

Sorusch, N., Wunderlich, K., Bauss, K., Nagel-Wolfrum, K. and Wolfrum, U. 2014. Usher syndrome protein network functions in the retina and their relation to other retinal ciliopathies. Adv. Exp. Med. Biol. 801: 527–533.

Souter, M. and Forge, A. 1998. Intercellular junctional maturation in the stria vascularis: possible association with onset and rise of endocochlear potential. Hear. Res. 119: 81–95.

Stooke-Vaughan, G. A., Huang, P., Hammond, K. L., Schier, A. F. and Whitfield, T. T. 2012. The role of hair cells, cilia and ciliary motility in otolith formation in the zebrafish otic vesicle. Development 139: 1777–87.

Stooke-Vaughan, G. A., Obholzer, N. D., Baxendale, S., Megason, S. G. and Whitfield, T. T. 2015. Otolith tethering in the zebrafish otic vesicle requires Otogelin and -Tectorin. Development 142: 1137–1145.

Torban, E., Patenaude, A. M., Leclerc, S., Rakowiecki, S., Gauthier, S., Andelfinger, G., Epstein, D. J. and Gros, P. 2008. Genetic interaction between members of the Vangl family causes neural tube defects in mice. Proc. Natl. Acad. Sci. USA 105: 3449–3454.

Wallingford, J. B. and Mitchell, B. 2011. Strange as it may seem: the many links between Wnt signaling, planar cell polarity, and cilia. Genes Dev. 25: 201–213.

Wan, G., Corfas, G. and Stone, J. S. 2013. Inner ear supporting cells: rethinking the silent majority. Semin. Cell Dev. Biol. 24: 448–59.

Wang, Y., Guo, N. and Nathans, J. 2006. The role of Frizzled3 and Frizzled6 in neural tube closure and in the planar polarity of inner-ear sensory hair cells. J. Neurosci. 26: 2147–2156.

Whitfield, T. T. 2015. Development of the inner ear. Curr. Opin. Genet. Dev. 32: 112–8.

Wiederhold, M. L. 1976. Mechanosensory transduction in "sensory" and "motile" cilia. Annu. Rev. Biophys. Bioeng. 5: 39–62.

Wolfrum, U., Liu, X., Schmitt, A., Udovichenko, I. P. and Williams, D. S. 1998. Myosin VIIa as a common component of cilia and microvilli. Cell Motil. Cytoskeleton 40: 261–71.

Zheng, J., Shen, W., He, D. Z., Long, K. B., Madison, L. D. and Dallos, P. 2000. Prestin is the motor protein of cochlear outer hair cells. Nature 405: 149–155.

Zuo, X., Fogelgren, B. and Lipschutz, J. H. 2011. The small GTPase Cdc42 is necessary for primary ciliogenesis in renal tubular epithelial cells. J. Biol. Chem.

Conclusion

Cilia have been an exciting, albeit challenging, field of research for the past twenty years. The number of established and candidate ciliopathy-related genes is now around two hundred and the number of diseases that have been re-classified as Ciliopathies is exceeding thirty. This poses a significant development in cellular biology, since for some years the eukaryotic cilium was considered to be of limited importance. Nowadays, this is no longer believed to be the case and eukaryotic cilia are constantly being connected to more human diseases and are being demonstrated to affect numerous signalling pathways.

What becomes obvious from the chapters of this book is the range of organs that carry a eukaryotic cilium and the diverse impact that the absence or malfunction of this organelle has on tissue function and human disease. Strikingly, cilia defects can either affect structures associated with development, such as the embryonic node or adult organs, such as the human eye and heart. Further, numerous signalling pathways have now been shown to require the eukaryotic cilium for optimal signalling and downstream cascade responses. Significantly, the number of genes that when mutated cause severe human diseases, whose protein products have been associated with ciliary structures, is constantly increasing. Further evidence is also accumulating on the importance of the transition zone, the ciliary axoneme and the BBSome in correct cellular function.

Our knowledge on the complex network of interactions that require the eukaryotic cilium is vastly expanding. Alongside it, a re-classification of human diseases and a better understanding of their molecular mechanisms are arising. In some cases, different approaches have already been adopted in the diagnosis and prognosis of human conditions. What now remains to be seen is whether our knowledge of eukaryotic cilia and the molecular pathways related to them, can serve a definite therapeutic purpose in assisting in the discovery of novel treatments for ciliopathies. The future is full of exciting possibilities.

Index